MULTIVARIABLE SYSTEM CONTROL

ANDRÉ FOSSARD
Centre d'Etudes et de Recherches de Toulouse, France

with

C. GUÉGUEN

Translated by
P.A. Cook
UMIST, Manchester, U.K.

1977

NORTH-HOLLAND PUBLISHING COMPANY
AMSTERDAM · NEW YORK · OXFORD

North-Holland ISBN: 0 7204 0745 1

Published by:
NORTH-HOLLAND PUBLISHING COMPANY
AMSTERDAM · OXFORD · NEW YORK

Distributors for the U.S.A. and Canada:
Elsevier/North-Holland, Inc.
52 Vanderbilt Avenue
New York, N.Y. 10017

Translation and revised edition of:
Commande des systèmes multidimensionnels

Library of Congress Cataloging in Publication Data

Fossard, André.
Multivariable system control.

A translation with revisions of the author's Commande des systèmes multidimensionnels.
Bibliography: p.
1. Control theory. I. Guéguen, C., joint author. II. Title.
QA402.3.F6613 1977 629.8'312 77-6423
ISBN 0-7204-0745-1

PRINTED IN THE NETHERLANDS

MULTIVARIABLE SYSTEM CONTROL

Foreword to the French edition

Jean-Charles Gille

The work by Messrs. Fossard and Guéguen, which I have the pleasure of introducing, is the most important book currently available on multivariable systems and their control.

This work, accessible to any reader who has grasped the elementary concepts of linear systems and state variables, is constructed around two principal ideas, both recent in comparison with the stage (not long past) when a multivariable system was thought of as characterised simply by its transfer function matrix, just as a single-input, single-output system could be described by a transfer function.

The first of these ideas is the non-equivalence of different representations of a system; this is made the subject of a full treatment, based on the concepts of controllability and observability, in the initial chapters. The second idea is the utilisation of the structure of a system, and of the subsystems into which one may be led to decompose it, in order to solve the control problem, as developed in the subsequent chapters.

With regard to these problems (the choice of a representation and the discovery of the internal structure), the book has the merit of gathering up and unifying a great many methods and results hitherto scattered through the literature in the form of papers and contributions to conferences. Several of these methods are due to the authors and are here making their first appearance. The book thus replaces a great number of diverse works, differing in terminology, notation, and level of difficulty.

The senior author of the book, who studied at the Higher National School of Aeronautics and subsequently at California Institute of Technology, has long been well-known. He was one of the founders of C.E.R.A., where he soon outstripped his former mentors. He worked there on problems of both theoretical and applied research (the flight and landing of aircraft, stabilisation of satellites, control of large industrial systems) in the course of which he was led to a deep study of the problems of representation and control of multivariable systems. He created and sustained a research group in this area, consisting of engineers from various French institutions, during the years when he presided over the automatic control section of A.F.C.E.T.

Professor Fossard was moreover the founder of the first teaching group on multivariable systems in France, an activity which he established at the Higher National School of Aeronautics as early as 1967. He is currently involved in teaching, both at this School and also at I.N.S.A. and at E.N.S.E.E.H. in Toulouse.

The work of synthesis here presented is the fruit of his labour. It only remains to wish him the success he deserves with the French-speaking scientific and technological community.

Preface to the French edition

Despite the considerable progress made in automatic control during recent decades, at both the theoretical and the practical level, and although the subject is now reaching maturity, the control of multivariable systems (i.e. those with several inputs and outputs) is still not well-developed or widely known. This situation in fact, although understandable, constitutes a serious deficiency, since the control of complex industrial processes, of an essentially multivariable nature, is now opening up a vast field of application for the control engineer.

As with every technique at its beginnings, the methods of multivariable control were conceived somewhat erroneously and gave rise to a number of false ideas, whose consequences are still being felt today. There was a tendency for the engineer to extend to this new case the theories and methods whose efficacy had been demonstrated in single-variable problems, without always applying the necessary effort to the adaptation; in particular, the use of operational methods based on the Laplace transform, and the straightforward transposition of transfer function concepts, brought about many errors, some due to lack of appreciation of the ideas of controllability and observability, others arising because the transfer function matrix gave insufficient indication of the system structure. We have emphasised these difficulties, and given a fairly deep treatment of the fundamental theoretical points whose misunderstanding can lead to an inadequate, and often fallacious, view of such problems.

However, a theory is of interest to the engineer only in so far as it leads to methods of practical application. Consequently, we illustrate each theoretical point by examples and give in an appendix the computer programs required for the automatic implementation of all the techniques presented.

The first chapter is devoted to a review of the properties of the state-space representation of a dynamical system, including the concepts of controllability, observability and reproducibility. Although these ideas are fundamental for the design of multivariable control systems, as will subsequently become clear, the review presented here has been strictly limited. In any case, this chapter should not be regarded as an introduction to state-space methods, which are assumed to be already known, at least for the single-variable case.

The second chapter is concerned with problems of representation, that is to say, the formulation of various mathematical models representing the system to be controlled (state-space equations, transfer function matrix, differential operators). The emphasis is placed on the conditions of validity of these different representations, and the formulation of simple and effective algorithms to pass from one to another. Most of these methods have been implemented at the Centre for Study and Research in Automatic Control by Messrs. Guéguen, Gauvrit and Fossard. With the conceptual bases thus established in the first two chapters, the third applies them to the actual problems of control whose solution forms the subject of the following chapters, concerned with the technical implications of the foregoing concepts and the importance of structural representations adapted to the problems to be solved.

The last two chapters are entirely devoted to the practical solution of control problems, which have been treated under two headings, on account of the dichotomy among the questions which arise and the nature of the methods employed, namely: problems of interactive control (chapter 4) and of non-interactive control (chapter 5). Although the techniques proposed rely mainly on state-space methods (in particular, the algorithms of Luenberger and Gilbert will be utilised), we have endeavoured, whenever possible, to bridge the gap between these and frequency-domain methods, emphasising particularly, in the latter case, the precautions required and the difficulties of realisation.

The book ends with an important appendix where we have gathered together the programs for numerical calculation relating to those methods whose manual implementation can be laborious (decomposition into sub-systems, determination of decoupling compensators, etc).

This work derives, for the most part, from research reports drawn up at the Centre for Study and Research in Automatic Control, under the aegis of the General Delegation for Scientific and Technical Research. It should be within reach of any reader familiar with the ideas of classical control (single-variable) and of state-space representation, for which we refer especially to the study of systems using state-space methods, by J.C. Gille, published in the same series.

In conclusion, we should like to thank all those who have helped us, directly and indirectly, in the preparation of this work, either by their contributions to certain chapters or through the comments which they have made to us, in particular Messrs. M. Gauvrit, M. Clique, N. Imbert, C. Foulard, J.J. Eltgen and E. Toumire. Our gratitude is especially due to Messrs. J.C. Gille, M. Pélegrin and Y. Sevely, who encouraged us to write this book and without whom it would undoubtedly never have seen the light of day.

Translator's note

It has been said that translations, like women, may be either beautiful or faithful but not both. For the present one I claim neither quality, but, as befits a writer on control system design, I have tried to achieve the best possible compromise.

Aside from the linguistic aspect, the present edition differs in some respects from the French original. Thus, chapter 1 has been revised and expanded to include sections on standard forms for state-space matrices and on discrete-time systems, while chapter 2 has been almost completely rewritten although the content is substantially retained. In the course of translation, a number of minor errors were discovered and these have been corrected as far as possible. I can only hope that I have not introduced any more.

Finally, it is my pleasure to thank my colleagues Professor H.H. Rosenbrock and Dr. P.E. Wellstead, through whose mediation and encouragement I came to undertake this task, and especially Mrs. E.R. Brooks, who laboured to make the appearance of the typescript compensate for the infelicities of style and inaccuracies of expression which it undoubtedly contains and for which I, of course, accept full responsibility.

Control Systems Centre,
UMIST,
Manchester, England.
P.A. Cook

Contents

Chapter 1

State-space representation

Although the problems of automatic control were at first treated as problems of rational mechanics, and, as such,were tackled by time-domain methods (being then most often concerned with second-order servo-mechanisms characterised by high inertia and considerable friction), they soon came, particularly under the influence of electrical and electronic engineers, to be treated as filtering problems, i.e. by working in the frequency domain. The developments which were to be made along these lines, associated with the concept of a transfer function, are well known, and it is no doubt unnecessary to mention here all the single-variable methods and techniques which thereby arose, since these methods are still used, almost exclusively, by the majority of design teams.

With a further swing of the pendulum, however, these frequential methods began, some years ago, gradually to give place to a theory of control in state-space, where the system to be controlled is represented by a vector differential equation of first order, in place of an operational transmittance. The advantages of such a representation (1), already

(1) Let us mention in particular its advantage in optimal control and, more generally, in discrete form, its adaptation to the numerical solution of system equations.

revealed within the framework of single-variable systems (i.e. with only one input and one output), were to become still more marked when the increasing complexity of the systems to be controlled, and the necessity of representing them better, led control engineers to consider the case of multivariable systems (i.e. with several inputs and outputs). The use of frequency-domain methods in this type of problem was, indeed, to reveal itself as a source of apparently inexplicable contradictions between the solutions on paper and the experimental results.

As we shall see, such errors are most often due to a lack of appreciation of the ideas of controllability and observability, and one of the essential benefits of state-space methods will be to make this fact clear. It is,indeed, important, once and for all, to convince oneself that these concepts are neither obvious trivialities (as is sometimes the case in single-variable theory) nor useless mathematical speculations (as some believe). On the contrary, at least as regards multivariable systems, we shall see that these notions are fundamental, in practice as well as in theory, both in problems of representation and structure and in control problems.

Since the state-space representation is regarded, in this book, only as a working tool, and the concepts of controllability and observability, for their part, only in respect of their implications, the present chapter will be restricted to a review of only those aspects which are essential for the understanding of the later chapters. The interested reader can refer, if he wishes, to the excellent books recently published on the subject. The only parts somewhat more developed will be the paragraphs concerning certain little-known criteria which are simpler to apply than the general classical ones, assuming that the reader is already familiar with the problems of representation of single-variable systems.

1.1 CONCEPTS AND FORMS OF STATE-SPACE

1.1.1 The notion of state

We know that the properties of a physical process depend on a number of quantities which we can, in general, identify either as input variables u or as output variables y. The relations defining the process, deduced from the laws of physics, are then of the form:

$$G(u, y) = 0 \tag{1.1}$$

We also know that these input-output relations do not, in general, enable us to determine uniquely the behaviour of the system. In particular, for a differential system, the evolution of the output in a time interval $[t_0, t_1]$, denoted by $y(t_0, t_1)$, depends not only on the segment $u(t_0, t_1)$ of the input applied, but also on the initial conditions existing at the moment of application of the input. It is thus important, in order to ensure the indispensable uniqueness of the causal relations $u \to y$, to adjoin to equation (1.1) a vector parameter $x(t_0)$ (1), which we shall call the "state" of the process at the instant t_0 provided that:

1) the equation

$$y(t_0, t) = A\left[x(t_0), u(t_0, t)\right] \tag{1.2}$$

holds for every input-output pair, whatever may be t_0 and $t > t_0$, for all x_0 belonging to the space X (called "state-space");

2) $y(t_0, t)$ is uniquely determined, given x_0 and $u(t_0, t)$;

3) for every pair of segments $[y(t_0, t_1), u(t_0, t_1)]$ satisfying equation (1.2), the segment pair $[y(t_0, t_2), u(t_0, t_2)]$ also satisfies it, for all $t_2 < t_1$;

4) for all input segments $u(t_0, t)$ and $u(t, t_1)$, to which correspond the respective output segments $y(t_0, t)$ and $y(t, t_1)$, we have

$$y(t_0, t, t_1) = A\left[x(t_0), u(t_0, t, t_1)\right].$$

(1) To simplify the notation, x_0 will often be written for $x(t_0)$.

The coherence properties expressed by conditions 2) to 4) reveal the fact that the state of a system constitutes, in a certain sense, a partition between the past and the future, and represents the minimal recollection of the past required for the determination of the future.

The above interpretation of the state of a system in terms of a parametrisation of its input-output relations is not the only one which could be given. It is also possible, in a complementary fashion, to start from a priori concepts of state and transition.

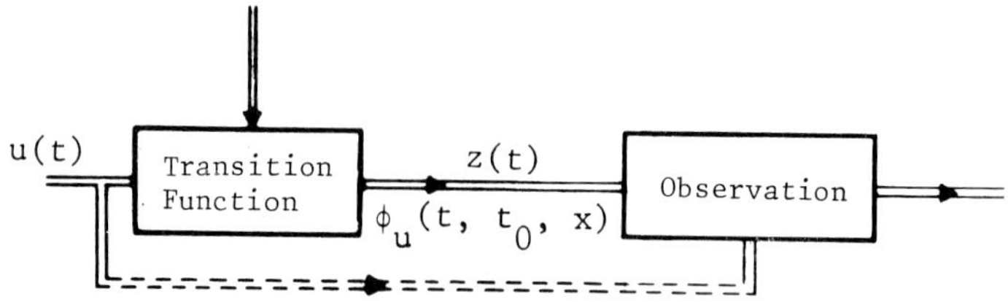

Fig. 1.1 Schematic representation of a system in terms of its transition function.

Here, the inputs enter as the means of operation on the state, and the outputs appear as quantities which depend on it (cf. figure 1.1). We thus have a mathematical definition of a dynamical system in the following axiomatic form:

1) given a space X (called state-space), an ordered set $\tau \subset R$ of time-values t, and a space of real vector functions of time representing the admissible inputs for the system;

2) given $t_0 \in \tau$, $x_0 \in X$ and $u \in \mathcal{U}$, the state of the system is defined, for all $t \geq t_0$, by the transition function

$$x(t) = \phi_u(t, x_0, t_0)$$

such that $\phi_u(t_0, x_0, t_0) = x_0$ for all x_0, t_0, u, belonging to X, τ, $\mathcal{U}$, respectively, and

$$\phi_u(t_2, x_0, t_0) = \phi_u[t_2, \phi_u(t_1, x_0, t_0), t_1],$$

$$\forall\, x_0, t_0, t_1, t_2 \text{ with } t_2 \geq t_1 \geq t_0.$$

3) every output y(t) of the system is given by the value of a real vector function

$$y(t) = \psi(t, x, u)$$

If, in particular, we make the following supplementary hypotheses (which will be retained throughout most of this book):

- X is of finite dimension;
- the time is continuous;
- ψ and ϕ are linear in u and x;
- u and y are of respective dimensions m and p;

it can be shown that the transition function is the solution of a vector differential equation

$$\dot{x} = A(t)\ x(t) + B(t)\ u(t)$$
$$y = C(t)\ x(t) + D(t)\ u(t) \qquad (1.3)$$

where x is a vector of dimension n, and A, B, C, D are real matrices of respective dimensions (n,n), (n,m), (p,n), (p,m), which we shall refer to as the evolution, control, observation, and direct transmission matrices, respectively (cf. figure 1.2). A general solution of equation (1.3) for arbitrary initial conditions x_0 can easily be found. If X(t) is a fundamental solution of the equation

$$\dot{X} = AX \quad \text{with} \quad |X(t_0)| \neq 0,$$

the first equation (1.3) admits a unique solution of the form

$$x(t) = X(t)\ X^{-1}(t_0)\ x(t_0) + \int_{t_0}^{t} X(t)\ X^{-1}(\tau)\ B(\tau)\ u(\tau)\ d\tau$$

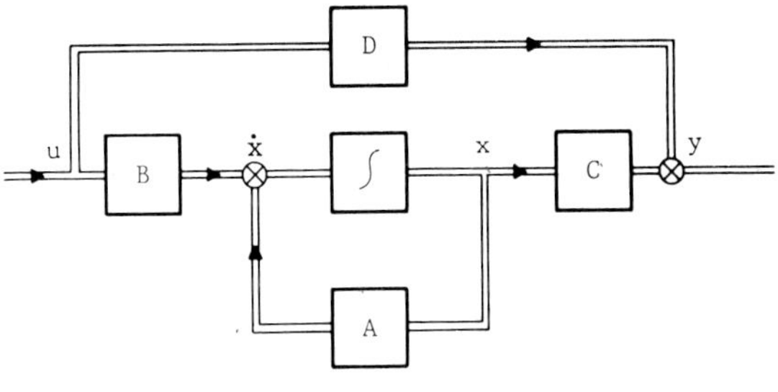

Fig. 1.2 State-space representation of a linear system

If we then define the matrix $\Phi(t,\tau)$ by

$$\Phi(t, \tau) = X(t)\, X^{-1}(\tau), \qquad \Phi(t, t) = I$$

we have

$$x(t) = \Phi(t, t_0)\, x_0 + \int_{t_0}^{t} \Phi(t, \tau)\, B(\tau)\, u(\tau)\, d\tau \qquad (1.4)$$

This last equation defines a transition function satisfying all the preceding axioms.

1.1.2 State-space in continuous time

We note now, a point which will be most important in what follows, that a state-space representation is not unique. In fact, if we have one representation

$$\dot{x} = Ax + Bu, \; y = Cx,$$

we can derive an infinity of them, by making a change of variables $x = M\tilde{x}$ where M is a nonsingular matrix. Under this transformation, equations (1.3) become

$$M\dot{\tilde{x}} = AM\tilde{x} + M^{-1}Bu, \quad y = CM\tilde{x} + Du,$$

i.e.

$$\dot{\tilde{x}} = M^{-1}AM\tilde{x} + M^{-1}Bu, \quad y = CM\tilde{x} + Du,$$

taking the analogous form to (1.3) if we set

$$\tilde{A} = M^{-1}AM$$

$$\tilde{B} = M^{-1}B \qquad (1.5)$$

$$\tilde{C} = CM$$

Hence, we do not speak of <u>the</u> state-space representation of a system, but of <u>a</u> representation, all triples (A,B,C), $(\tilde{A},\tilde{B},\tilde{C})$, related by equations (1.5), being equivalent (1).

(1) In particular, they lead to the same transfer function matrix, $\tilde{Z}(s) = \tilde{C}(sI-\tilde{A})^{-1}\tilde{B} = CM(sI-M^{-1}AM)^{-1}M^{-1}B = CM[M^{-1}(sI-A)M]^{-1}\, M^{-1}B = C(sI-A)^{-1}B = Z(s)$

This multiplicity of forms should not be regarded as a drawback, but rather as an advantage. If certain triples (A,B,C) lend themselves better to calculation, or show more clearly the properties of a given system, it will often be beneficial to use them:

- either, to find directly a state-space representation where the matrices A, B, C have special forms (cf. chapter 2);
- or, if a representation is already available, to find a transformation matrix M which puts $\tilde{A}$, $\tilde{B}$, $\tilde{C}$ into particular simple forms (cf. chapters 3, 4, 5).

We can, in particular, seek special forms for:

- the matrix A alone;
- the matrices A and B simultaneously;
- the matrices A and C simultaneously;
- the set of three matrices, A, B, C.

These special forms of state-space representation, called canonical forms, are easy to obtain in the case of single-input, single-output systems. In the multivariable case, the problem is more delicate and will be treated in chapters 3 and 5.

A - Canonical forms for the evolution matrix

We recall the following mathematical definitions and properties:-

1) An eigenvalue of a matrix A is any root of its characteristic equation

$$\det(\lambda I-A) = 0 \tag{1.6}$$

2) An eigenvector, associated with an eigenvalue λ, is any vector x satisfying the equation

$$(\lambda I-A)x = 0. \tag{1.7}$$

An eigenvector is defined only to within a constant of proportionality; to a simple eigenvalue λ, there corresponds a unique eigenvector, while, to an eigenvalue λ of order ν, there may correspond from 1 to ν eigenvectors.

3) Every matrix satisfies its own characteristic equation (Cayley -

Hamilton Theorem).

Diagonal form - If all the roots of the characteristic equation are distinct, it is possible to put $\tilde{A}$ in diagonal form

$$\tilde{A} = \text{diag}[\lambda_1, \lambda_2, \ldots \lambda_n] \ , \quad \lambda_i \neq \lambda_j \ ;$$

the transformation matrix M is the matrix of eigenvectors.

Example:

$$A = \begin{bmatrix} 0 & 2 & -1 \\ 2 & 0 & -1 \\ 0 & 3 & -2 \end{bmatrix} , \ \det(\lambda I - A) = (\lambda+1)(\lambda+2)(\lambda-1);$$

with

$$M = \begin{bmatrix} 1 & 1 & 1 \\ 1 & 0 & 1 \\ 3 & 2 & 1 \end{bmatrix} , \ \tilde{A} = M^{-1}AM = \text{diag}(-1, -2, 1).$$

Jordan form - In the case of a multiple eigenvalue λ, the form obtained depends on the number of independent eigenvectors which can be associated with λ, that is to say, the rank of the matrix $\lambda I - A$. If ρ is the rank of this matrix, of dimension n, the number of Jordan blocks in which λ will appear is $n-\rho$.

We note, however, that, although the rank of $A-\lambda I$ determines the number of blocks appearing in $\tilde{A}$, we cannot in general, by these means, either specify its form or completely determine the transformation matrix. The simplest approach is to use the equation $AM = M\tilde{A}$, while making assumptions about the structure of $\tilde{A}$.

Examples: Let the matrix A be

$$A = \begin{bmatrix} 6 & 3 & -8 \\ 1 & -2 & 1 \\ 6 & 2 & -7 \end{bmatrix}$$

We have $\det(\lambda I - A) = (\lambda+1)^3$ and

$$\text{rank}(A+I) = \text{rank}\begin{bmatrix} 7 & 3 & -8 \\ 1 & -1 & 1 \\ 6 & 2 & -6 \end{bmatrix} = 2.$$

The three eigenvalues $\lambda = -1$ will appear in a single Jordan block (since $n-\rho = 1$):

$$\tilde{A} = \begin{bmatrix} -1 & 1 & 0 \\ 0 & -1 & 1 \\ 0 & 0 & -1 \end{bmatrix}, \quad M = \begin{bmatrix} 1 & 1 & 0 \\ 3 & 0 & 3 \\ 2 & -2 & 1 \end{bmatrix}.$$

Let us now consider

$$A = \begin{bmatrix} 0 & 0 & 0 & 1 \\ 0 & 0 & 1 & 0 \\ 0 & 0 & 0 & 1 \\ 0 & 0 & 0 & 0 \end{bmatrix}; \ \det(\lambda I - A) = \lambda^4.$$

There is a zero eigenvalue of order 4, and rank $(A+\lambda I)$ = rank $A = 2$. Hence, we are certain to make two Jordan blocks appear, but, a priori, we may have:

- either a) one block of order 1 and one of order 3;

- or b) two blocks of order 2:

$$\text{a)} \begin{bmatrix} 0 & 0 & 0 & 0 \\ 0 & 0 & 1 & 0 \\ 0 & 0 & 0 & 1 \\ 0 & 0 & 0 & 0 \end{bmatrix} \qquad \text{b)} \begin{bmatrix} 0 & 1 & 0 & 0 \\ 0 & 0 & 0 & 0 \\ 0 & 0 & 0 & 1 \\ 0 & 0 & 0 & 0 \end{bmatrix}.$$

Using the matrix equation $AM = M\tilde{A}$, we find that form a) is the one required, and that

$$M = \begin{bmatrix} 1 & 0 & 1 & 1 \\ 0 & 1 & 1 & 0 \\ 0 & 0 & 1 & 1 \\ 0 & 0 & 0 & 1 \end{bmatrix}.$$

If we had taken

$$A = \begin{bmatrix} 0 & 1 & -1 & 2 \\ 0 & 0 & 0 & 1 \\ 0 & 0 & 0 & 1 \\ 0 & 0 & 0 & 0 \end{bmatrix},$$

which has the same characteristic equation and the same rank as the previous one, we should, on the other hand, have ended up with form b) (using the same transformation matrix M).

<u>Companion form</u> - Another canonical form, which will be frequently used in what follows, is the "companion" form. This term should be understood in the sense of "companion to the characteristic equation". In fact, the coefficients which appear in the last row (or in the first column, depending on the form chosen) of a companion matrix are, within an overall sign, the coefficients of the characteristic equation:

$$A_1 = \begin{bmatrix} 0 & 1 & 0 & \cdots & 0 \\ \vdots & & \ddots & \ddots & \vdots \\ \vdots & & & \ddots & 0 \\ 0 & \cdots & \cdots & 0 & 1 \\ -a_0 & -a_1 & \cdots & \cdots & -a_{n-1} \end{bmatrix}, \quad A_2 = \begin{bmatrix} -a_{n-1} & 1 & 0 & \cdots & 0 \\ \vdots & 0 & \ddots & \ddots & 0 \\ -a_1 & \vdots & & \ddots & 1 \\ -a_0 & 0 & \cdots & \cdots & 0 \end{bmatrix} \tag{1.8}$$

where

$$\det(sI-A_1) = \det(sI-A_2) = s^n+a_{n-1}s^{n-1}+\ldots+a_1s+a_0 = \psi(s).$$

<u>Note</u>: These matrices have other advantageous properties which will be utilised later and which concern the determination of the inverse matrix $(sI-A)^{-1}$ (cf. paragraph 2.1.3.A).

Let us first consider the case where we wish to put an arbitrary matrix A into the horizontal companion form A_1, by a transformation M. The equation $M^{-1}AM = A_1$, rewritten in the form $AM = MA_1$, is solved, column by column, by exhibiting the columns of M in the form

$$M = [{}^1M, {}^2M, \ldots {}^nM].$$

We have, in fact,

$$[{}^1M,\ldots{}^nM]\begin{bmatrix}0 & 1 & 0 & \cdots & 0\\ \vdots & & \ddots & & 0\\ & & & \ddots & \\ 0 & \cdots & & 0 & 1\\ -a_0 & \cdots & & \cdots & -a_{n-1}\end{bmatrix} = A[{}^1M,\ldots{}^nM] \tag{1.9}$$

i.e., by equating columns, starting with the last one,

$$\begin{aligned} {}^{n-1}M - a_{n-1}{}^nM &= A\,{}^nM \\ &\vdots \\ {}^1M - a_1\,{}^nM &= A\,{}^2M \end{aligned}$$

whence

$$\begin{aligned} {}^{n-1}M &= (A+a_{n-1}I)\,{}^nM \\ &\vdots \\ {}^1M &= (A^{n-1}+a_{n-1}A^{n-2}+\ldots+a_1I)\,{}^nM \end{aligned} \tag{1.10}$$

The first n-1 columns of M are thus determined in terms of the last one, which can be chosen arbitrarily.

Note 1: The equation corresponding to the first column on each side of (1.9) is automatically satisfied because of the Cayley-Hamilton Theorem.

Note 2: It is possible to utilise the remaining degree of freedom in the choice of M in order, at the same time as A is put into companion form, to put B in a special form. In particular, in the case of a single-input system, where B is a column vector, if we take ${}^nM = B$, the matrix $\tilde{B}$, which is the transform of B, will have the form $[0,\ldots 0,1]^T$. Indeed, from $M^{-1}B = \tilde{B}$, we have (1)

$$B = M\tilde{B} = [{}^1M,\ldots{}^nM]\begin{bmatrix}0\\ \vdots\\ 0\\ 1\end{bmatrix}.$$

(1) As we shall see later, since the matrix M has to be nonsingular, this can in fact only be done if (A,B) is controllable.

In the case where the companion form sought is of the vertical type (coefficients of the characteristic equation in the first column), it is preferable to calculate, not M, but the matrix $(M^{-1})^T = P$. From the equation $M^{-1}AM = A_2$, we have in fact $M^{-1}A = A_2M^{-1}$, whence, after transposition,

$$A^TP = PA_2^T . \tag{1.11}$$

Displaying as before the columns of P in the form $[{}^1P, \ldots {}^nP]$, we have

$$A^T[{}^1P, \ldots {}^nP] = [{}^1P, \ldots {}^nP] \begin{bmatrix} -a_{n-1} & \cdots & \cdots & -a_0 \\ 1 & 0 & \cdots & 0 \\ 0 & \ddots & & \vdots \\ 0 & \cdots 0 & 1 & 0 \end{bmatrix}$$

and hence, equating columns, starting with the first,

$$\begin{aligned} {}^2P &= (A^T + a_{n-1}I)\,{}^1P \\ &\vdots \\ {}^nP &= [(A^T)^{n-1} + a_{n-1}(A^T)^{n-2} + \ldots + a_1I]\,{}^1P \end{aligned} \tag{1.12}$$

The last n-1 columns of P are thus expressed in terms of the first one, which can be chosen arbitrarily (with the proviso that P should be invertible).

Note: The same remarks as were made previously remain valid in this case. In particular, if, instead of taking ${}^1P$ arbitrarily, we take ${}^1P = C^T$, in the case of a single-output system, the transformation $x = M\tilde{x}$, with $M = (P^{-1})^T$, simultaneously takes A into the form $\tilde{A} = A_2$ and C to the form $\tilde{C} = [1, 0, \ldots 0]$.(1)

(1) As we shall see, this is possible only if the system (A,C) is observable.

B - Canonical forms for the control and observation matrices

In the case of systems with several inputs and outputs, the problem of putting the matrices A, (A,B) or (A,C) into canonical forms is more delicate. In particular, in the latter two cases, the matrix $\tilde{A}$ will appear in block diagonal form with companion blocks, the number of which, moreover, is not fixed a priori. This problem will be treated in chapter 3, where we shall exhibit forms such as:

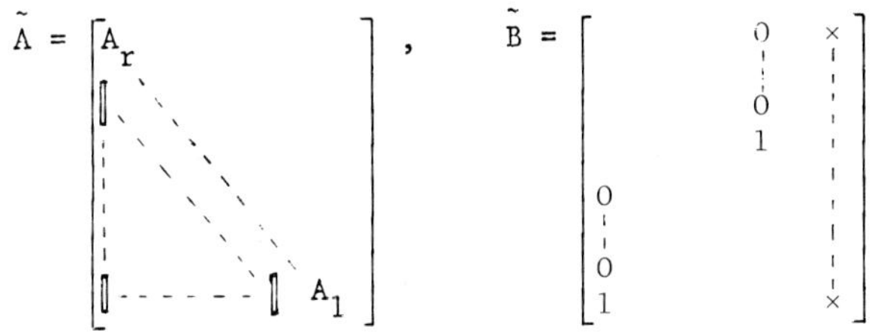

$$\tilde{A} = \begin{bmatrix} A_r & & \\ & \ddots & \\ & & A_1 \end{bmatrix}, \quad \tilde{B} = \begin{bmatrix} & 0 & \times \\ & \vdots & \\ & 0 & \\ & 1 & \\ 0 & & \\ \vdots & & \\ 0 & & \\ 1 & & \times \end{bmatrix}$$

or

$$\tilde{A} = \begin{bmatrix} A_1^* & \cdots & \\ & \ddots & \\ & & A_m^* \end{bmatrix}, \quad \tilde{B} = \begin{bmatrix} 0 & 0 & 0 \\ \vdots & \vdots & \\ 0 & & \\ 1 & 0 & \\ \times & 1 & \\ & & 0 \\ & & \vdots \\ \times & \cdots \times & 1 \end{bmatrix}$$

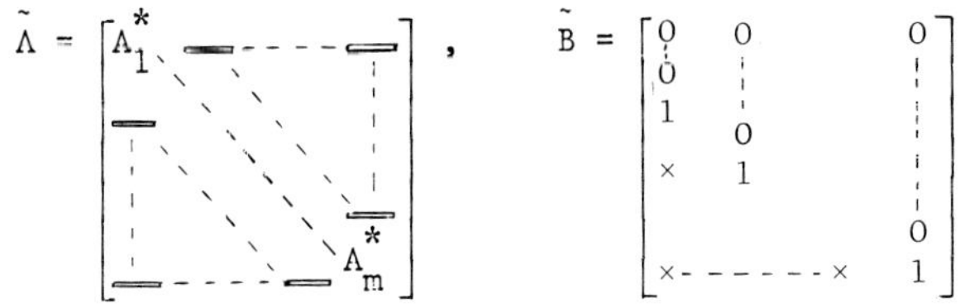

where A_i, A_i^* are horizontal companion blocks, of dimensions (n_i, n_i), and $\Longrightarrow$, $\mathbf{|}$ denote respectively a row and column of dimension n_i.

1.1.3 State-space in discrete time. Discretisation of the continuous state-equations

The preceding, continuous-time, state-space models are perfectly adapted to the simulation of continuous systems on analogue computers (in fact, it seems quite convincing that all analogue simulations should be carried out on the basis of a state-space model). Very often, however, and for a variety of reasons, we are led to simulate a continuous system on a numerical calculator, which requires the evolution equations to be set up in discrete form.

The problem of discretising the continuous-time state-equations

$$\dot{x} = Ax + Bu, \quad y = Cx,$$

consists of finding a discrete-time model

$$x_{k+1} = Mx_k + Nu_k, \quad y_k = Cx_k, \tag{1.13}$$

such that, if a discrete input u_k is applied to it, the state-vector x_k will take, at each sampling instant, the same values as would the continuous state-vector x if the system were submitted to the same input through the intermediary of a hold element (cf. figure 1.3).

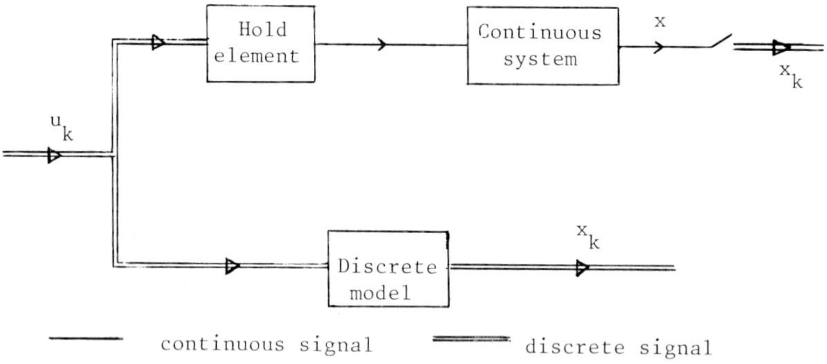

Fig. 1.3 Discretisation of a continuous system

The problem is thus to find how M and N are related to A and B. This question can be resolved in various ways, either by discrete integration of the continuous state-equations or by utilising the properties of Laplace transforms or z-transforms. We remark that x_k represents the state-vector at the instant kT and x_{k+1} at (k+1)T.

A - Discrete integration of the continuous equations

We know that the integral solution of the system (1.3) for initial conditions x_0 is, in general, given by

$$x(t) = \Phi(t, t_0)x_0 + \int_{t_0}^{t} \Phi(t,\tau)B(\tau)u(\tau)d\tau \tag{1.4}$$

where $\Phi(t,\tau)$ is the transition matrix which, for time-invariant systems, is the exponential function $e^{A(t-\tau)}$. If we agree to take $t_0 = 0$ (which makes no difference to the generality of the problem and simplifes the expressions), we thus have, at the instants kT and (k+1)T,

$$x_k = x(kT) = e^{AkT}x_0 + \int_0^{kT} e^{A(kT-\tau)}Bu(\tau)d\tau$$

and

$$x_{k+1} = e^{A(k+1)T}x_0 + \int_0^{(k+1)T} e^{A[(k+1)T-\tau]}Bu(\tau)d\tau$$

so that

$$x_{k+1} = e^{AT}\, e^{AkT}x_0 + \int_0^{(k+1)T} e^{AT}\, e^{A(kT-\tau)}Bu(\tau)d\tau$$

$$= e^{AT}\left[e^{AkT}\, x_0 + \int_0^{kT} e^{A(kT-\tau)}Bu(\tau)d\tau + \int_{kT}^{(k+1)T} e^{A(kT-\tau)}Bu(\tau)d\tau\right].$$

We recognise in the first two terms inside the brackets, the expression for x_k. It follows that

$$x_{k+1} = e^{AT}\, x_k + e^{AT}\int_{kT}^{(k+1)T} e^{A(kT-\tau)}Bu(\tau)d\tau.$$

By identification with the required form

$$x_{k+1} = Mx_k + Nu_k$$

we thus deduce that

$$M = e^{AT} \tag{1.14}$$

and

$$Nu_k = e^{AT}\int_{kT}^{(k+1)T} e^{A(kT-\tau)}Bu(\tau)d\tau.$$

Although the expression for M can thus be calculated independently of any hypothesis about the nature of the hold element used, we see by contrast that the determination of N requires an integration involving assumptions about the hold element . We shall consider here the case of a zero-order hold.

If a zero-order hold element is used, the control $u(t)$ during the interval $[kT, (k+1)T]$ will be constant, equal to the value $u(kT)$ taken at the beginning of the sampling period. We have then

$$Nu_k = \int_{kT}^{(k+1)T} e^{A[(k+1)T-\tau]} Bu_k d\tau$$

and so

$$N = \int_{kT}^{(k+1)T} e^{A[(k+1)T-\tau]} d\tau B.$$

The calculation of this integral is easily performed by making the change of variable $v = (k+1)T-\tau$, so that

$$N = \int_0^T e^{Av} dv\ B. \tag{1.15}$$

Note 1: If the matrix A is nonsingular, N can be written explicitly

$$N = A^{-1}(e^{AT}-I)B. \tag{1.16}$$

We have generally, in fact,

$$A\int_0^T e^{Av} dv = e^{AT}-I.$$

Note 2: The calculation of N can also be performed by application of Sylvester's Theorem. If all the eigenvalues are distinct and nonzero, we have

$$N = \sum_{i=1}^{n} \left[\left(\frac{e^{\lambda_i T}-1}{\lambda_i}\right) \prod_{j\neq i} \left(\frac{A-\lambda_j I}{\lambda_i-\lambda_j}\right)\right] B.$$

The determination of the discrete model thus requires the calculation of e^{AT}. Several methods can be utilised.

i) Calculation of the series

$$e^{AT} = I + AT + \frac{A^2T^2}{2!} + \dots + \frac{A^nT^n}{n!} + \dots$$

This method is not, in general, convenient for practical application and has, in particular, the disadvantage of not usually admitting a term-by-term interpretation.

ii) Special case where the matrix A is diagonal. The only simple case is that where A is diagonal. Indeed, in this case, e^{AT} is also diagonal and it follows from the preceding paragraph that, if

$$A = \text{diag}(\lambda_1, \lambda_2, \dots \lambda_n),$$

then

$$e^{AT} = \text{diag}\left[e^{\lambda_1 T}, e^{\lambda_2 T}, \dots e^{\lambda_n T}\right].$$

The simplicity of this case encourages us to reduce other cases to it by a change of basis. In fact, if A is not diagonal and M is its diagonalising matrix (matrix of eigenvectors), we have

$$\tilde{A} = M^{-1}AM, \; A = M\tilde{A}M^{-1}$$

and

$$e^{AT} = Me^{\tilde{A}T}M^{-1}.$$

iii) Application of Sylvester's formula. The application of Sylvester's theorem to the calculation of the matrix function e^{AT} can also be envisaged. In particular, if all the eigenvalues are distinct, we have

$$e^{AT} = \sum_{i=1}^{n} \left[e^{\lambda_i T} \prod_{j \neq i} \left(\frac{A-\lambda_j I}{\lambda_i-\lambda_j}\right)\right].$$

Example: Suppose we are seeking a discrete state-space model corresponding to the transfer function

$$F(s) = \frac{s+4}{(s+1)(s+2)}.$$

We note, to begin with, that the poles of F(s) are simple and real, and hence, in view of our previous remarks, we shall use a continuous state-space representation in which A has diagonal form. The calculation of e^{AT} is then immediate. We have

$$\begin{bmatrix} \dot{x}_1 \\ \dot{x}_2 \end{bmatrix} = \begin{bmatrix} -1 & 0 \\ 0 & -2 \end{bmatrix} \begin{bmatrix} x_1 \\ x_2 \end{bmatrix} + \begin{bmatrix} 1 \\ 1 \end{bmatrix} u,$$

$$y = [3, -2] \begin{bmatrix} x_1 \\ x_2 \end{bmatrix},$$

whence

$$M = e^{AT} = \begin{bmatrix} e^{-T} & 0 \\ 0 & e^{-2T} \end{bmatrix}, \quad N = A^{-1}(e^{AT}-I)B = \begin{bmatrix} 1-e^{-T} \\ \frac{1}{2}(1-e^{-2T}) \end{bmatrix},$$

giving the discrete-time representation

$$x_{k+1} = \begin{bmatrix} e^{-T} & 0 \\ 0 & e^{-2T} \end{bmatrix} x_k + \begin{bmatrix} 1-e^{-T} \\ \frac{1}{2}(1-e^{-2T}) \end{bmatrix} u_k,$$

$$y_k = [3, -2] x_k .$$

This method is by far the simplest. By way of illustration, let us suppose that the same transfer function had been associated with a different representation, for example, with the companion form

$$\begin{bmatrix} \dot{x}_1 \\ \dot{x}_2 \end{bmatrix} = \begin{bmatrix} -3 & 1 \\ -2 & 0 \end{bmatrix} \begin{bmatrix} x_1 \\ x_2 \end{bmatrix} + \begin{bmatrix} 1 \\ 4 \end{bmatrix} u,$$

$$y = [1, 0] \begin{bmatrix} x_1 \\ x_2 \end{bmatrix}.$$

The calculation of e^{AT} could then be conducted:-

- <u>By development in series</u>

$$e^{AT} = I + AT + \frac{A^2T^2}{2!} + \ldots$$

$$= \begin{bmatrix} 1 & 0 \\ 0 & 1 \end{bmatrix} + \begin{bmatrix} -3 & 1 \\ -2 & 0 \end{bmatrix} T + \begin{bmatrix} 7 & -3 \\ 6 & -2 \end{bmatrix} \frac{T^2}{2!} + \begin{bmatrix} -15 & 7 \\ -14 & 6 \end{bmatrix} \frac{T^3}{3!} + \ldots$$

The series development of the entry in position (1,1) of e^{AT} thus begins

$$1 - 3T + \frac{7T^2}{2} - \frac{15T^3}{6} + \frac{31T^4}{24} + \ldots$$

If this series is calculated numerically, it is by no means obvious that it corresponds to $2e^{-2T} - e^{-T}$!

- <u>By a diagonalising transformation</u>

The transformation matrix M which converts the companion matrix A into diagonal form $\tilde{A}$ is

$$M = \begin{bmatrix} 1 & 1 \\ 2 & 1 \end{bmatrix}, \text{ whence } M^{-1} = \begin{bmatrix} -1 & 1 \\ 2 & -1 \end{bmatrix}.$$

We thus have

$$e^{AT} = Me^{\tilde{A}T} M^{-1} = \begin{bmatrix} 1 & 1 \\ 2 & 1 \end{bmatrix} \begin{bmatrix} e^{-T} & 0 \\ 0 & e^{-2T} \end{bmatrix} \begin{bmatrix} -1 & 1 \\ 2 & -1 \end{bmatrix}$$

$$= \begin{bmatrix} 2e^{-2T} - e^{-T} & e^{-T} - e^{-2T} \\ 2e^{-2T} - 2e^{-T} & 2e^{-T} - e^{-2T} \end{bmatrix}.$$

- <u>By the use of Sylvester's formula</u>

We then have, since the eigenvalues are −1 and −2,

$$e^{AT} = e^{-T} \left(\frac{A+2I}{-1-(-2)} \right) + e^{-2T} \left(\frac{A+I}{-2-(-1)} \right)$$

$$= e^{-T} \begin{bmatrix} -1 & 1 \\ -2 & 2 \end{bmatrix} + e^{-2T} \begin{bmatrix} 2 & -1 \\ 2 & -1 \end{bmatrix}.$$

We evidently recover the previous result.

B - Utilisation of Laplace transforms

One of the essential drawbacks of the above method is the need to calculate the function e^{AT}. It is posisble to avoid this by using Laplace transformations, whose utility in the integration of differential equations is well known.

If we take the time origin at the instant kT, the Laplace transform of the continuous equation

$$\dot{x} = Ax + Bu$$

is

$$X(s) = (sI-A)^{-1}\left[x_k + \frac{Bu_k}{s}\right]$$

where x_k is the initial condition for x (at the instant kT) and u_k/s is the transform of the step input of magnitude u_k (cf. figure 1.4).

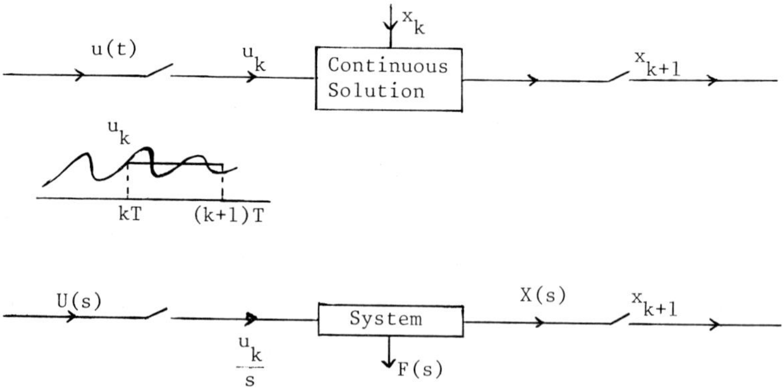

Fig. 1.4 Laplace transformation of continuous system

We have then

$$x(t) = {}^{-1}X(s) = {}^{-1}(sI-A)^{-1}x_k + {}^{-1}(sI-A)^{-1}\frac{Bu_k}{s}$$

$$= M(t)x_k + N(t)u_k.$$

In particular, for $t = T$ [i.e. at the instant $(k+1)T$], we have

$$x_{k+1} = M(T)x_k + N(T)u_k,$$

whence

$$\left.\begin{aligned} M(T) &= {}^{-1}(sI-A)^{-1}\Big|_{t=T}, \\ N(T) &= {}^{-1}(sI-A)^{-1}\frac{B}{s}\Big|_{t=T}. \end{aligned}\right. \qquad (1.17)$$

Example: Let us take again the previous example,

$$\begin{bmatrix} \dot{x}_1 \\ \dot{x}_2 \end{bmatrix} = \begin{bmatrix} -3 & 1 \\ -2 & 0 \end{bmatrix} \begin{bmatrix} x_1 \\ x_2 \end{bmatrix} + \begin{bmatrix} 1 \\ 4 \end{bmatrix} u, \quad y = \begin{bmatrix} 1, & 0 \end{bmatrix} \begin{bmatrix} x_1 \\ x_2 \end{bmatrix}.$$

We have

$$(sI-A)^{-1} = \frac{1}{(s+1)(s+2)} \begin{bmatrix} s & 1 \\ -2 & s+3 \end{bmatrix}, \quad (sI-A)^{-1}\frac{B}{s} = \frac{1}{s(s+1)(s+2)} \begin{bmatrix} s+4 \\ 4s+10 \end{bmatrix},$$

i.e.

$$(sI-A)^{-1} = \begin{bmatrix} -\frac{1}{s+1} + \frac{2}{s+2}, & \frac{1}{s+1} - \frac{1}{s+2} \\ -\frac{2}{s+1} + \frac{2}{s+2}, & \frac{2}{s+1} - \frac{1}{s+2} \end{bmatrix}, \quad (sI-A)^{-1}\frac{B}{s} = \begin{bmatrix} \frac{2}{s} - \frac{3}{s+1} + \frac{1}{s+2} \\ \frac{5}{s} - \frac{6}{s+1} + \frac{1}{s+2} \end{bmatrix},$$

whence

$$M = \begin{bmatrix} -e^{-T} + 2e^{-2T}, & e^{-T} - e^{-2T} \\ -2e^{-T} + 2e^{-2T}, & 2e^{-T} - e^{-2T} \end{bmatrix}, \quad N = \begin{bmatrix} 2 - 3e^{-T} + e^{-2T} \\ 5 - 6e^{-T} + e^{-2T} \end{bmatrix},$$

and

$$x_{k+1} = Mx_k + Nu_k,$$

$$y_k = [1,\ 0]\ x_k.$$

Note: It will be noticed that this method requires us to calculate the matrices $(sI-A)^{-1}$ and $(sI-A)^{-1}B/s$, and then to decompose them into partial fractions so as to return to the corresponding time-functions. The advantage of this method thus depends on the ease with which these operations can be performed [the use of Leverrier's algorithm for the calculation of $(sI-A)^{-1}$, decomposition routines, tables of transforms, etc.].

C - Utilisation of z-transforms

We may also be able to make use of z-transforms, since we are working in discrete time. The calculations to be performed can be deduced directly from the diagram of figure 1.5.

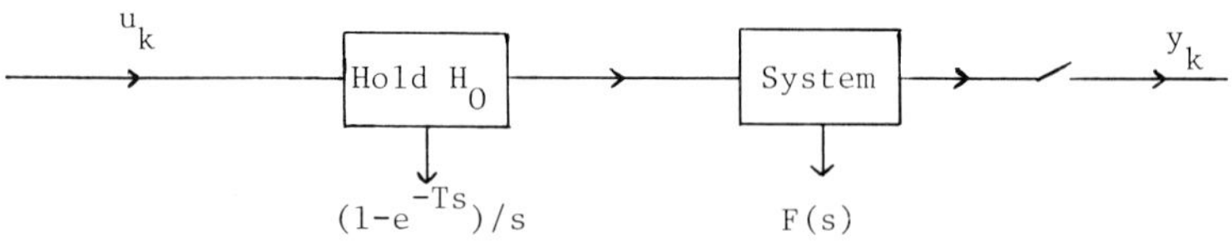

Fig. 1.5 Calculation of z-domain transfer function.

The transfer function, in the z-domain, between u_k and y_k, is

$$\left[\frac{1-e^{-Ts}}{s}\right] F(s) = (1-z^{-1})\ \frac{F(s)}{s}$$

which corresponds to a recurrence relation (the counterpart of a differential equation in continuous time) between y_{k+1}, y_k, y_{k-1},... and u_k, u_{k-1},... . Putting this relation in state-space form then gives the desired discrete-time model.

Example: Let us retain the same example with

$$F(s) = \frac{s+4}{(s+1)(s+2)} \quad .$$

We have

$$\left(\frac{1 - e^{-Ts}}{s}\right) \frac{(s+4)}{(s+1)(s+2)} = (1 - z^{-1}) \left(\frac{2}{s} - \frac{3}{s+1} + \frac{1}{s+2}\right)$$

$$= (1 - z^{-1}) \left[\frac{2}{1-z^{-1}} - \frac{3}{1-e^{-T}z^{-1}} + \frac{1}{1-e^{-2T}z^{-1}} \right]$$

whence, after regrouping the terms, we get the transfer function

$$\frac{Y(z)}{U(z)} = \frac{z^{-1}(2-3e^{-T}+e^{-2T}) + z^{-2}(e^{-T}-3e^{-2T}+2e^{-3T})}{1-z^{-1}(e^{-T}+e^{-2T}) + z^{-2}e^{-3T}} \, ,$$

corresponding to the recurrence relation

$$y_{k+1}-(e^{-T}+e^{-2T})y_k+e^{-3T}y_{k-1} = (2-3e^{-T}+e^{-2T})u_k+(e^{-T}-3e^{-2T}+2e^{-3T})u_{k-1}$$

which has the state-space representation

$$\begin{bmatrix} x_1 \\ x_2 \end{bmatrix}_{k+1} = \begin{bmatrix} e^{-T}+e^{-2T} & 1 \\ -e^{-3T} & 0 \end{bmatrix} \begin{bmatrix} x_1 \\ x_2 \end{bmatrix}_k + \begin{bmatrix} 2-3e^{-T}+e^{-2T} \\ e^{-T}-3e^{-2T}+2e^{-3T} \end{bmatrix} u_k ,$$

$$y_k = [1, \ 0] \begin{bmatrix} x_1 \\ x_2 \end{bmatrix}_k \quad .$$

1.2 CONTROLLABILITY AND OBSERVABILITY

1.2.1 Introduction to the concepts of controllability and observability

If we regard a state-space representation as a parametrisation of input-output relations, we may ask if the choice of parameters, i.e. state variables, is appropriate. Since the inputs are the means available for acting on the state of the system, and the outputs are the quantities via which the state is observed, two basic questions come to mind:

- can we, by acting on the inputs, transfer the system from an arbitrary state $x(t_0)$ to another arbitrary state $x(t_1)$?
- can we, by observing the outputs over a sufficiently long time-intercal $[t_0, t_1]$, determine the initial state $x(t_0)$ of the system?

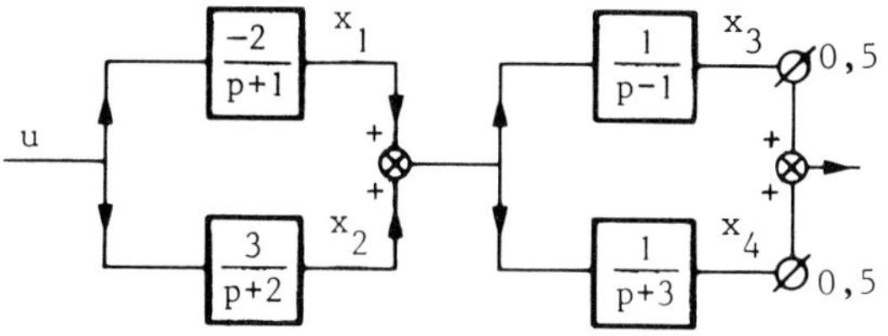

Fig. 1.6 Example of an uncontrollable and unobservable system.

Let us consider for example the case of the single-variable system illustrated in figure 1.6, and let us take the state-variables $x_1, \dots x_4$ indicated. We then have the equations:

$$\begin{bmatrix} \dot{x}_1 \\ \dot{x}_2 \\ \dot{x}_3 \\ \dot{x}_4 \end{bmatrix} = \begin{bmatrix} -1 & & & \\ 0 & -2 & & \\ 1 & 1 & 1 & \\ 1 & 1 & 0 & -3 \end{bmatrix} \begin{bmatrix} x_1 \\ x_2 \\ x_3 \\ x_4 \end{bmatrix} + \begin{bmatrix} -2 \\ 3 \\ 0 \\ 0 \end{bmatrix} u$$

$$[y] = [0 \quad 0 \quad 0{,}5 \quad 0{,}5] \, [x_1 \quad x_2 \quad x_3 \quad x_4]^T$$

Making the change of variables

$$\begin{aligned} x_1 &= 2\,\tilde{x}_2 \\ x_2 &= 3\,\tilde{x}_3 \\ x_3 &= \tilde{x}_1 - \tilde{x}_2 - \tilde{x}_3 \\ x_4 &= \tilde{x}_2 + 3\,\tilde{x}_3 + \tilde{x}_4 \end{aligned}$$

the preceding equations become

$$\dot{\tilde{x}}_1 = \tilde{x}_1$$

$$\dot{\tilde{x}}_2 = -\tilde{x}_2 - u$$

$$\dot{\tilde{x}}_3 = -2\tilde{x}_3 + u$$

$$\dot{\tilde{x}}_4 = -3\tilde{x}_4 - 2u$$

$$y = \tilde{x}_1 + \tilde{x}_3 + 0{,}5\,\tilde{x}_4$$

which are represented diagrammatically in figure 1.7.

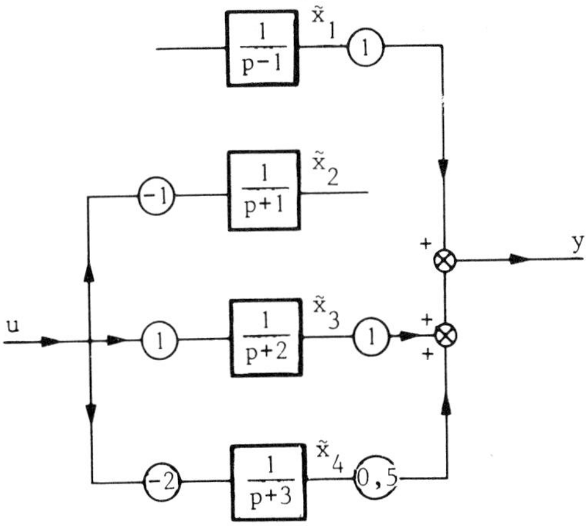

Fig. 1.7 Manifestation of the uncontrollable and unobservable modes.

In this form, it is apparent that:

- $\tilde{x}_1$ is not connected to the input and its evolution cannot be affected by it. Such a state will be said to be uncontrollable.

- $\tilde{x}_2$ is not connected to the output. From observation of the output, it will thus not be possible to determine $\tilde{x}_2$. Such a state will be said to be unobservable.

1.2.2 Definitions and mathematical review

A - Definitions and theorems concerning controllability

A phase $[x_0, t_0]$ will be said to be <u>controllable</u> if there can be found a final time t_1, later than t_0, and an input u(t) during the interval $[t_0, t_1]$, which transfers the phase $[x_0, t_0]$ to $[0, t_1]$.

If every state x belonging to X gives a controllable phase at time t_0, we shall say that the system is "completely controllable at the instant t_0". If the system is completely controllable in the above sense whatever t_0 may be, we shall say that it is "completely controllable".

In terms of the transition matrix, it follows from equation (1.4) that a phase $[x_0, t_0]$ is controllable if and only if there exists a function u(t) and a time $t_1 > t_0$ such that

$$x_0 = -\int_{t_0}^{t_1} \Phi(t_0, t)\, B(t)\, u(t)\, dt \qquad (1.18)$$

We recall below, without proof, the fundamental theorems of which we shall make use in the remainder of this book (1):-

1) The set of states controllable at an instant t_0 forms a linear subspace of X, which will be denoted by $C(t_0)$.

2) If x_0 and x_1 belong to $C(t_0)$, it is possible, by starting from each of these initial states, to reach the same state $x \in X$ at the instant t.

3) If x_0 does not belong to $C(t_0)$, the whole evolution starting from this phase remains outside C(t). On the other hand, if x_0 belongs to $C(t_0)$, it can always be transferred to an arbitrary state $x \in C(t)$ for any t later than t_0. This theorem expresses the fact that an uncontrollable state can never become controllable, but that it is always possible, starting from a controllable state, to reach another arbitrary controllable state.

4) A necessary and sufficient condition that $[x_0, t_0]$ can be transferred to $[x_1, t_1]$ is that

(1) For the proofs of these theorems, see in particular refs. 1.8 and 1.10 to 1.12.

$$\Phi(t_0, t_1) - x_0 \in \mathcal{R}\left[W(t_0, t_1)\right] \qquad (1.19)$$

where (W) is the range-space of the transformation (1)

$$W(t_0, t_1) = \int_{t_0}^{t_1} \Phi(t_0, t)\, B(t)\, B^T(t)\, \Phi^T(t_0, t)\, dt \,. \qquad (1.20)$$

B - Decomposition of a system (2)

Let us consider the following decomposition of a system

$$X = X_a(t) \oplus X_b(t)$$

where the subspaces X_a and X_b are defined by

$$X_a = \Phi(t, t_0)\, \mathcal{R}\left[W(t_0, t_1(t_0))\right]$$
$$X_b = \Phi^T(t_0, t)\, \mathcal{N}\left[W(t_0, t_1(t_0))\right]$$

The orthogonality of this decomposition enables us to decompose every state x in X into two components x_a and x_b, such that

$$x = x_a \oplus x_b \,.$$

It can then be shown that equations (1.3), with D = 0, take the form

$$\begin{bmatrix} \dot{x}_a \\ \dot{x}_b \end{bmatrix} = \begin{bmatrix} A_{11} & A_{12} \\ 0 & A_{22} \end{bmatrix} \begin{bmatrix} x_a \\ x_b \end{bmatrix} + \begin{bmatrix} B_1 \\ 0 \end{bmatrix} u \qquad (1.21)$$

$$y = [C_1, C_2] \begin{bmatrix} x_a \\ x_b \end{bmatrix}$$

(1) The nullspace of this transformation will subsequently be denoted by $[W(t_0, t_1)]$.

(2) We are concerned here with decomposition in the sense of Kalman.

It can be proved (cf. refs. 1.8, 1.12) that, if A(t) and B(t) are analytic functions of time, then so are $\Phi(t_0, t)$ and $W(t_0, t)$, and the dimension of the controllable subspace C(t) is constant, i.e. independent of time. The decomposition into subspaces is thus fixed and can be represented diagrammatically as in figure 1.8.

The system will then be completely controllable if and only if a time t_1 can be found such that the n rows of the matrix

$$\Phi(t_0, t)\, B(t)$$

are linearly independent functions of time on the interval $[t_0, t_1]$.

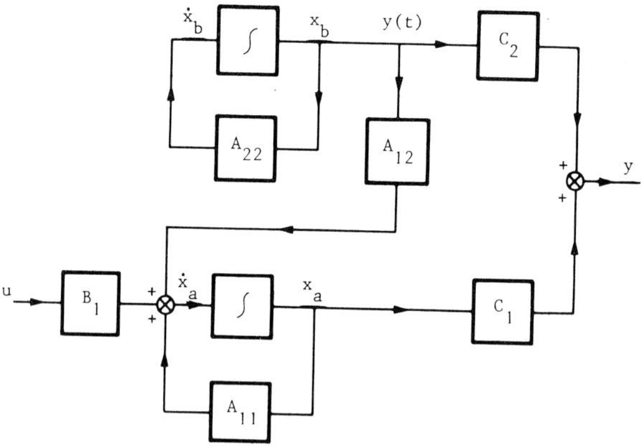

Fig. 1.8 Decomposition of a system with regard to controllability.

In particular, for constant systems, where A and B are time-invariant, a necessary and sufficient condition for controllability is that the matrix (1)

$$Q = [B, AB, \dots A^{n-1}B] \tag{1.22}$$

should have rank n.

(1) This controllability criterion is often known as Kalman's Criterion.

Note: This criterion is not "minimal". More often than not, it will turn out that the matrix

$$Q_c = [B,\ AB, \ldots\ A^{\nu-1}B]$$

is of rank n for some ν less than n. The smallest such ν will then be called the controllability index.

C - Observability and duality

The notion of observability, introduced in an intuitive fashion in subsection 1.2.1, can, just like that of controllability, be specified more rigorously.

We shall say that a phase $[x_0, t_0]$ is observable if the initial state x_0 can be determined from a knowledge of the output y(t) and the input u(t) over an interval $[t_0, t_1]$.

In the same way as in paragraph A, we shall say that the system is completely observable at t_0 if every phase $[x_0, t_0]$ is itself observable.

It can be shown that the system is completely observable at t_0 if and only if a time t_1 can be found such that the n columns of the matrix

$$C(t)\,\Phi\,(t, t_0)$$

are linearly independent on the interval $[t_0, t_1]$.

The evident analogy between this theorem and that of paragraph B is connected with the idea of duality. We shall call two systems dual if they are defined respectively by the equations

$$S : \begin{cases} \dot{x} = A(t)\ x(t) + B(t)\ u(t) \\ y = C(t)\ x(t) \end{cases} \qquad S^* : \begin{cases} \dot{z} = A^T(t)\ z(t) + C^T(t)\ v(t) \\ s = B^T(t)\ z(t) \end{cases} \tag{1.23}$$

These systems are such that, if S is controllable at t_0, S^* is observable at t_0, and vice versa. It is thus possible to test the observability of a system by examining the controllability of the dual system.

It follows in particular that, with regard to observability, the state-space can be decomposed into two complementary subspaces X_c and X_d, respectively observable and unobservable, such that:

- X_d is invariant;
- the components of X_d are not connected to the output (cf. figure 1.9);
- the dimensions of these subspaces are constant if the matrices A(t) and C(t) are analytic;
- <u>for constant systems</u>, an observability criterion analogous to the controllability criterion (1.22) can be written in the form (1)

$$\text{rank}\ Q_0 = \text{rank}\ \left[C^T,\ A^TC^T, \ldots\ (A^{n-1})^TC^T\right] = n. \qquad (1.24)$$

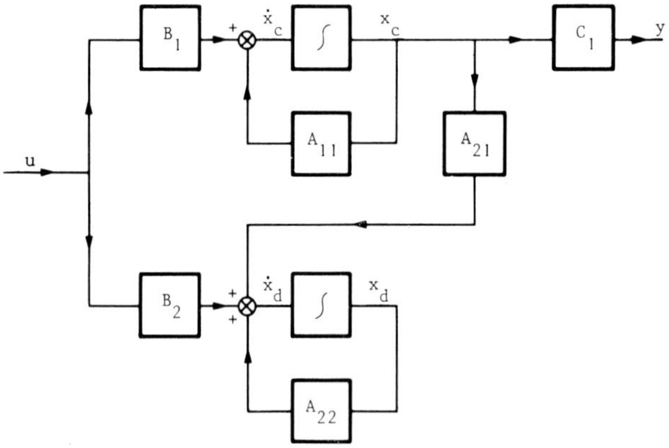

Fig. 1.9 Decomposition of a system with regard to observability

(1) As in the case of controllability, if

$$\text{rank}\ Q_{0\nu} = \text{rank}\ \left[C^T, \ldots\ (A^{\nu-1})^TC^T\right] = n,$$

the smallest such ν will be called the observability index.

D - Canonical decomposition of a system

It follows from paragraphs B and C that, from an overall viewpoint, the state-space can be decomposed into four subspaces X_1 to X_4 such that

$$X = X_1 \oplus X_2 \oplus X_3 \oplus X_4$$

the spaces X_i being defined by

$X_1 = X_a \cap X_d$ (controllable and unobservable subspace)

$X_2 = X_a \cap X_c$ (controllable and observable subspace)

$X_3 = X_b \cap X_d$ (uncontrollable and unobservable subspace)

$X_4 = X_b \cap X_c$ (uncontrollable and observable subspace)

In view of the properties recalled above, the state equations can be put in the form

$$\begin{bmatrix} x_1 \\ x_2 \\ x_3 \\ x_4 \end{bmatrix} = \begin{bmatrix} A_{11} & A_{12} & A_{13} & A_{14} \\ 0 & A_{22} & 0 & A_{24} \\ 0 & 0 & A_{33} & A_{34} \\ 0 & 0 & 0 & A_{44} \end{bmatrix} \begin{bmatrix} x_1 \\ x_2 \\ x_3 \\ x_4 \end{bmatrix} + \begin{bmatrix} B_1 \\ B_2 \\ 0 \\ 0 \end{bmatrix} u \qquad (1.25)$$
$$y = \begin{bmatrix} 0 & C_2 & 0 & C_4 \end{bmatrix} \begin{bmatrix} x_1 & x_2 & x_3 & x_4 \end{bmatrix}^T$$

as illustrated in figure 1.10.

<ins>Note</ins>: Although everything we have just said concerns only linear systems, it can be shown that part of the results, in particular the form (1.25), can be extended to bilinear systems; cf., for example, C. Bruni, G. Di Pillo and G. Koch, "Bilinear systems : an appealing class of nearly linear systems", IEEE. Trans. Autom. Control, AC-19 no. 4 (August 1974).

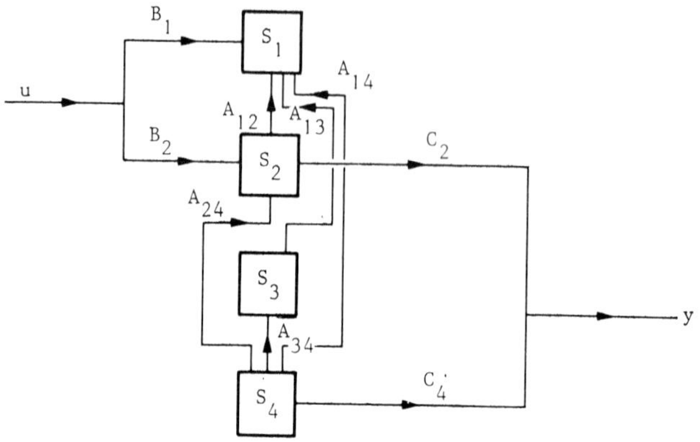

Fig. 1.10 Canonical decomposition of a system

1.2.3 Particular criteria

The controllability criterion which has just been given for the case of constant systems can be quite difficult to apply in practice because of the computational problems involved. Although in the general case, where A is arbitrary, we shall be obliged to have recourse to it (the subroutines for testing controllability will be particularly useful), it is possible, for special forms of A, to derive criteria whose application is much easier. The purpose of this subsection is to exhibit a certain number of these special simplified criteria.

A - A theorem about truncated functions

We shall understand by the term "truncated function" a function of time which vanishes outside a certain interval $[0, T]$, and by an "entire function" a function of the variable s having no poles. It can then be shown that the Laplace transform of a truncated function is an entire function. Although the converse of this theorem is true only under certain conditions, the consequences which will be drawn from it will remain general on account of the algebraic nature of the linear systems concerned.

B - Controllability criteria when the evolution matrix is in Jordan form

Let us suppose that the A matrix of the system takes the form of a number of Jordan blocks A(i) associated with the various eigenvalues, according to the following arrangement:

$$A = \begin{bmatrix} A(1) & & & & \\ & A(2) & & & \\ & & \ddots & & \\ & & & A(i) & \\ & & & & \ddots \\ & & & & & A(r) \end{bmatrix}$$

Various cases can arise:-

i) With an eigenvalue λ_i, there is associated one and only one Jordan block $A(\lambda_i)$, of dimension n_i:

$$A(i) = A(\lambda_i) = \begin{bmatrix} \lambda_i & 1 & & & \\ & \lambda_i & 1 & & \\ & & \ddots & & \\ & & & \lambda_i & 1 \\ & & & & \lambda_i \end{bmatrix}$$

If we Laplace-transform the equation

$$\dot{x} = Ax + Bu$$

with arbitrary initial conditions x_0, we get

$$X(s) = (sI-A)^{-1}[x_0 + BU(s)]$$

and the controllability condition can be expressed, in view of the theorem of paragraph A, by demanding that X(s) should be an entire function. Since the eigenvalues associated with the blocks A(i) are distinct, the "entire" character of X will hold, on account of the block diagonal form of A, separately for each of the expressions

$$^{i}X(s) = [sI-A(\lambda_i)]^{-1}\ [x_0+B(i)U(s)]$$

where B(i) is the block of n_i rows of B corresponding to the block $A(\lambda_i)$. The separation of the criterion with respect to each of the eigenvalues enables us to deal with the elements associated with a particular eigenvalue λ_i alone, and we shall, in the following, omit the index i so as to simplify the notation. We then have

$$X(s) = (sI - A)^{-1} [x_0 + BU(s)]$$

$$= \begin{bmatrix} \frac{1}{s-\lambda} & \frac{1}{(s-\lambda)^2} & \cdots & \frac{1}{(s-\lambda)^n} \\ 0 & \frac{1}{s-\lambda} & \cdots & \frac{1}{(s-\lambda)^{n-1}} \\ \vdots & & \ddots & \vdots \\ 0 & \cdots & 0 & \frac{1}{s-\lambda} \end{bmatrix} \begin{bmatrix} x_1(0) + {}_1BU(s) \\ x_2(0) + {}_2BU(s) \\ \vdots \\ x_n(0) + {}_nBU(s) \end{bmatrix} .$$

The state x_n will be controllable if

$$X_n(s) = \frac{1}{s-\lambda} [x_n(0) + {}_nBU(s)]$$

is an entire function, i.e. if there exists a U such that

$$x_n(0) + {}_nBU(\lambda) = 0$$

for any $x_n(0)$. This condition will be satisfied if, and only if, ${}_nB \neq 0$.

Similarly, the state x_{n-1} will be controllable if

$$X_{n-1}(s) = \frac{1}{s-\lambda} \left[x_{n-1}(0) + {}_{n-1}BU(s)\right] + \frac{1}{(s-\lambda)^2} \left[x_n(0) + {}_nBU(s)\right]$$

is an entire function, that is to say, if the expression

$$x_n(0) + {}_nBU(s) + (s-\lambda) \left[x_{n-1}(0) + {}_{n-1}BU(s)\right]$$

has $s = \lambda$ as a double root, i.e. if it is possible to satisfy the two conditions, on this expression and its derivative,

$$x_n(0) + {}_nBU(\lambda) = 0,$$

$${}_nB\dot{U}(\lambda) + x_{n-1}(0) + {}_{n-1}BU(\lambda) = 0.$$

With $U(\lambda)$ already determined, this last equation enables us to obtain $U(\lambda)$ if and only if ${}_nB \neq 0$.

Hence, proceeding step by step to x_1, we see that the sole condition for complete controllability is

$$ {}_nB \neq 0. $$

A possible control function is given by

$$ U(s) = U(\lambda) + (s-\lambda)\dot{U}(\lambda) + \dots + \frac{(s-\lambda)^{n-1}}{(n-1)!} U^{(n-1)}(\lambda). $$

<u>Theorem</u>: If the matrix A is decomposed into Jordan form

$$ A = \text{diag}\left[A(\lambda_i)\right] $$

with one and only one block corresponding to each eigenvalue λ_i, the system is controllable if and only if the last rows ${}_{n_i}B$, of the blocks of B corresponding to the various $A(\lambda_i)$, are non-null.

2) An eigenvalue λ_i may be associated, no longer with only a single block, but with s_i elementary blocks, according to the following arrangement:

$$ A(\lambda_i) = \begin{bmatrix} A(\lambda_i, n_1) & & & \\ & A(\lambda_i, n_2) & & \\ & & \ddots & \\ & & & A(\lambda_i, n_{s_i}) \end{bmatrix} $$

where $A(\lambda_i, n_k)$ is a Jordan sub-block of dimension n_k associated with the eigenvalue λ_i. We shall then have, with the same notation as before, i.e. omitting the index i,

$$ X(s) = (sI - A)^{-1}\left[x_0 + BU(s)\right] $$

$$= \begin{bmatrix} [sI-A(\lambda_i, n_1)]^{-1} [{}^1x(0) + B(1)U(s)] \\ \vdots \\ [sI-A(\lambda_i, n_{s_i})]^{-1} [{}^{s_i}x(0) + B(s_i)U(s)] \end{bmatrix}$$

with

$$B = \begin{bmatrix} B(1) \\ \vdots \\ B(s_i) \end{bmatrix} , \quad x_0 = \begin{bmatrix} {}^1x(0) \\ \vdots \\ {}^{s_i}x(0) \end{bmatrix}$$

where ${}^kx$ has dimension n_k and B(k) has dimensions (n_k, m), the form of $[sI-A(\lambda_i, n_k)]^{-1}$ being the same as previously indicated. Since the various equations corresponding to the blocks $A(\lambda_i, n_k)$ associated with the same eigenvalue λ_i are no longer independent, the existence of a function U(s), such that X(s) is entire, depends on the matrix formed by the last rows $_{n_k}B(k)$ of each sub-block being of full rank.

<u>Theorem:</u> If the matrix A is decomposed into Jordan blocks in the form

$$A = \text{diag}\,[A(\lambda_i, n_k)]$$

where s_i Jordan sub-blocks are associated with the eigenvalue λ_i, the system is controllable if and only if, for each i, the matrix composed of the last rows of the sub-blocks of B associated with the sub-blocks $A(\lambda_i, n_k)$ is of rank s_i.

<u>Note 1:</u> It follows, in particular, that, if the number of sub-blocks associated with the same eigenvalue is greater than m, the system is uncontrollable.

<u>Note 2:</u> If all the eigenvalues of A are distinct, i.e. if A is in diagonal form

$$A = \text{diag}\,(\lambda_1, \dots \lambda_n)$$

we have

$$X_i(s) = \frac{1}{s-\lambda_i}\left[x_i(0) + {}_iBU(s)\right]$$

and the truncation conditions become simply

which will hold if ${}_iB \neq 0$. The controllability condition thus reduces in this case to the requirement that no row of B should be null.

Examples:

1) Let the system be defined by the following A and B matrices:

$$A = \left[\begin{array}{ccc|cc} -1 & 1 & & & \\ & -1 & 1 & & \\ & & -1 & & \\ \hline & & & -2 & 1 \\ & & & & -2 \end{array}\right], \qquad B = \begin{bmatrix} {}_1B \\ {}_2B \\ \alpha, \ \beta \\ {}_4B \\ \gamma, \ \delta \end{bmatrix}$$

The system is controllable if $[\alpha, \beta] \neq 0$ and $[\gamma, \delta] \neq 0$. It will be noticed that there is no need to specify the rows ${}_1B$, ${}_2B$ and ${}_4B$.

2) Let the system be defined by the matrices:

$$A = \left[\begin{array}{ccc|cc} -1 & 1 & & & \\ & -1 & 1 & & \\ & & -1 & & \\ \hline & & & -1 & 1 \\ & & & & -1 \end{array}\right] \qquad B = \begin{bmatrix} {}_1B \\ {}_2B \\ \alpha, \ \beta \\ {}_4B \\ \gamma, \ \delta \end{bmatrix}$$

As there are two Jordan blocks associated with the same eigenvalue, the controllability condition becomes:

$$\text{rank} \begin{bmatrix} \alpha & \beta \\ \gamma & \delta \end{bmatrix} = 2$$

i.e.

$$\alpha\delta - \beta\gamma \neq 0.$$

C - Controllability criteria when the evolution matrix is in companion form

We recall that this form may appear, in general, with A as a block diagonal matrix:

$$A = \text{diag}\left[A(i)\right]$$

where A(i) is a companion block which can be in either "vertical" (a) or "horizontal" (b) form:

$$A(i) = \begin{bmatrix} -a^i_{n_i-1} & 1 & 0 & \cdots & 0 \\ \vdots & 0 & \ddots & \ddots & \vdots \\ \vdots & \vdots & & \ddots & 0 \\ \vdots & \vdots & & & 1 \\ -a^i_0 & 0 & \cdots & \cdots & 0 \end{bmatrix} \quad (a) \qquad A(i) = \begin{bmatrix} 0 & 1 & 0 & \cdots & 0 \\ \vdots & & \ddots & \ddots & \vdots \\ \vdots & & & \ddots & 0 \\ 0 & \cdots & \cdots & 0 & 1 \\ -a^i_0 & \cdots & \cdots & \cdots & -a^i_{n_i-1} \end{bmatrix} \quad (b)$$

These forms have special properties which will be widely used in chapter 2:-

i) The characteristic polynomial of A(i) can be written directly

$$\psi_i(s) = \det\left[sI-A(i)\right] = s^{n_i} + a^i_{n_i-1}s^{n_i-1} + \ldots + a^i_1 s + a^i_0 \qquad (1.26)$$

ii) The inverse matrix $\left[sI-A(i)\right]^{-1}$ can be calculated from the expressions:

with A(i) in form a),

$$\left[sI-A(i)\right]^{-1} = \frac{1}{\psi_i(s)} \mathscr{A}_i \begin{bmatrix} 1 \\ s \\ \vdots \\ s^{n_i-1} \end{bmatrix} \left[s^{n_i-1}, \ldots s, 1\right] \text{ modulo } \psi_i(s); \qquad (1.27)$$

with A(i) in form b),

$$[sI-A(i)]^{-1} = \frac{1}{\psi_i(s)} \begin{bmatrix} 1 \\ s \\ \vdots \\ s^{n_i-1} \end{bmatrix} [s^{n_i-1}, \dots s, 1] \mathscr{A}_i \text{ modulo } \psi_i(s); \quad (1.28)$$

where $\mathscr{A}_i$ is a triangular matrix with unity elements on the diagonal (and hence invertible) and with the elements below the diagonal being coefficients of the characteristic polynomial:

$$\mathscr{A}_i = \begin{bmatrix} 1 & 0 & \cdots & 0 \\ a^i_{n_i-1} & \ddots & \ddots & \vdots \\ \vdots & \ddots & \ddots & 0 \\ a^i_1 & \cdots & a^i_{n_i-1} & 1 \end{bmatrix} \quad (1.29)$$

These expressions for $(sI-A)^{-1}$, and the consequent ones for $(sI-A)^{-1}B$, enable us to obtain simple forms of the controllability criteria by using the truncated-function theorem.

a) We shall consider first the case where the companion blocks are in vertical form. Two situations can arise:-

1) The space X is cyclic with respect to A, i.e. there is only a single companion block. We then have

$$X(s) = (sI-A)^{-1} [x_0 + BU(s)]$$

$$= \frac{\mathscr{A}}{\psi(s)} \begin{bmatrix} 1 \\ \vdots \\ s^{n-1} \end{bmatrix} [s^{n-1} \dots 1] [x_0 + BU(s)] \text{ mod } \psi(s)$$

and, for every root λ of ψ(s), X(s) will have a pole unless U(λ) can be chosen so as to cancel it. Thus, the system is controllable in respect of the mode $s = \lambda_i$ if it is possible to find U such that

$$[\lambda_i^{n-1}, \ldots \lambda_i, 1] \begin{bmatrix} x_{01} + {}_1BU(\lambda_i) \\ \vdots \\ x_{0n} + {}_nBU(\lambda_i) \end{bmatrix} = 0$$

where ${}_iB$ denotes the ith row of B. This last equation can be put in the form

$$\mathcal{X}_0(\lambda_i) + \mathcal{B}(\lambda_i)U(\lambda_i) = 0, \qquad (1.30)$$

with

$$\mathcal{X}_0(\lambda_i) = x_{0n} + \ldots + \lambda_i^{n-1} x_{01}$$

and

$$\mathcal{B}(\lambda_i) = {}_nB + \ldots + \lambda_i^{n-1}\, {}_1B,$$

and it will be possible to find a $U(\lambda_i)$ if $\mathcal{B}(\lambda_i) \neq 0$.

Note: In the case that there is a multiple root λ of order ν, it must also be a multiple root, of the same order, of $x_0 + BU(s)$. The successive derivatives of this expression, however, do not bring about any further conditions.

Theorem: A necessary and sufficient condition for the system

$$\dot{x} = \begin{bmatrix} -a_{n-1} & 1 & 0 & \cdots & 0 \\ \vdots & 0 & \ddots & & 0 \\ \vdots & \vdots & & \ddots & 1 \\ -a_0 & 0 & \cdots & & 0 \end{bmatrix} x + \begin{bmatrix} {}_1B \\ \vdots \\ {}_nB \end{bmatrix} u$$

to be controllable is that

$$\mathcal{B}(\lambda) = {}_nB + \lambda\, {}_{n-1}B + \ldots + \lambda^{n-1}\, {}_1B = [\lambda^{n-1}, \ldots 1] \begin{bmatrix} {}_1B \\ \vdots \\ {}_nB \end{bmatrix}$$

should be non-null for every root λ of the characteristic equation $\psi(s) = 0$.

2) In general, the matrix A can appear in block diagonal form, having companion blocks A(i) associated with polynomials $\psi_i(s)$. If these polynomials have no common roots, controllability can be tested as before within each block. If, on the contrary, a root λ_i appears in r blocks, $k_1, k_2, \ldots k_r$, these blocks must be considered simultaneously, and the same argument as in 1) shows that, if $\mathcal{B}_{k_j}(\lambda_i)$ denotes the polynomial row vector associated with the block k_j, the system

$$\dot{x} = \begin{bmatrix} A(k_1) & & \\ & \ddots & \\ & & A(k_r) \end{bmatrix} x + \begin{bmatrix} B(k_1) \\ \vdots \\ B(k_r) \end{bmatrix} u$$

will be controllable if and only if

$$\text{rank} \begin{bmatrix} \mathcal{B}_{k_1}(\lambda_i) \\ \vdots \\ \mathcal{B}_{k_r}(\lambda_i) \end{bmatrix} = r$$

Example: Let the system be defined by the matrices

$$A = \begin{bmatrix} -7 & 1 & 0 & 0 \\ -14 & 0 & 1 & 0 \\ -8 & 0 & 0 & 0 \\ 0 & 0 & 0 & -2 \end{bmatrix}, \quad B = \begin{bmatrix} 1 & 3 \\ 1 & 4 \\ 0 & 1 \\ \alpha & \beta \end{bmatrix}$$

The matrix A contains two companion blocks associated with the characteristic polynomials:

$$\psi_1(s) = s^3 + 7s^2 + 14s + 8 = (s+1)(s+2)(s+4),$$

$$\psi_2(s) = s+2 .$$

It will be noticed in particular that the mode s = −2 appears in both blocks. We have:

$$\mathcal{B}_1(s) = [0,1] + [1,4]s + [1,3]s^2 = [s+s^2,\ 1+4s+3s^2],$$

$$\mathcal{B}_2(s) = [\alpha,\beta].$$

The modes s = −1 and s = −4 appear only in the first block, and so it is sufficient to examine $\mathcal{B}_1(-1)$ and $\mathcal{B}_1(-4)$. We have:

$$\mathcal{B}_1(-1) = [0,0],\ \mathcal{B}_1(-4) = [12,33] \neq 0.$$

Thus, the first mode is uncontrollable, the second controllable.

For s = −2, we have to examine the whole matrix

$$\begin{bmatrix} \mathcal{B}_1(-2) \\ \mathcal{B}_2(-2) \end{bmatrix} = \begin{bmatrix} 2 & 5 \\ \alpha & \beta \end{bmatrix}$$

Hence, if $2\beta - 5\alpha = 0$, the mode s = −2 is uncontrollable; if $2\beta - 5\alpha \neq 0$, it is controllable.

b) We now consider the case where the companion blocks are in horizontal form. In this case, the previous criteria must be modified to take account of the transfer of $\;_i$ in the expression for $[sI-A(i)]^{-1}$. Allowing for this, the calculations can be carried out in a completely analogous manner.

In the case that there is only one companion block, the system is controllable if U can be found such that

$$\mathcal{X}_0(\lambda_i) + \mathcal{B}^*(\lambda_i)U(\lambda_i) = 0$$

for every root λ_i of $\psi(s)$, i.e. if

$$\mathcal{B}^*(\lambda_i) \neq 0.$$

In this case, however, $\mathcal{B}^*(\lambda_i)$ is defined by

$$\mathcal{B}^*(\lambda) = [{}_n\mathcal{A} + \lambda\ {}_{n-1}\mathcal{A} + \ldots + \lambda^n\ {}_1\mathcal{A}\ ]B$$

where ${}_i\mathcal{A}$ is the ith row of the matrix $\mathcal{A}$.

If A contains more than one companion block, and λ_i is a common root of the characteristic polynomials of r blocks, the system is controllable if and only if the matrix composed of the r rows ${}_k\mathcal{B}^*(\lambda_i)$ associated with these blocks is of rank r.

Example: Let

$$\dot{x} = \begin{bmatrix} 0 & 1 & 0 & 0 \\ 0 & 0 & 1 & 0 \\ -6 & -11 & -6 & 0 \\ 0 & 0 & 0 & -1 \end{bmatrix} x + \begin{bmatrix} 1 & 1 \\ -1 & 0 \\ 1 & -3 \\ 1 & 1.5 \end{bmatrix} u$$

We have

$$\psi_1(s) = s^3 + 6s^2 + 11s + 6 = (s+1)(s+2)(s+3),$$

$$\psi_2(s) = s + 1,$$

$$\mathcal{A}_1 = \begin{bmatrix} 1 & 0 & 0 \\ 6 & 1 & 0 \\ 11 & 6 & 1 \end{bmatrix}, \quad \mathcal{A}_2 = [1],$$

$${}_1\mathcal{B}^*(\lambda) = [\lambda^2+6\lambda+11,\ \lambda+6, 1] \begin{bmatrix} 1 & 1 \\ -1 & 0 \\ 1 & -3 \end{bmatrix} = [\lambda^2+5\lambda+6,\ \lambda^2+6\lambda+8].$$

Since the eigenvalues $\lambda = -2$ and $\lambda = -3$ appear only in $\psi_1(s)$, we deduce from

$${}_1\mathcal{B}^*(-2) = [0, 0]$$

the controllability of $s = -2$, and from

$${}_1\mathcal{B}^*(-3) = [0,-1] \neq 0$$

the controllability of s = -3. On the other hand, $\lambda = -1$ is a root of both ψ_1 and ψ_2, and we have

$$\text{rank}\begin{bmatrix} {}_1\mathcal{B}^*(-1) \\ {}_2\mathcal{B}^*(-1) \end{bmatrix} = \text{rank}\begin{bmatrix} 2 & 3 \\ 1 & 1.5 \end{bmatrix} = 1 < 2,$$

whence the mode s = -1 is uncontrollable.

D - Observability criteria

In consequence of the ideas of duality discussed in paragraph 1.2.2.C, we are freed from the necessity of deriving particular criteria for observability. Since, however, such criteria are often useful in practice, we recall here the observability criteria corresponding to those for controllability given in paragraphs B and C above, for the cases where A appears in Jordan or companion form.

1) If A is decomposed into Jordan blocks, in the form

$$A = \text{diag}[A(i)],$$

then:

- if, to each eigenvalue λ_i, there corresponds one and only one Jordan block, the system is observable if, and only if, the first columns ${}^1C(i)$, of the blocks of C corresponding to the blocks of A, are non-null;

- if there are s_i Jordan blocks corresponding to an eigenvalue λ_i, the system is observable if, and only if, the matrix composed of the first column of each of the corresponding sub-blocks of C is of rank s_i.

2) If the matrix A has the form of a companion block, a necessary and sufficient condition for observability is that the polynomial column vector

$$\mathcal{C}(s) = C\mathcal{A}\begin{bmatrix} 1 \\ s \\ \vdots \\ s^{n-1} \end{bmatrix}$$

should be non-null for every root λ of the characteristic polynomial $\psi(s)$ of A, if the companion form used is the vertical one. In the case of the horizontal companion form, the above expression should be replaced by

$$\mathcal{C}(s) = C\begin{bmatrix} 1 \\ s \\ \vdots \\ s^{n-1} \end{bmatrix}$$

If A is in block diagonal form, consisting of companion blocks A(k) associated with polynomials $\psi_k(s)$, and s_i is the number of polynomials possessing the common root λ_i, the mode $s = \lambda_i$ is observable if and only if

$$\text{rank}\left[{}^{1}\mathcal{C}(\lambda_i), \dots {}^{s_i}\mathcal{C}(\lambda_i)\right] = s_i$$

where ${}^{k}\mathcal{C}(\lambda_i)$ is the polynomial column vector derived from the block C(k) associated with A(k) by the previous relations.

1.2.4 Appearance of uncontrollable or unobservable modes, and their physical significance

A - Appearance of uncontrollable and unobservable modes

The study of the system of figure 1.6, which was used as an introductory example, revealed the existence of two particular modes, one at $s = 1$, which was uncontrollable and the other at $s = -1$ which was unobservable. In fact, for this simple example of a single-variable system formed of two subsystems in cascade, the mechanism of uncontrcllability and unobservability appears clearly if the system is rewritten in the equivalent form of figure 1.11, by replacing the subsystems in parallel with equivalent single subsystems.

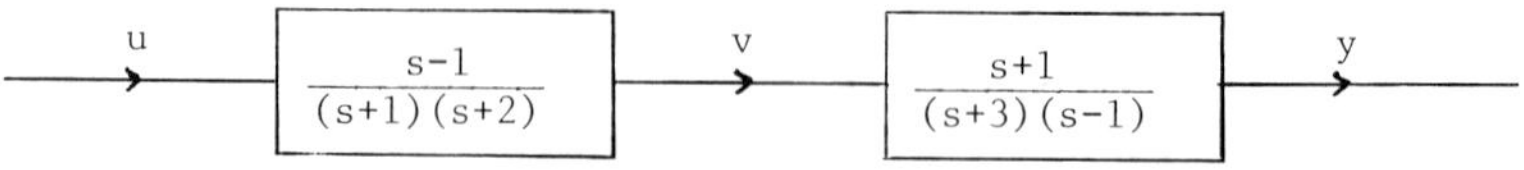

Fig. 1.11 Re-expression of the system of figure 1.6

The uncontrollability and unobservability result, in fact, from the cancellation of a pole of one transfer function by a zero of the other and vice versa. More precisely, if the cancellation is of the type "zero upstream with pole downstream", as for $s = 1$, uncontrollability occurs, while, if it is of the type "pole upstream with zero downstream", as for $s = -1$, unobservability arises.

Note: It follows in particular that, if the two systems are interchanged, uncontrollability is replaced by unobservability and vice versa. The ordering of the systems, with regard to the direction of signal flow, is thus significant.

The mechanism is more delicate in the case of multi-input, multi-output systems, where cancellations can take place as a result of the properties of the determinants, as we shall see in the following paragraph. In all cases, however, the appearance of an uncontrollable or unobservable mode results from some kind of cancellation between different subsystems. This means that, if we cascade two subsystems which "have nothing to do with one another", the probability of creating uncontrollable or unobservable modes is practically zero. It is, however, different in control problems where a process is connected to a compensator designed, as a function of the process, to produce certain open-loop or closed-loop properties.

If Z is the transfer function matrix of the process, and C is that of the compensator, designed so that the open loop should have a given transmittance N, we see from figure 1.12 that the two systems connected are far from "having nothing to do with one another".

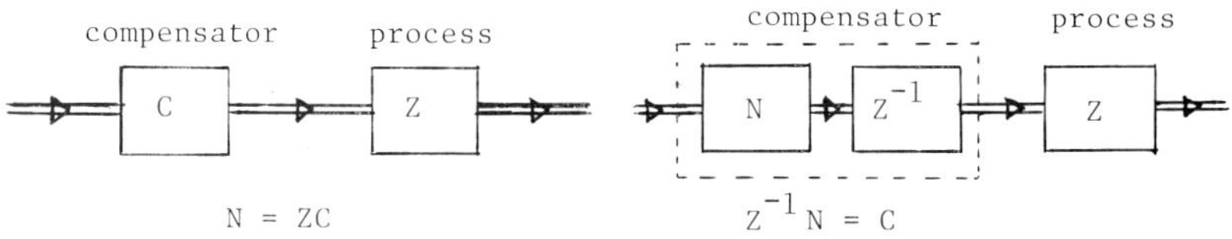

Fig. 1.12 Cascade of a process and a compensator

In such a case, if we want to avoid the risk of producing an uncontrollability or unobservability when the compensator is cascaded with the process, we shall see in chapter 3 that certain constraints must be imposed on N. For example, if we write the rational function matrices Z, C, N, in the form of polynomial matrices $\mathcal{Z}$, $\mathcal{C}$, $\mathcal{N}$, divided by polynomials z, c, n, respectively, we shall be led to impose conditions of the form:

$$[\operatorname{adj}\mathcal{Z}]\,\mathcal{N} \text{ contains the factors of } \det\mathcal{Z},$$
$$n \text{ contains the factors of } z.$$

B - Practical importance of the concepts of controllability and observability

i) An uncontrollable mode, as we have seen, is not connected to the inputs. The response associated with such a mode will thus evolve in time in a manner depending only on its dynamics and the corresponding initial conditions, and hence independent of the control inputs applied, whether in open-loop or closed-loop configuration. The fact that, in particular, feedback control is incapable of improving matters, which is physically obvious from what has gone before, follows easily from the equations.

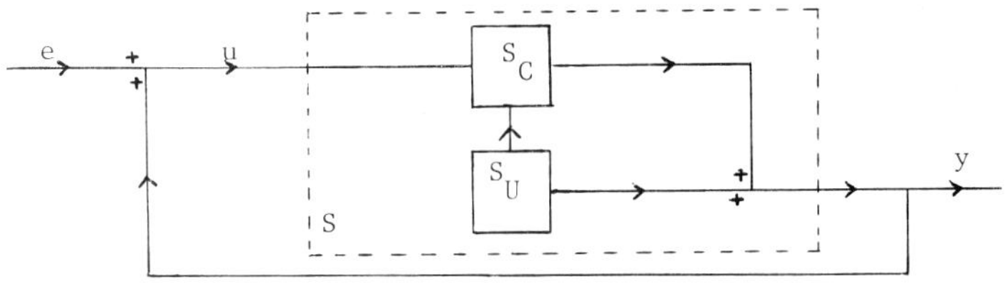

Fig. 1.13 System containing a controllable part and an uncontrollable part

Let us consider the system S of figure 1.13, which comprises a controllable part S_C and an uncontrollable part S_U. The state equations of the open-loop system are of the form:

$$\dot{x}_C = A_C x_C + B_C u + K_U x_U ,$$

$$\dot{x}_U = A_U x_U ,$$

$$y = C_C x_C + C_U x_U .$$

The feedback law being u = e + y, the closed-loop equations become

$$\begin{bmatrix} \dot{x}_C \\ \dot{x}_U \end{bmatrix} = \begin{bmatrix} A_C + B_C C_C & K_U + B_C C_U \\ 0 & A_U \end{bmatrix} \begin{bmatrix} x_C \\ x_U \end{bmatrix} + \begin{bmatrix} B_C \\ 0 \end{bmatrix} e$$

$$y = \begin{bmatrix} C_C & C_U \end{bmatrix} \begin{bmatrix} x_C \\ x_U \end{bmatrix} .$$

The characteristic equation of the closed-loop system is

$$\det \left[sI - A_C - B_C C_C\right] \det \left[sI - A_U\right] = 0$$

and the corresponding modes consist of :

a) the modes of the controllable part of the closed-loop system (first factor;

b) the modes of the uncontrollable part of the open-loop system (second factor).

Analogous conclusions could be reached in the same way with regard to unobservable modes.

(ii) From a practical point of view, it is worthwhile to make a distinction between uncontrollable (or unobservable) modes which are stable and those which are not.

The exact cancellation of a pole with a zero is a purely theoretical matter. In practice, when we say that a pole cancels with a zero, what actually happens is the formation of a dipole:

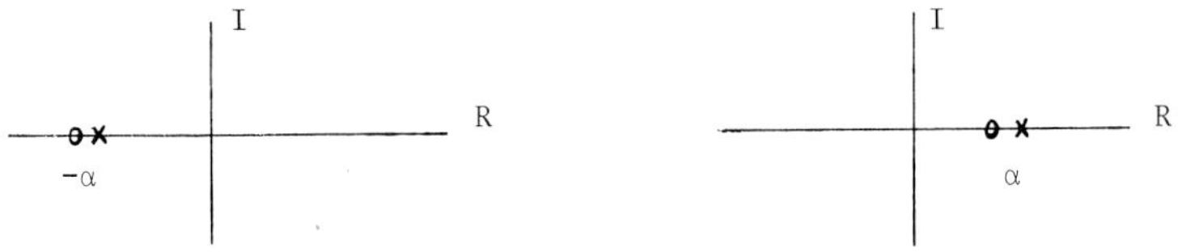

If this dipole is in the left half-plane (around $-\alpha$), it will contribute to the response a term of the form $\varepsilon\, e^{-\alpha t}$, where ε is the residue associated with the pole, very small because of the existence of the nearby zero. This term will, more often than not, be negligible. On the other hand, if the dipole is in the right half-plane (around α), there will be a term $\varepsilon\, e^{\alpha t}$, unstable however small ε may be.

(iii) Suppose that, in a system, there is an unstable mode which is observable but uncontrollable. Since it is observable, the output is unstable. Thus, there is no risk of "forgetting" this mode. On the contrary, the fact that it is uncontrollable excludes all possibility of stabilising the system. What is required in this case is not a control law but a modification of the structure of the system.

(iv) Let us suppose now that the system has an unstable mode which is controllable but not observable. Since this mode is not connected to the outputs, these may be observed to be stable. Nevertheless, the internal instability of the system will still entail one or other of the consequences:

- either, a break-up of the system when the unstable variable reaches a certain amplitude;

- or, the appearance of a nonlinear function arising, e.g., from saturation effects, so that the model linearised around the operating point is no longer valid.

1.3 CONTROLLABILITY IN THE OUTPUT SPACE

1.3.1 Introduction

The ideas of controllability and observability defined in the preceding section were concerned witn concepts relating to the state of the system, controllability for example being connected with the possibility of bringing the system from a given initial state x_0 to a given final state x_1. Since, however, the purpose of a control system is to generate desired outputs by appropriate manipulation of the input vector, we see the necessity of defining a notion of controllability with respect to the outputs; is it possible, by the use of a suitable control input, to bring the output vector from an initial value y_0 at time t_0 to the value y_1 at time t_1? It is easy to appreciate that these two ideas are different, the first involving only the matrices A and B while the second depends on A,B,C and possibly the direct transmission matrix D. It is also clear that state controllability is not necessary for output controllability, as the following examples show.

Example 1: Let us consider a system which is not state controllable,

$$\begin{bmatrix} \dot{x}_1 \\ \dot{x}_2 \end{bmatrix} = \begin{bmatrix} -\alpha & 0 \\ 0 & -\alpha \end{bmatrix} \begin{bmatrix} x_1 \\ x_2 \end{bmatrix} + \begin{bmatrix} 1 \\ 0 \end{bmatrix} u$$

$$y = \begin{bmatrix} 0 & 2\alpha \end{bmatrix} \begin{bmatrix} x_1 \\ x_2 \end{bmatrix} + u$$

where the state x_2 is uncontrollable. Suppose we wish to bring y from y_0 at t_0 to y_1. We have

$$y(t_0) = y_0 = 2\alpha x_{20} + u(t_0)$$

so that

$$u(t_0) = y_0 - 2\alpha x_{20}$$

Since x_2 is not controllable, its time development follows the law, independent of the control input,

$$x_2 = x_{20}\, e^{-\alpha(t-t_0)} \;;$$

at the instant t_1, we require

$$y(t_1) = y_1 = 2\alpha x_2(t_1) + u(t_1)$$

$$y_1 = 2\alpha x_{20}\, e^{-\alpha(t_1-t_0)} + u(t_1)$$

whence

$$u(t_1) = y_1 - 2\,\alpha x_{2_0}\, e^{-\alpha(t_1-t_0)}$$

For $t > t_1$, we shall be able to hold y at y_1. It suffices to take u(t) such that:

$$y(t) = y_1 = 2\,\alpha x_{2_0}\, e^{-\alpha(t-t_0)} + u(t) \qquad \forall t > t_1$$

so that

$$u(t) = y_1 - 2\,\alpha x_{2_0}\, e^{-\alpha(t-t_0)}$$

The system is thus controllable with respect to the outputs, without being so in regard to the states, thanks to the influence of D (here equal to 1).

Example 2: The same thing would evidently happen if we were to consider the case of an uncontrollable system whose uncontrollable states were also unobservable, e.g.

$$\begin{bmatrix}\dot{x}_1\\ \dot{x}_2\\ \dot{x}_3\end{bmatrix} = \begin{bmatrix}-1 & 0 & 0\\ 0 & -2 & 0\\ 0 & 0 & -3\end{bmatrix}\begin{bmatrix}x_1\\ x_2\\ x_3\end{bmatrix} + \begin{bmatrix}1 & 2\\ 0 & 0\\ 3 & 1\end{bmatrix}\begin{bmatrix}u_1\\ u_2\end{bmatrix}$$

$$\begin{bmatrix}y_1\\ y_2\end{bmatrix} = \begin{bmatrix}3 & 0 & 1\\ 2 & 0 & 4\end{bmatrix}\begin{bmatrix}x_1\\ x_2\\ x_3\end{bmatrix}$$

where the state x_2 is uncontrollable but does not appear in the outputs. We can thus bring y_1 and y_2 to desired values at time t_1. Furthermore, since the uncontrollable state is stable, the presence of this uncontrollability will not be embarassing in practice.

After these few examples, let us consider the general problem; let the system be

$$\dot{x} = Ax + Bu$$
$$y = Cx + Du$$

whose solution is given by:

$$y(t) = C(t)\Phi(t,t_0)x_0 + \int_{t_0}^{t} S(t,\tau)u(\tau)d\tau + D(t)u(t) \qquad (1.31)$$

with

$$S(t,\tau) = C(t)\Phi(t,\tau)B(\tau). \qquad (1.32)$$

The elements of the matrix $S(t,\tau)$ can be interpreted as impulse responses of the system, for $t > \tau$. In particular, with null initial conditions, the input-output relation becomes:

$$y(t) = \int_{t_0}^{t} S(t, \tau)\ u(\tau)\ d\tau + Du(t)$$

with

$$\begin{aligned} S(t,\tau) &= C(t)\Phi(t,\tau)B(\tau) \\ &= C(t)\Phi(t,0)\Phi(0,\tau)B(\tau) \\ &= \psi(t)\ \theta(\tau) \qquad (1.33) \end{aligned}$$

so that $S(t,\tau)$ can be expressed as the product of two matrices, one of dimensions (p,n) depending only on the variable t, the other of dimensions (n,m) depending only on τ.

1.3.2 Condition for controllability of the outputs

Definition: A system is (completely) controllable with respect to the outputs at the instant t_0 if there exists an input which transfers an arbitrary initial output $y(t_0)$ to an arbitrary vector y_1 in a finite time. If we consider only processes without direct transmission ($D \equiv 0$) and set:

$$\tilde{y}(t) = y(t) - C(t)\ \Phi(t,\ t_0)\ x_0$$

equation (1.31) gives:

$$\tilde{y}(t) = \int_{t_0}^{t} S(t,\ \tau)\ u(\tau)\ d\tau$$

It may then be shown (1) that: a necessary and sufficient condition for controllability with respect to the outputs is that there exists a time t_1 such that the p rows of the matrix $S(t_1,\tau)$ are linearly independent functions on the interval $[t_0,t_1]$.

Equivalently, this condition can be expressed by the requirement that the matrix

$$P(t_0,\ t_1) = \int_{t_0}^{t_1} S(t_1,\ \tau)\ S^T(t_1,\ \tau)\ d\tau$$

is nonsingular.

In particular, for the case of constant systems, it can be shown that a necessary and sufficient condition for controllability with respect to the outputs is that

$$\text{rank}\ [CB,\ldots CA^{n-1}B,D] = p.$$

1.4 REPRODUCIBILITY

The problem of reproducibility can be formulated as follows: given a linear dynamical system, defined by matrices (A,B,C), a space of desired outputs S_y, and a space of admissible inputs S_u, does there exist, for $\bar{y}$ belonging to S_y, a control function u belonging to S_u, such that:

$$y = \phi_u(t,\ x_0,\ t_0) = \bar{y}\ .$$

This is a complicated problem which has, in particular, been studied in various forms by Brockett and Mesarovic (cf. ref. 1.3).

We shall limit the present discussion to the case of reproducibility in a class of functions (2), taking S_y to be the space

$$C_p^r = C^r \times C^r \ldots \times C^r$$

of functions with r continuous derivatives on the interval $[0,T]$ and seeking conditions under which it is possible to generate a y belonging to C_p^r by means of a control input belonging to the space $C_p^{r-\alpha}$ of

(1) Cf. ref. 1.3.
(2) Cf. refs. 1.2,1.5.

functions continuously differentiable $r-\alpha$ times. The corresponding set of inputs and initial conditions will be designated, following Birta's notation, by L^r_α. It can then be shown that, for constant systems:-

1) The necessary and sufficient condition for L^r_1 not to be empty (i.e. that every $\hat{y}$ in C^r_p, $r \geq 1$, can be generated by a u in C^{r-1}_m) is that

$$\text{rank }(CB) = p.$$

A possible control input can then be defined by

$$\hat{u} = \begin{bmatrix} \hat{u}_1 \\ \hat{u}_2 \end{bmatrix} \quad \text{with} \quad \begin{cases} \hat{u}_1 = 0 \\ \hat{u}_2 = Z_2^{-1}\left[\dot{\hat{y}} - CAx\right] \end{cases}$$

where Z_2 is the matrix of dimensions (p,p) obtained from:

$$CB = \begin{bmatrix} Z_1 & Z_2 \end{bmatrix}$$

and x is a solution of the equation

$$\dot{x} = Ax + B\hat{u}$$

for initial conditions x_0 such that

$$\hat{y}_0 = Cx_0.$$

2) The necessary and sufficient condition that the space L^r_2 is not empty (i.e. that every $\hat{y}$ in C^r_p, $r \geq 2$, can be generated by a u in C^{r-2}_m) is that

$$\text{rank } \mathcal{G}_1 = \text{rank} \begin{bmatrix} CB \\ VCAB \end{bmatrix} = p,$$

where V is the matrix formed from the set of row vectors which annihilate CB. A possible control is defined by

$$\hat{u} = \begin{bmatrix} \hat{u}_1 \\ \hat{u}_2 \end{bmatrix} \text{ with } \hat{u}_1 = 0,\ \hat{u}_2 = Z_2^{-1}\left[\begin{bmatrix} \langle\dot{\hat{y}}\rangle_\sigma \\ V\ddot{\hat{y}} \end{bmatrix} - \begin{bmatrix} \langle CA\rangle_\sigma \\ VCA^2 \end{bmatrix} x\right],$$

where the notation is as follows. If rank $(\mathcal{G}_1)$ = p and rank (CB) = p-s, it is possible to eliminate s suitably chosen rows from $\mathcal{G}_1$, forming a sequence σ $(\sigma_1, \sigma_2, \dots \sigma_s)$, so that the matrix $\langle\mathcal{G}_1\rangle_\sigma$ thus obtained remains of rank p. Similarly, $\langle\hat{\dot{y}}\rangle_\sigma$ is the vector of dimension p-s obtained by removing from $\hat{\dot{y}}$ the s components corresponding to σ. Also, Z_2 is the matrix of dimensions (p,p) defined by:

$$\langle \mathcal{G}_1 \rangle_\sigma = \begin{bmatrix} Z_1 & Z_2 \end{bmatrix}$$

and x is a solution of the equation

$$\dot{x} = Ax + B\hat{u}$$

for initial conditions x_0 such that

$$\hat{y}_0 = Cx_0 \qquad V\hat{y}_0 = VCAx_0$$

3) The necessary and sufficient condition for the space L_3^r not to be empty is that the matrix

$$\mathcal{G}_2 = \begin{bmatrix} CB \\ VCAB \\ W \begin{bmatrix} CAB \\ VCA^2\,B \end{bmatrix} \end{bmatrix}$$

should be of rank p, where W is the matrix of annihilators of $\mathcal{G}_1$. If we have

$$\text{rank } (\mathcal{G}_2) = p, \text{ rank } (\mathcal{G}_1) = p-\rho, \text{ rank (CB)} = p-s,$$

we can find a sequence σ_1 corresponding to the elimination of s rows of $\mathcal{G}_1$ and a sequence σ_2 corresponding to the elimination of ρ rows of $\mathcal{G}_2$ such that

$$\text{rank}\langle \mathcal{G}_1\rangle_{\sigma_1} = p - \rho \qquad \text{rank}\langle \mathcal{G}_2\rangle_\sigma = \text{rank}\left\langle \begin{matrix} \langle \mathcal{G}_1\rangle_{\sigma_1} \\ W_3\ CAB + W_4\ VCA^2\ B \end{matrix} \right\rangle_{\sigma_2} = p$$

with

$$W = \begin{bmatrix} W_3, & W_4 \end{bmatrix}$$

where the matrices V, W_3, W_4 have respective dimensions (s,p), (ρ,p), (ρ,s). We can then take for $\hat{u}$ the control defined by

$$\hat{u} = \begin{bmatrix} 0 \\ \hat{u}_2 \end{bmatrix}$$

with

$$\hat{u}_2 = Z_2^{-1} \left[\begin{bmatrix} \left\langle \begin{matrix} \langle \dot{\hat{y}} \rangle_{\sigma_1} \\ V\ddot{\hat{y}} \end{matrix} \right\rangle_{\sigma_2} \\ \begin{bmatrix} W_3 & W_4 \end{bmatrix} \begin{bmatrix} \ddot{\hat{y}} \\ V\dddot{\hat{y}} \end{bmatrix} \end{bmatrix} - \begin{bmatrix} \left\langle \begin{matrix} \langle CA \rangle_{\sigma_1} \\ VCA^2 \end{matrix} \right\rangle_{\sigma_2} \\ \begin{bmatrix} W_3 & W_4 \end{bmatrix} \begin{bmatrix} CA^2 \\ VCA^3 \end{bmatrix} \end{bmatrix} x \right]$$

where x is a solution of the equation

$$\dot{x} = Ax + B\hat{u}$$

for initial conditions such that

$$\begin{bmatrix} \hat{y}_0 \\ V\dot{\hat{y}}_0 \\ [W_3, W_4] \begin{bmatrix} \dot{\hat{y}}_0 \\ V\ddot{\hat{y}}_0 \end{bmatrix} \end{bmatrix} = \begin{bmatrix} C \\ VCA \\ [W_3, W_4] \begin{bmatrix} CA \\ VCA^2 \end{bmatrix} \end{bmatrix} x_0$$

and Z_2 is obtained from

$$\langle \mathcal{G}_2 \rangle_\sigma = [Z_1 \quad Z_2]$$

This latter theorem generalises easily for arbitrary classes L_α^r. The proofs of the first two theorems can be found in ref. 1.2, those of the generalised theorems in ref. 1.5 (1).

(1) Or better in the report "Identifiable structures of multivariable systems" (CERA/ENSA, December, 1968), where the details of the calculations are given.

Example: (1) Let the system be

$$\dot{x} = \begin{bmatrix} -1 & & & \\ & -1 & & \\ & & -2 & \\ & & & -3 \end{bmatrix} x + \begin{bmatrix} 0 & 1 \\ 1 & 1 \\ 1 & 1 \\ 1 & 2 \end{bmatrix} u, \qquad y = \begin{bmatrix} 1 & 0 & 2 & 1 \\ 2 & 2 & 1 & 0 \end{bmatrix} x$$

and suppose an input $\hat{u}$ is to be found such that the output is

$$\hat{y} = \begin{bmatrix} \frac{27}{440} (1-t)^{11/3} \\ 0 \end{bmatrix}.$$

It is easily verified that

$$CB = \begin{bmatrix} 3 & 5 \\ 3 & 5 \end{bmatrix}, \quad V = [1, -1], \quad {}_1 = \begin{bmatrix} 3 & 5 \\ 3 & 5 \\ -3 & -5 \end{bmatrix},$$

$$W = \begin{bmatrix} 1 & -1 & 0 \\ 0 & 1 & 1 \end{bmatrix},$$

and that consequently

$$\mathcal{G}_2 = \begin{bmatrix} 3 & 5 \\ 3 & 5 \\ -3 & -5 \\ -3 & -5 \\ 7 & 13 \end{bmatrix}$$

is of rank 2. Although the spaces L_1^3, L_2^3 are empty, we can thus find a $\hat{u}$ in the class C_2^0 to generate $\hat{y}$. Following the notation of the preceding theorem, let us take:

$$\sigma_1 = (3),$$

$$\sigma_2 = (1,3),$$

(1) We shall see in chapter 5 the connections between this problem and that of non-interaction.

so that

$$\left\langle \begin{matrix} \langle g_1 \rangle_{\sigma_1} \\ W_3\ CAB + W_4\ VCA^2\ B \end{matrix} \right\rangle_{\sigma_2} = \begin{bmatrix} \boxed{3 \quad 5} \\ 3 \quad 5 \\ \boxed{-3 \quad -5} \\ 7 \quad 13 \end{bmatrix} = \begin{bmatrix} 3 & 5 \\ 7 & 13 \end{bmatrix} = Z_2$$

and then we have

$$\hat{u}_2 = \frac{1}{4}\begin{bmatrix} 13 & -5 \\ -7 & 3 \end{bmatrix} \left[\left[\left\langle \begin{matrix} \left\langle \begin{matrix} \frac{9}{40}(1-t)^{8/3} \\ 0 \end{matrix} \right\rangle_{\sigma_1} \\ \frac{3}{5}(1-t)^{5/3} \end{matrix} \right\rangle_{\sigma_2} \atop (1-t)^{2/3} \right] - \begin{bmatrix} -1 & 0 & -4 & -3 \\ -2 & -2 & -2 & 0 \\ -1 & -2 & -4 & 9 \\ 3 & +4 & -4 & -27 \end{bmatrix} x \right]$$

whence

$$\hat{u}_2 = \frac{1}{4}\begin{bmatrix} -5 \\ 3 \end{bmatrix} (1-t)^{2/3} - \frac{1}{4}\begin{bmatrix} -41 & -46 & -6 & 135 \\ 23 & 26 & 2 & -81 \end{bmatrix} x$$

with the initial conditions

$$\begin{bmatrix} \frac{27}{440} \\ 0 \\ \frac{9}{40} \\ \frac{3}{5} \end{bmatrix} = \begin{bmatrix} 1 & 0 & 2 & 1 \\ 2 & 2 & 1 & 0 \\ -1 & 2 & -2 & -3 \\ -3 & -4 & 2 & 9 \end{bmatrix} \begin{bmatrix} x_{10} \\ x_{20} \\ x_{30} \\ x_{40} \end{bmatrix}$$

The corresponding system is illustrated in figure 1.14.

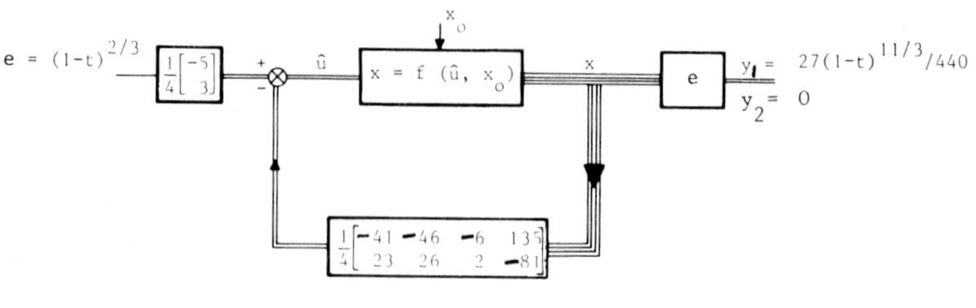

Fig. 1.14 Example of reproducibility in a function space.

Chapter 2

Representations

In the first chapter, we considered mathematical models giving accurate descriptions of linear dynamical systems. Such models, called "state-space models" and defined by equations of the form

$$\begin{aligned} \dot{x} &= Ax + Bu \\ y &= Cx + Du \end{aligned} \qquad (2.1)$$

are, from a theoretical viewpoint, of paramount importance. They do, however, correspond to a rather abstract conception of a system, being constructed in accordance with the axioms satisfied by the transition function; they do not necessarily express the physical structure of the system, and the state variables may not even be physically measurable quantities. Hence, such a representation should be considered as a mathematical tool which, though useful and even necessary for the correct design of many complex control systems, does not correspond, in general, to our basic knowledge of the system involved.

In many cases, it is not possible to write down the theoretical equations of a complex system, either because the number of equations is too large, or because of the impossibility of measuring some internal variables, or else because the underlying laws are not exactly known. In such a case, the only knowledge available is that of the external behaviour as expressed by input-output relations. If the inputs and outputs have been suitably chosen, such a set of transmittances may give a satisfactory representation of the system. In the case of linear time-invariant systems, this corresponds to the well-known transfer function matrix representation, relating the Laplace transforms of input and output vectors by an expression such as

$$Y(s) = Z(s)\,U(s) \tag{2.2}$$

We shall suppose that such transfer function matrices, of dimensions (p, m) where m is the number of inputs and p the number of outputs, are physically realisable, i.e. each element is a rational function of s with the degree of the numerator less than or equal to that of the denominator (when the inequality is strict for each element, the matrix will be said to be proper).

In other cases, especially when dealing with electrical or mechanical systems, it is possible, by applying the laws of Physics, to obtain, at least in first approximation, a set of differential equations relating the inputs u and outputs y via intermediate variables $\tilde{y}$. In the case of linear systems, with D denoting the operator of differentiation with respect to time, we get equations of the form

$$L^*(D)y + K(D)\tilde{y} = M^*(D)u$$

where L^*, K, M^* are polynomial matrices in D. By eliminating the intermediate variables, this equation can be transformed into

$$L(D)y = M(D)u \tag{2.3}$$

where L, M are matrices of respective dimensions (p, p), (p, m).

In the following, we shall consider only "physical systems", for which:

i) it is possible to associate a representation of the form (2.2) with the representation (2.3), which means that the matrix L(D) is invertible;

ii) the associated transfer function matrix Z(s) is proper;

and we shall refer to relation (2.3) as a "representation by differential operators" or "differential representation".

Contrary to what may be thought by the engineer used to working only with frequency-domain representations, such as Z(s), the various representations defined by equations (2.1) to (2.3) may not "represent" a given system with the same accuracy, when some information about the structure of the system is available. As an example, to illustrate what is meant, consider the single-input, single-output system of figure 2.1, which consists of four subsystems connected two-by-two in parallel and then in cascade.

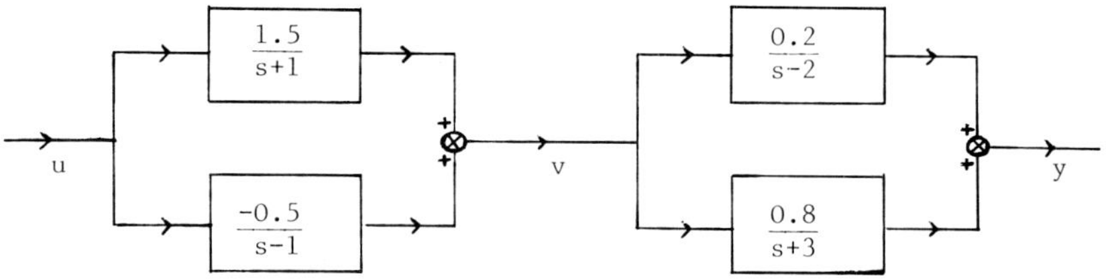

Fig. 2.1 A system comprising four subsystems

A state representation can readily be obtained by taking as state variables the outputs of the various first-order subsystems:

$$\begin{bmatrix} x_1 \\ x_2 \\ x_3 \\ x_4 \end{bmatrix} = \begin{bmatrix} -1 & 0 & 0 & 0 \\ 0 & 1 & 0 & 0 \\ 1 & 1 & 2 & 0 \\ 1 & 1 & 0 & -3 \end{bmatrix} \begin{bmatrix} x_1 \\ x_2 \\ x_3 \\ x_4 \end{bmatrix} + \begin{bmatrix} 1.5 \\ -0.5 \\ 0 \\ 0 \end{bmatrix} u \qquad (2.4)$$

$$y = [0,\ 0,\ 0.2,\ 0.8]x^T$$

The characteristic polynomial (s+1)(s-1)(s-2)(s+3) is of order 4 and exhibits the four modes indicated in the diagram.

To get the input-output differential equation, we can eliminate x_1 and x_2 from the equations for the first two subsystems, giving

$$\ddot{v} - v = \dot{u} - 2u,$$

and then eliminate v to obtain the desired equation

$$\dddot{y} + 2\ddot{y} - 5\dot{y} - 6y = \dot{u} - 2u \qquad (2.5)$$

This equation is only of order 3, its characteristic polynomial being

$s^3 + 2s^2 - 5s - 6 = (s-2)(s+3)(s+1)$. The mode $s = 1$ has been "lost" by this representation.

If we now look for the transfer function, it is easily seen to be given by

$$\frac{Y(s)}{U(s)} = \frac{1}{(s+1)(s+3)} \tag{2.6}$$

Two modes, those at $s = 1$ and $s = 2$, have been lost in this representation, and the forms (2.4), (2.5), (2.6) are clearly not equivalent.

The purpose of this chapter is twofold:
first, to specify the properties of the transfer function matrix representation (2.2) and the differential operator representation (2.3), and, by comparison with the state-space model (2.1), to determine the conditions under which they give an accurate picture of a given system;

second, to develop algorithms enabling us to convert the transfer function matrix and differential operator representations into an equivalent state-space form.

2.1 TRANSFER FUNCTION MATRICES AND STATE REPRESENTATIONS

2.1.1 The transfer function matrix, controllability and observability

We saw in chapter 1 that a time invariant (1) linear dynamical system, defined by its state equations, could be decomposed, in the most general case, into four subsystems. If the state vector x is decomposed in the form

$$x = \begin{bmatrix} x_1 \\ x_2 \\ x_3 \\ x_4 \end{bmatrix}$$

(1) In fact, this property still holds for time-varying systems and has been extended to certain "nearly linear" nonlinear systems, such as bilinear systems, cf. P. Alessandro (refs. 2.2, 2.3).

where the components represent subsets of states as follows:

x_1 - controllable and unobservable,
x_2 - controllable and observable,
x_3 - uncontrollable and unobservable,
x_4 - uncontrollable and observable,

then the state equations can be written (cf. refs. 2.14, 2.15)

$$\begin{bmatrix} \dot{x}_1 \\ \dot{x}_2 \\ \dot{x}_3 \\ \dot{x}_4 \end{bmatrix} = \begin{bmatrix} A_{11} & A_{12} & A_{13} & A_{14} \\ 0 & A_{22} & 0 & A_{24} \\ 0 & 0 & A_{33} & A_{34} \\ 0 & 0 & 0 & A_{44} \end{bmatrix} \begin{bmatrix} x_1 \\ x_2 \\ x_3 \\ x_4 \end{bmatrix} + \begin{bmatrix} B_1 \\ B_2 \\ 0 \\ 0 \end{bmatrix} u,$$

$$y = \begin{bmatrix} 0 & C_2 & 0 & C_4 \end{bmatrix} \begin{bmatrix} x_1 \\ x_2 \\ x_3 \\ x_4 \end{bmatrix}. \qquad (2.7)$$

The transfer function matrix associated with the system given by equations (2.7) is obtained by taking the Laplace transforms of these first-order differential equations, with null initial conditions:

$$Z(s) = C(sI - A)^{-1}B$$

i.e. on account of the special structure of the A, B and C matrices:

$$Z(s) = \begin{bmatrix} 0 & C_2 & 0 & C_4 \end{bmatrix} \begin{bmatrix} (sI-A_{11})^{-1} & \times & \times & \times \\ 0 & (sI-A_{22})^{-1} & \times & \times \\ 0 & 0 & (sI-A_{33})^{-1} & \times \\ 0 & 0 & 0 & (sI-A_{44})^{-1} \end{bmatrix} \begin{bmatrix} B_1 \\ B_2 \\ 0 \\ 0 \end{bmatrix}$$

$$= C_2(sI-A_{22})^{-1}B_2.$$

It is thus identical to the transfer function matrix obtained by considering, instead of equations (2.7), the system defined by the equations

$$\dot{x}_2 = A_{22}x_2 + B_2u,$$
$$y = C_2x_2.$$

This means that the transfer function matrix represents only the controllable and observable part of a system, and loses all information about the uncontrollable and unobservable parts. This point will be of fundamental importance in all the following, not only in regard to representations, as we shall see in this chapter, but also when dealing with control problems, as will be made clear in chapters 4 and 5.

This explains the form (2.6) previously found for the system of figure 2.1. In fact, using a transformation matrix constructed from the eigenvectors of the A matrix, the state equations (2.4) can be rewritten in the form

$$\begin{bmatrix} \dot{\xi}_1 \\ \dot{\xi}_2 \\ \dot{\xi}_3 \\ \dot{\xi}_4 \end{bmatrix} = \begin{bmatrix} -1 & 0 & 0 & 0 \\ 0 & 1 & 0 & 0 \\ 0 & 0 & 2 & 0 \\ 0 & 0 & 0 & -3 \end{bmatrix} \begin{bmatrix} \xi_1 \\ \xi_2 \\ \xi_3 \\ \xi_4 \end{bmatrix} + \begin{bmatrix} 0.75 \\ -0.25 \\ 0 \\ -0.625 \end{bmatrix} u,$$

$$y = \left[\frac{2}{3}, \; 0, \; \frac{1}{5}, \; \frac{4}{5} \right] \xi,$$

as illustrated in figure 2.2.

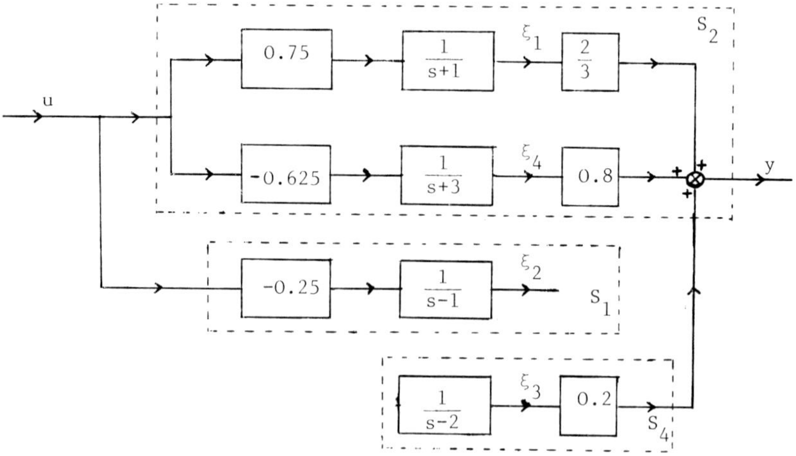

Fig. 2.2 Canonical decomposition of the system of fig. 2.1.

Only the controllable and observable modes, $s = -1$ and $s = -3$, appear in the transfer function representation.

2.1.2 The minimal realisation problem

The inverse problem, namely that of passing from a given transfer function matrix $Z(s)$ to an "equivalent" state-space form, will consequently require some care. Before dealing with practical techniques, it will be worthwhile to make the following remarks.

1) Part of the problem is to find a matrix triple (A, B, C), of appropriate dimensions, such that

$$Z(s) = C(sI-A)^{-1}B. \tag{2.8}$$

However, this equation can be satisfied by an infinity of triples and not all of them are solutions to our problem. The dimension of the state vector is, in fact, not fixed by equation (2.8), since any number of extra state variables can be adjoined to it without affecting $Z(s)$, provided that they correspond to uncontrollable or unobservable modes. It should be noticed that these additional states are arbitrary, and may be introduced by the particular method in use, but have no physical significance.

Example: Consider for example the controllable and observable system

$$\dot{x} = \begin{bmatrix} -1 & 0 & 0 \\ 0 & -2 & 0 \\ 0 & 0 & -3 \end{bmatrix} x + \begin{bmatrix} 1 & 0 \\ 1 & 2 \\ 2 & 1 \end{bmatrix} u,$$

$$y = \begin{bmatrix} 1 & 1 & 2 \\ 3 & 1 & 5 \end{bmatrix} x,$$

whose transfer function matrix is

$$Z(s) = \frac{1}{(s+1)(s+2)(s+3)} \begin{bmatrix} 6s^2 + 21s + 17 & 4s^2 + 14s + 10 \\ 14s^2 + 49s + 41 & 7s^2 + 23s + 16 \end{bmatrix}.$$

The system defined by the state equations

$$\dot{x} = \begin{bmatrix} -1 & 0 & 0 & 0 & 0 & 0 \\ 0 & -2 & 0 & 0 & 0 & 0 \\ 0 & 0 & -3 & 0 & 0 & 0 \\ 0 & 0 & 0 & -1 & 0 & 0 \\ 0 & 0 & 0 & 0 & -2 & 0 \\ 0 & 0 & 0 & 0 & 0 & 4 \end{bmatrix} x + \begin{bmatrix} 1 & 0 \\ 1 & 2 \\ 2 & 1 \\ 2 & 1 \\ 1 & 2 \\ 1 & 2 \end{bmatrix} u,$$

$$y = \begin{bmatrix} 1 & 1 & 2 & 2 & 2 & 0 \\ 3 & 1 & 5 & 6 & 1 & 0 \end{bmatrix} x,$$

has the same transfer function matrix, since the extra modes are either uncontrollable or unobservable. Clearly, however, such a form cannot be regarded as a suitable representation of the system described by this transfer function matrix.

2) It follows that, when we want to find a state representation corresponding to Z(s), there are two problems to be solved:

first, to obtain a triple (A, B, C) satisfying

$$Z(s) = C(sI-A)^{-1}B;$$

second, to ensure that the triple is controllable and observable, i.e. that A is of minimal dimension.

We shall see that the first problem is always quite easy to solve, not to say trivial. The second will in general be more difficult. Consequently, two kinds of method may be considered (cf. figure 2.3):-

i) Solve both problems in a single step. In a few cases this will be easy, for instance, when all the poles of the transfer function matrix are real and simple (cf. paragraph 2.1.3.B). In the general case, it will necessitate, in one form or another, the determination of the invariants of the system, by a rather cumbersome procedure.

ii) Solve the problems, one after the other. This means that, as a first step, a triple (A, B, C) satisfying equation (2.8) is found, and the state representation obtained is tested to see whether or not it is controllable and observable . If not, it will be necessary to eliminate, by a reduction process, all the uncontrollable or unobservable modes. We shall see (cf. paragraph 2.1.3.C) that this procedure can be advantageous.

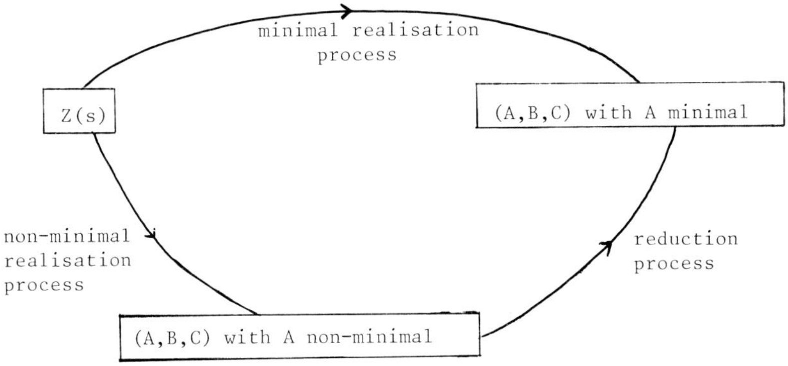

Fig. 2.3 Methods of obtaining a minimal realisation.

Note: It should also be remembered that, even with the condition that A is minimal, a state-space representation is not unique, since it is defined only to within a linear transformation. Thus, if (A, B, C) is a solution, any triple ($\tilde{A}$, $\tilde{B}$, $\tilde{C}$) will also be a solution if, for some non-singular matrix M,

$$\tilde{A} = M^{-1}AM, \quad \tilde{B} = M^{-1}B, \quad \tilde{C} = CM,$$

since then

$$\begin{aligned}\tilde{C}(sI-\tilde{A})^{-1}\tilde{B} &= CM[sI-M^{-1}AM]^{-1}M^{-1}B\\ &= C(sI-A)^{-1}B\\ &= Z(s).\end{aligned}$$

This property is important to the extent that it allows us, when searching for a solution, to specify the structure of the A matrix.

2.1.3 The methods

A - General case: use of the invariants

a) Traiangularisation and diagonalisation of matrices. Let us first recall a few theorems concerning polynomial matrices (cf. ref. 2.6).

Theorem 1. Given a polynomial matrix M(s), there exist two polynomial matrices V(s) and W(s), whose determinants are constant (i.e. independent of s), such that

$$M(s) = V(s)\,\Gamma^*(s)\,W(s) \qquad (2.9)$$

where Γ^* is a diagonal matrix

$$\Gamma^*(s) = \text{diag}\left[\gamma_1,\ldots\gamma_r,\ 0,\ldots\ 0\right], \qquad (2.10)$$

whose nonzero elements are the "invariants" of the matrix M. These have the property that γ_i divides γ_{i+1}. They can be obtained either directly or through a series of elementary row and column operations (1).

Directly, the invariants are obtained as follows. Define $\Delta_0 = 1$, and let Δ_i be the greatest common divisor of all the minors of order i formed from M(s). Then, $\gamma_i = \Delta_i/\Delta_{i-1}$.

<u>Example</u>: Let

$$M(s) = \begin{bmatrix} s+2 & 5s+7 \\ 2s+3 & 2(3s+4) \end{bmatrix}.$$

Then

$\Delta_0 = 1$, $\Delta_1 = 1$, $\Delta_2 = \det M(s) = -(s+1)(s+\frac{5}{4})$, and M(s) can be written in the form

$$M(s) = V(s)\begin{bmatrix} 1 & 0 \\ 0 & -(s+1)(s+\frac{5}{4}) \end{bmatrix}W(s).$$

<u>Theorem 2</u>. Given a rational matrix $Z(s) = M(s)/\psi(s)$, there exist two polynomial matrices V(s) and W(s), with constant determinants, such that

$$Z(s) = V(s)\,\Gamma(s)\,W(s) \qquad (2.11)$$

where $\Gamma(s)$ is a diagonal matrix

$$\Gamma(s) = \text{diag}\left[\frac{\varepsilon_1}{\psi_1},\ldots\frac{\varepsilon_q}{\psi_q},\varepsilon_{q+1},\ldots\ \varepsilon_r,\ 0,\ldots\ 0\right] \qquad (2.12)$$

with ε_i and ψ_i being relatively prime polynomials.

(1) See paragraph 2.2.2.C.b).

Indeed, from Theorem 1,

$$Z(s) = \frac{M(s)}{\psi(s)} = V(s)\,\frac{\Gamma^*(s)}{\psi(s)}\,W(s)$$

$$= V(s)\ \mathrm{diag}\left[\frac{\gamma_1}{\psi},\ldots \frac{\gamma_r}{\psi},\ldots 0,\ldots 0\right]W(s)$$

and the desired form follows by writing

$$\frac{\gamma_i(s)}{\psi(s)} = \frac{\varepsilon_i(s)}{\psi_i(s)}$$

with ε_i and ψ_i relatively prime.

Example: Let $Z(s) = M(s)/(s+1)^2$, where $M(s)$ is as in the previous example. Then,

$$\frac{\gamma_1}{\psi} = \frac{1}{(s+1)^2} = \frac{\varepsilon_1}{\psi_1},$$

$$\frac{\gamma_2}{\psi} = \frac{-(s+1)(s+\frac{5}{4})}{(s+1)^2},$$

whence, after cancellation of the common factor s + 1,

$$\frac{\varepsilon_2}{\psi_2} = \frac{-(s+\frac{5}{4})}{s+1}$$

Once a transfer function matrix $Z(s)$ has been written in the form (2.11), called "Smith-McMillan" form, it is easy to find a minimal state representation whose order is

$$n = \sum_{i=1}^{q} \text{degree } (\psi_i).$$

Thus, let ${}^iV$ denote the ith column of V and ${}_iW$ denote the ith row of W, so that

$$V = [{}^1V,\ldots {}^pV]\ ,\quad W = \begin{bmatrix} {}_1W \\ \vdots \\ {}_mW \end{bmatrix}.$$

Then,

$$Z(s) = [{}^1V, \dots {}^pV] \operatorname{diag}\left[\frac{\varepsilon_1}{\psi_1}, \dots \frac{\varepsilon_q}{\psi_q}, \varepsilon_{q+1}, \dots \varepsilon_r, 0, \dots 0\right] \begin{bmatrix} {}_1W \\ \vdots \\ {}_mW \end{bmatrix}$$

$$= \sum_{i=1}^{q} {}^iV \frac{\varepsilon_i}{\psi_i} {}_iW + \sum_{i=q+1}^{r} {}^iV \varepsilon_i {}_iW$$

$$= \sum_{i=1}^{q} Z_i + \sum_{i=q+1}^{r} Z_i^* . \quad (2.13)$$

The matrix Z(s) has thus been written as a sum of elementary matrices, each of which, being the product of a column by a row, is of rank 1. Of these matrices, q are rational, associated with the various ψ_i, and the remaining r - q are polynomial matrices contributing to the non-proper part of Z(s). Since physical transfer function matrices are proper, the Z_i^* will not be considered further, as they must necessarily cancel, either among themselves or with the non-proper parts of the Z_i. Hence, if we associate with the proper part of each Z_i a state representation of order n_i = degree (ψ_i), then a state representation of Z can be obtained by combining these "partial" representations.

b) Companion state-space form corresponding to an elementary transfer function matrix. For a given elementary matrix (1)

$$Z_i = \frac{{}^iV \; {}_iW}{\psi_i} \quad (2.14)$$

where ψ_i is a polynomial of degree n_i, we shall consider two cases, depending on the nature of the roots of ψ_i. Further, for convenience, we shall, in this paragraph, drop the subscripts i and write

$$Z = \frac{VW}{\psi} , \quad (2.14a)$$

leaving it to be understood that Z is a matrix of rank one, V a column vector and W a row vector.

(1) In this expression, ε_i has been absorbed into either ${}^iV$ or ${}_iW$.

General case: Let

$$\psi(s) = s^n + a_{n-1}s^{n-1} + \ldots + a_1 s + a_0.$$

The problem is to find a triple (A, B, C) such that:

(i) $C(sI-A)^{-1}B = Z(s)$,

(ii) dimension of A = n.

Let us fix the structure of A, specifically by taking a companion form

$$A = \begin{bmatrix} 0 & 1 & 0 & \cdots & 0 \\ 0 & & \ddots & \ddots & \vdots \\ \vdots & \ddots & & \ddots & 0 \\ 0 & \cdots & 0 & 0 & 1 \\ -a_0 & \cdots & \cdots & \cdots & -a_{n-1} \end{bmatrix} \qquad (2.15)$$

We recall that this form has two interesting and important properties:

(i) $\det(sI-A) = \psi(s)$, (2.16)

(ii) $$\operatorname{adj}(sI-A) = \begin{bmatrix} 1 \\ \vdots \\ s^{n-1} \end{bmatrix} \begin{bmatrix} s^{n-1}, & \ldots & 1 \end{bmatrix} \begin{bmatrix} 1 & 0 & \cdots & 0 \\ a_{n-1} & \ddots & \ddots & 0 \\ \vdots & \ddots & \ddots & \\ a_1 & \cdots & a_{n-1} & 1 \end{bmatrix} \bmod \psi(s) \qquad (2.17)$$

The latter expression (1) will often be written in the abbreviated form

$$\operatorname{adj}(sI-A) = \mathrm{I} \longmapsto \mathscr{A} \bmod \psi(s).$$

The problem is now to determine the B and C matrices so that

$$Z(s) = \frac{V(s)\, W(s)}{\psi(s)} = \frac{C \operatorname{adj}(sI-A)B}{\psi(s)}$$

(1) We recall that $a \bmod \psi = b$ means $a = b + k\psi$ with degree (b) < degree(ψ). Other properties are: $(a+b) \bmod \psi = a \bmod \psi + b \bmod \psi$, $(a\psi_1) \bmod (\psi_1\psi_2) = \psi_1(a \bmod \psi_2)$, $(ab) \bmod \psi = [(a \bmod \psi)\ (b \bmod \psi)] \bmod \psi$.

where B and C are constant matrices of respective dimensions (n, m) and (p, n). In view of equation (2.17), we can write

$$C\,\mathrm{adj}(sI-A)B = C\Big[\longmapsto \mathcal{A}\,B \bmod \psi.$$

The product of the first two matrices on the right-hand side is a polynomial column vector $\mathcal{C}(s)$. In fact, if the ith row of C has components

$${}_iC = [c_{oi}, \ldots c_{n-1,i}],$$

the ith element of $\mathcal{C}(s)$ is

$$c_{0i} + c_{1i}s + \ldots + c_{n-1,i}\,s^{n-1}.$$

Similarly, the product $\longmapsto \mathcal{A}\,B$ is a polynomial row vector $\mathcal{B}(s)$, and so we have

$$\psi(s)\,Z(s) = V(s)W(s) = \mathcal{C}(s)\mathcal{B}(s) \bmod \psi(s). \qquad (2.18)$$

To satisfy this equation, we can take

$$V(s) = \mathcal{C}(s),$$

$$W(s) = \mathcal{B}(s) \bmod \psi(s),$$

so that

$$C = \left[V(0),\ \dot{V}(0), \ldots \frac{V^{n-1}(0)}{(n-1)!}\right] \qquad (2.19)$$

and

$$B = \mathcal{A}^{-1}\begin{bmatrix} \dfrac{W^{n-1}(0)}{(n-1)!} \\ \vdots \\ \dot{W}(0) \\ W(0) \end{bmatrix}. \qquad (2.20)$$

Example: Let

$$Z(s) = \frac{1}{s^3+2s^2+2s+1}\begin{bmatrix} s+2 & s^2+4s+3 \\ 2s+3 & 2s^2+7s+5 \end{bmatrix}$$

which can be written in Smith-McMillan form as

$$Z(s) = \begin{bmatrix} s+2 & -1 \\ 2s+3 & -2 \end{bmatrix} \begin{bmatrix} \dfrac{1}{s^3+2s^2+2s+1} & 0 \\ 0 & \dfrac{1}{s^2+s+1} \end{bmatrix} \begin{bmatrix} 1 & s+1 \\ 0 & -1 \end{bmatrix}$$

$$= \begin{bmatrix} s+2 \\ 2s+3 \end{bmatrix} \frac{1}{s^3+2s^2+2s+1} \begin{bmatrix} 1, & s+1 \end{bmatrix} + \begin{bmatrix} -1 \\ -2 \end{bmatrix} \frac{1}{s^2+s+1} \begin{bmatrix} 0, & -1 \end{bmatrix}$$

$$= Z_1 + Z_2 \;.$$

Using equations (2.15), (2.19) and (2.20), the first elementary matrix Z_1 is associated with the state representation

$$\dot{x}_1 = \begin{bmatrix} 0 & 1 & 0 \\ 0 & 0 & 1 \\ -1 & -2 & -2 \end{bmatrix} x_1 + \begin{bmatrix} 1 & 0 & 0 \\ 2 & 1 & 0 \\ 2 & 2 & 1 \end{bmatrix}^{-1} \begin{bmatrix} 0 & 0 \\ 0 & 1 \\ 1 & 1 \end{bmatrix} u,$$

$$v_1 = \begin{bmatrix} 2 & 1 & 0 \\ 3 & 2 & 0 \end{bmatrix} x_1,$$

and similarly Z_2 is associated with

$$\dot{x}_2 = \begin{bmatrix} 0 & 1 \\ -1 & -1 \end{bmatrix} x_2 + \begin{bmatrix} 1 & 0 \\ 1 & 1 \end{bmatrix}^{-1} \begin{bmatrix} 0 & 0 \\ 0 & -1 \end{bmatrix} u,$$

$$v_2 = \begin{bmatrix} -1 & 0 \\ -2 & 0 \end{bmatrix} x_2,$$

so that a state representation of Z is given by

$$\dot{x} = \begin{bmatrix} 0 & 1 & 0 & 0 & 0 \\ 0 & 0 & 1 & 0 & 0 \\ -1 & -2 & -2 & 0 & 0 \\ 0 & 0 & 0 & 0 & 1 \\ 0 & 0 & 0 & -1 & -1 \end{bmatrix} x + \begin{bmatrix} 0 & 0 \\ 0 & 1 \\ 1 & -1 \\ 0 & 0 \\ 0 & -1 \end{bmatrix} u,$$

$$y = \begin{bmatrix} 2 & 1 & 0 & -1 & 0 \\ 3 & 2 & 1 & -2 & 0 \end{bmatrix} x.$$

Note 1: On account of equation (2.20), the method requires, in general, the inversion of the matrix $\mathcal{A}$. It should be noticed that, whatever the dimension of this matrix is, its inversion is straightforward. It is, indeed, a very special matrix; not only is it triangular with ones on the diagonal, but every subdiagonal is made up of identical elements. Further, the same properties hold for $\mathcal{A}^{-1}$, and consequently only n - 1 elements out of the n^2 need to be calculated (they can be computed by setting the products of successive rows of $\mathcal{A}^{-1}$ by the first column of $\mathcal{A}$ equal to zero).

Note 2: Instead of choosing the companion matrix A in the form (2.15), where the coefficients of the characteristic polynomial appear in the last row, we could have taken the dual form

$$A = \begin{bmatrix} -a_{n-1} & 1 & 0 & \cdots & 0 \\ \vdots & 0 & \ddots & \ddots & \vdots \\ \vdots & \vdots & & \ddots & 0 \\ -a_1 & \vdots & & & 1 \\ -a_0 & 0 & \cdots & \cdots & 0 \end{bmatrix} \qquad (2.21)$$

Then, property i) would remain, but property ii) would be replaced by:

$$\operatorname{adj}(sI-A) = \begin{bmatrix} 1 & 0 & \cdots & \cdots & 0 \\ a_{n-1} & \ddots & \ddots & & \vdots \\ \vdots & \ddots & \ddots & \ddots & 0 \\ a_1 & \cdots & \cdots & a_{n-1} & 1 \end{bmatrix} \begin{bmatrix} 1 \\ \vdots \\ \vdots \\ s^{n-1} \end{bmatrix} [s^{n-1}, \ldots 1] \bmod \psi(s). \qquad (2.22)$$

Also, proceeding in a manner completely analogous to the previous development, we should obtain

$$B = \begin{bmatrix} \dfrac{W^{n-1}(0)}{(n-1)!} \\ \vdots \\ \dot{W}(0) \\ \dot{W}(0) \end{bmatrix}, \quad C = \left[V(0),\ \dot{V}(0), \ldots \frac{V^{n-1}(0)}{(n-1)!}\right] \mathcal{A}^{-1}. \qquad (2.23)$$

Example: For the previous example, where Z(s) was decomposed into

$$Z(s) = \begin{bmatrix} s+2 \\ 2s+3 \end{bmatrix} \frac{1}{s^3+2s^2+2s+1} \begin{bmatrix}1, & s+1\end{bmatrix} + \begin{bmatrix} -1 \\ -2 \end{bmatrix} \frac{1}{s^2+s+1} \begin{bmatrix}0, & -1\end{bmatrix},$$

the corresponding state-space form would be

$$\dot{x} = \begin{bmatrix} -2 & 1 & 0 & 0 & 0 \\ -2 & 0 & 1 & 0 & 0 \\ -1 & 0 & 0 & 0 & 0 \\ 0 & 0 & 0 & -1 & 1 \\ 0 & 0 & 0 & -1 & 0 \end{bmatrix} x + \begin{bmatrix} 0 & 0 \\ 0 & 1 \\ 1 & 1 \\ 0 & 0 \\ 0 & -1 \end{bmatrix} u,$$

$$y = \begin{bmatrix} 2 & 1 & 0 \\ 3 & 2 & 0 \end{bmatrix} \begin{bmatrix} 1 & 0 & 0 \\ 2 & 1 & 0 \\ 2 & 2 & 1 \end{bmatrix}^{-1} x_1 + \begin{bmatrix} -1 & 0 \\ -2 & 0 \end{bmatrix} \begin{bmatrix} 1 & 0 \\ 1 & 1 \end{bmatrix}^{-1} x_2$$

$$= \begin{bmatrix} 0 & 1 & 0 & -1 & 0 \\ -1 & 2 & 0 & -2 & 0 \end{bmatrix} x,$$

with

$$x = \begin{bmatrix} x_1 \\ x_2 \end{bmatrix}.$$

c) Jordan state-space form corresponding to an elementary matrix with multiple poles. In the case that $\psi(s)$ has the form $(s+\lambda)^n$, it is better, instead of taking A as a companion matrix, to use a Jordan block form. This exhibits more clearly the presence of a multiple root. Hence, let A be a Jordan matrix of order n:

$$A = \begin{bmatrix} -\lambda & 1 & 0 & \cdots & 0 \\ 0 & -\lambda & 1 & \ddots & \vdots \\ \vdots & \ddots & \ddots & \ddots & 0 \\ \vdots & & \ddots & \ddots & 1 \\ 0 & \cdots & \cdots & 0 & -\lambda \end{bmatrix} . \tag{2.24}$$

This matrix has the properties

i) $$\det(sI-A) = (s+\lambda)^n, \tag{2.25}$$

ii) $$(sI-A)^{-1} = \begin{bmatrix} \frac{1}{s+\lambda} & \frac{1}{(s+\lambda)^2} & \cdots & \frac{1}{(s+\lambda)^n} \\ 0 & \frac{1}{s+\lambda} & \ddots & \vdots \\ \vdots & \ddots & \ddots & \frac{1}{(s+\lambda)^2} \\ 0 & \cdots & 0 & \frac{1}{s+\lambda} \end{bmatrix} , \tag{2.26}$$

and our problem is to find the B and C matrices, such that

$$C(sI-A)^{-1}B = Z(s) = \frac{V(s)W(s)}{\psi(s)} = \frac{M(s)}{\psi(s)} .$$

We remark first that $(sI-A)^{-1}$ can be written in the form

$$(sI-A)^{-1} = \frac{I_1}{s+\lambda} + \frac{I_2}{(s+\lambda)^2} + \cdots + \frac{I_n}{(s+\lambda)^n} ,$$

where I_1 is the unit matrix of order n, and $I_2, \ldots I_n$ are matrices whose only nonzero elements are on shifted diagonals made up of ones:

$$I_2 = \begin{bmatrix} 0 & 1 & 0 & \cdots & 0 \\ \vdots & \ddots & \ddots & \ddots & 0 \\ \vdots & & \ddots & \ddots & 1 \\ 0 & \cdots & \cdots & \cdots & 0 \end{bmatrix} , \quad \cdots\cdots\cdots\cdots \quad I_n = \begin{bmatrix} 0 & \cdots & 0 & 1 \\ \vdots & \ddots & & 0 \\ \vdots & & \ddots & \vdots \\ 0 & \cdots & \cdots & 0 \end{bmatrix} .$$

Hence,

$$Z(s) = \frac{CI_1B}{s+\lambda} + \frac{CI_2B}{(s+\lambda)^2} + \ldots + \frac{CI_nB}{(s+\lambda)^n} \qquad (2.27)$$

On the other hand, since $Z(s) = M(s)/(s+\lambda)^n$, where $M(s)$ is a polynomial matrix and consequently can be expanded in the form (1)

$$M(s) = M(-\lambda) + (s+\lambda)\dot{M}(-\lambda) + \ldots + (s+\lambda)^{n-1}\frac{\overset{n-1}{M}(-\lambda)}{(n-1)!} \;,$$

we have

$$Z(s) = \frac{M(-\lambda)}{(s+\lambda)^n} + \frac{\dot{M}(-\lambda)}{(s+\lambda)^{n-1}} + \ldots + \frac{\overset{n-1}{M}(-\lambda)}{(n-1)!(s+\lambda)} \,. \qquad (2.28)$$

Hence, comparing the expressions (2.27) and (2.28), the matrices B and C are to be found from the equations:

$$CB = \frac{\overset{n-1}{M}(-\lambda)}{(n-1)!}, \quad CI_2B = \frac{\overset{n-2}{M}(-\lambda)}{(n-2)!}, \ldots \quad CI_nB = M(-\lambda).$$

Now, let us make use of the properties of $I_2, \ldots I_n$. Denoting the columns of C by ${}^1C, \ldots {}^nC$, and the rows of B by ${}_1B, \ldots {}_nB$, we have:

$$\begin{aligned}
CB &= {}^1C\,{}_1B + \ldots + {}^nC\,{}_nB, \\
CI_2B &= {}^1C\,{}_2B + \ldots + {}^{n-1}C\,{}_nB, \\
&\;\vdots \\
CI_{n-1}B &= {}^1C\,{}_{n-1}B + {}^2C\,{}_nB, \\
CI_nB &= {}^1C\,{}_nB\,.
\end{aligned} \qquad (2.29)$$

Consequently, using the fact that $M(s)$ is a matrix of rank 1, equal to $V(s)W(s)$, we have, by equating coefficients of $(s+\lambda)^{-i}$ for $i = 1,\ldots n$:

$$CI_nB = {}^1C\,{}_nB = M(-\lambda) = V(-\lambda)W(-\lambda),$$

(1) Since $Z(s)$ is proper, all the elements of $M(s)$ have degree less than n.

so that

$$^{1}C = V(-\lambda), \quad {}_{n}B = W(-\lambda);$$

$$CI_{n-1}B = {}^{1}C\,{}_{n-1}B + {}^{2}C\,{}_{n}B = \dot{M}(-\lambda) = V(-\lambda)\dot{W}(-\lambda) + \dot{V}(-\lambda)W(-\lambda),$$

so that

$$^{2}C = \dot{V}(-\lambda), \quad {}_{n-1}B = \dot{W}(-\lambda);$$

and so on, giving generally

$$^{i}C = \frac{\overset{i}{V}(-\lambda)}{i\,!}, \quad {}_{n-i+1}B = \frac{\overset{i}{W}(-\lambda)}{i\,!}.$$

Thus, the B and C matrices are

$$B = \begin{bmatrix} \dfrac{\overset{n-1}{W}(-\lambda)}{(n-1)!} \\ \vdots \\ \ddot{W}(-\lambda) \\ W(-\lambda) \end{bmatrix}, \quad C = \left[V(-\lambda),\ \dot{V}(-\lambda), \ldots \dfrac{\overset{n-1}{V}(-\lambda)}{(n-1)!} \right]. \qquad (2.30)$$

<u>Example 1</u>: Let Z(s) be written in Smith-McMillan form as

$$Z(s) = \begin{bmatrix} s+2 & -1 \\ 2s+3 & -2 \end{bmatrix} \begin{bmatrix} \dfrac{1}{(s+1)^2} & 0 \\ 0 & \dfrac{1}{(s+1)^2} \end{bmatrix} \begin{bmatrix} 1 & 3s+5 \\ 0 & -1 \end{bmatrix}$$

or

$$Z(s) = Z_1 + Z_2$$

$$= \begin{bmatrix} s+2 \\ 2s+3 \end{bmatrix} \frac{1}{(s+1)^3} [1, \ 3s+5] + \begin{bmatrix} -1 \\ -2 \end{bmatrix} \frac{1}{(s+1)^2} [0, \ -1].$$

Using the expressions (2.30), it is straightforward to write down a state representation for each of Z_1 and Z_2. The complete representation is

$$\dot{x} = \begin{bmatrix} -1 & 1 & 0 & 0 & 0 \\ 0 & -1 & 1 & 0 & 0 \\ 0 & 0 & -1 & 0 & 0 \\ 0 & 0 & 0 & -1 & 1 \\ 0 & 0 & 0 & 0 & -1 \end{bmatrix} x + \begin{bmatrix} 0 & 0 \\ 0 & 3 \\ 1 & 2 \\ 0 & 0 \\ 0 & -1 \end{bmatrix} u,$$

$$y = \begin{bmatrix} 1 & 1 & 0 & -1 & 0 \\ 1 & 2 & 0 & -2 & 0 \end{bmatrix} x.$$

Note: We remarked in paragraph a) above that, if the Z matrix given was proper, there should be no need to consider any non-proper parts which might appear, since they must ultimately cancel. This is illustrated by the following example.

Example 2: Let

$$Z(s) = \frac{1}{(s+1)^2} \begin{bmatrix} s+2 & 5s+7 \\ 2s+3 & 2(3s+4) \end{bmatrix}$$

$$= \begin{bmatrix} s+2 & -1 \\ 2s+3 & -2 \end{bmatrix} \begin{bmatrix} \frac{1}{(s+1)^2} & 0 \\ 0 & \frac{-(4s+5)}{s+1} \end{bmatrix} \begin{bmatrix} 1 & 4s+6 \\ 0 & -1 \end{bmatrix}$$

$$= \begin{bmatrix} s+2 \\ 2s+3 \end{bmatrix} \frac{1}{(s+1)^2} [1, \; 4s+6] + \begin{bmatrix} 1 \\ 2 \end{bmatrix} \frac{4s+5}{s+1} [0, \; -1] \qquad (2.31)$$

$$= \begin{bmatrix} s+2 \\ 2s+3 \end{bmatrix} \frac{1}{(s+1)^2} [1, \; 4s+6] + \begin{bmatrix} 1 \\ 2 \end{bmatrix} \frac{1}{s+1} [0, \; -1] + \begin{bmatrix} 1 \\ 2 \end{bmatrix} 4 \, [0, \; -1].$$

The direct transmittance appearing in the last term is cancelled by a contribution from the first one; this first matrix is indeed non-proper since

$$\begin{bmatrix} s+2 \\ 2s+3 \end{bmatrix} \frac{1}{(s+1)^2} [1, \; 4s+6] = \frac{1}{(s+1)^2} \begin{bmatrix} s+2 & 4s^2+14s+12 \\ 2s+3 & 8s^2+24s+18 \end{bmatrix}$$

$$= \frac{1}{(s+1)^2}\begin{bmatrix} s+2 & 6s+8 \\ 2s+3 & 8s+10 \end{bmatrix} + \begin{bmatrix} 0 & 4 \\ 0 & 8 \end{bmatrix}.$$

Although, in expression (2.31), the direct transmittance in the second term must be isolated in order to apply the relations (2.30), this is not necessary for the first term. We have, finally, the representation:

$$\dot{x} = \begin{bmatrix} -1 & 1 & 0 \\ 0 & -1 & 0 \\ 0 & 0 & -1 \end{bmatrix} x + \begin{bmatrix} 0 & 4 \\ 1 & 2 \\ 0 & -1 \end{bmatrix} u,$$

$$y = \begin{bmatrix} 1 & 1 & 1 \\ 1 & 2 & 2 \end{bmatrix} x .$$

B - Gilbert's method for the case of simple real poles

For the case where all the poles of the transfer function matrix are real and of first order, Gilbert (ref. 2.7, theorem 7) has given a method which allows a very easy determination of the order of a minimal state representation and also the construction of one. This method, which is an extension of the classical modal method for single-variable systems, is so convenient as to be well worth using whenever possible.

If all the poles $-\lambda_i$ $(i = 1, \ldots r)$, of a proper transfer function matrix $Z(s)$, are real and of order 1, we can write

$$Z(s) = \sum_{i=1}^{r} \frac{M_i}{s+\lambda_i} \tag{2.32}$$

where M_i, the matrix of residues associated with $-\lambda_i$, is given by

$$M_i = \lim_{s \to -\lambda_i} (s+\lambda_i) Z(s). \tag{2.33}$$

Let n_i be the rank of M_i. Then, the order of a minimal state representation is

$$n = \sum_{i=1}^{r} n_i = \sum_{i=1}^{r} \text{rank}\,(M_i), \tag{2.34}$$

and the matrices (A, B, C) of such a representation can be straightforwardly obtained from the form (2.32) of Z(s).

Example: Let

$$Z(s) = \frac{1}{s(s+1)(s+2)}\begin{bmatrix} 2(2s^2+4s+1) & s(s+2) \\ 2s(3s+5) & (s+2)(3s+1) \end{bmatrix}$$

so that, by decomposition,

$$Z(s) = \frac{1}{s}\begin{bmatrix} 1 & 0 \\ 0 & 1 \end{bmatrix} + \frac{1}{s+1}\begin{bmatrix} 2 & 1 \\ 4 & 2 \end{bmatrix} + \frac{1}{s+2}\begin{bmatrix} 1 & 0 \\ 2 & 0 \end{bmatrix} . \qquad (2.35)$$

The matrices M_1, M_2, M_3, are respectively of ranks 2, 1, 1, so that n = 4. The block diagram corresponding to equation (2.35) is shown, for greater clarity, in figure 2.4, representing the Laplace-transform relations

$$Y_1 = \frac{U_1}{s} + \frac{2U_1+U_2}{s+1} + \frac{U_1}{s+2} ,$$

$$Y_2 = \frac{U_2}{s} + \frac{2(2U_1+U_2)}{s+1} + \frac{2U_1}{s+2} .$$

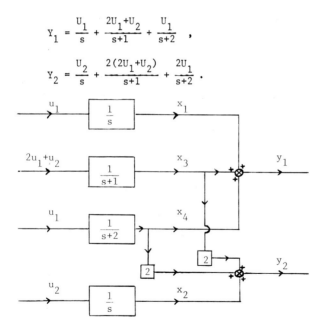

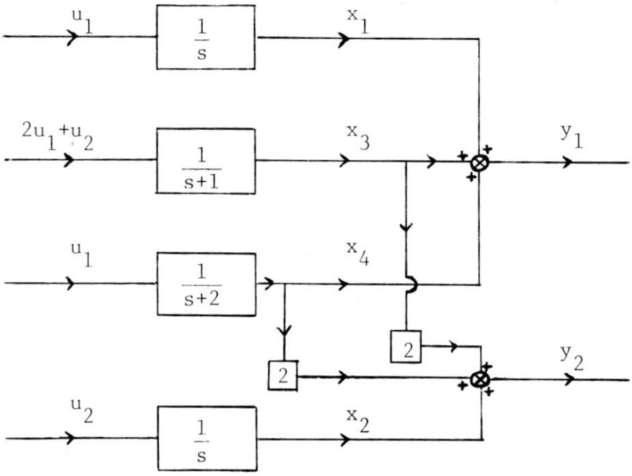

Fig. 2.4 Block diagram representation of equation (2.35).

If we take, as state-variables, the outputs of the first-order blocks in figure 2.4, we get directly:

$$\begin{bmatrix} \dot{x}_1 \\ \dot{x}_2 \\ \dot{x}_3 \\ \dot{x}_4 \end{bmatrix} = \begin{bmatrix} 0 & 0 & 0 & 0 \\ 0 & 0 & 0 & 0 \\ 0 & 0 & -1 & 0 \\ 0 & 0 & 0 & -2 \end{bmatrix} \begin{bmatrix} x_1 \\ x_2 \\ x_3 \\ x_4 \end{bmatrix} + \begin{bmatrix} 1 & 0 \\ 0 & 1 \\ 2 & 1 \\ 1 & 0 \end{bmatrix} \begin{bmatrix} u_1 \\ u_2 \end{bmatrix},$$

$$\begin{bmatrix} y_1 \\ y_2 \end{bmatrix} = \begin{bmatrix} 1 & 0 & 1 & 1 \\ 0 & 1 & 2 & 2 \end{bmatrix} \begin{bmatrix} x_1 \\ x_2 \\ x_3 \\ x_4 \end{bmatrix}.$$

Note: It is clear from the block diagram why the order of the representation is determined by the ranks of the M_i matrices. If one of the M_i has rank $\rho < p$, then $p - \rho$ of its rows can be expressed as linear combinations of the ρ others. The terms associated with $-\lambda_i$ in the expressions for $p - \rho$ of the outputs can then be generated from the blocks used for the ρ others. We shall, in fact, use this procedure systematically later, in the graph method (cf. paragraph 2.1.3.D).

C - Practical realisation by decomposition into a sum of rank-one matrices

a) Construction of a non-minimal state equation. Given a transfer function matrix $Z(s) = M(s)/(s+\lambda)^r$, we have seen, in paragraph 2.1.3.A, that it is possible, using Smith-McMillan form, to write it as a sum of elementary matrices. By an elementary matrix, we mean here a matrix of rank 1, which is the product of a column vector and a row vector, associated with an invariant of Z(s). While this method has the advantage of giving a minimal realisation directly as a combination of partial realisations related to the elementary matrices, it also has definite drawbacks with regard to the amount of computation required.

Another method is to use only the unit-rank property for the matrices in the decomposition. We can, indeed, immediately write Z(s) as a sum of rank-one matrices, simply by decomposing M(s) into its columns. Thus, let ${}^1M, \ldots {}^mM$ be the m columns of M(s). Then, Z(s) can be expanded as

$$Z(s) = \sum_{i=1}^{m} \frac{{}^iM[0,\ldots 0,1,0,\ldots 0]}{(s+\lambda)^r} = \sum_{i=1}^{m} Z_i. \tag{2.36}$$

Now, using the relations (2.30), a state representation, of order r, can be written directly for each Z_i:

$$\dot{x}_i = \begin{bmatrix} -\lambda & 1 & 0 & \cdots & 0 \\ 0 & \ddots & \ddots & & \vdots \\ \vdots & & \ddots & \ddots & 0 \\ & & & \ddots & 1 \\ 0 & \cdots & & 0 & -\lambda \end{bmatrix} x_i + \begin{bmatrix} 0 & \cdots & & \cdots & 0 \\ \vdots & & & & \vdots \\ 0 & \cdots & & \cdots & 0 \\ 0 \cdots 0 & & 1 & & 0 \cdots 0 \end{bmatrix} u = Ax_i + B_i u,$$

$$v_i = \left[{}^iM(-\lambda), \ldots \frac{{}^iM^{r-1}(-\lambda)}{(r-1)!} \right] x_i = C_i x_i \; . \tag{2.37}$$

The A matrix is not subscripted since it is the same for each Z_i, as all these matrices have the same denominator $(s+\lambda)^r$. By combining the equations (2.37) for i = 1,...m, we get the complete representation

$$\dot{x} = \begin{bmatrix} \dot{x}_1 \\ \vdots \\ \dot{x}_m \end{bmatrix} = \begin{bmatrix} A & 0 & \cdots & 0 \\ 0 & \ddots & & 0 \\ \vdots & & \ddots & \\ 0 & \cdots & 0 & A \end{bmatrix} x + \begin{bmatrix} B_1 \\ \vdots \\ B_m \end{bmatrix} u, \tag{2.38}$$

$$y = v_1 + \ldots + v_m = [C_1 \ldots C_m] x.$$

Example: Consider the matrix

$$Z(s) = \frac{1}{(s+1)^3} \begin{bmatrix} 2s^2+4s+7 & s^2+2s+4 \\ 4s+7 & s+3 \end{bmatrix}.$$

We have

$$Z(s) = \frac{1}{(s+1)^3} \begin{bmatrix} 2s^2+4s+7 \\ 4s+7 \end{bmatrix} [1, \; 0] + \frac{1}{(s+1)^3} \begin{bmatrix} s^2+2s+4 \\ s+3 \end{bmatrix} [0, \; 1].$$

For the first matrix, equations (2.30) give the state representation

$$\dot{x}_1 = Ax_1 + B_1u,$$
$$v_1 = C_1x_1,$$

with

$$A = \begin{bmatrix} -1 & 1 & 0 \\ 0 & -1 & 1 \\ 0 & 0 & -1 \end{bmatrix}, \quad B_1 = \begin{bmatrix} 0 & 0 \\ 0 & 0 \\ 1 & 0 \end{bmatrix}, \quad C_1 = \begin{bmatrix} 5 & 0 & 2 \\ 3 & 4 & 0 \end{bmatrix},$$

and similarly, for the second matrix,

$$\dot{x}_2 = Ax_2 + B_2u,$$
$$v_2 = C_2x_2,$$

where

$$B_2 = \begin{bmatrix} 0 & 0 \\ 0 & 0 \\ 0 & 1 \end{bmatrix}, \quad C_2 = \begin{bmatrix} 3 & 0 & 1 \\ 2 & 1 & 0 \end{bmatrix}.$$

Hence, for the overall representation, we get

$$\dot{x} = \begin{bmatrix} -1 & 1 & 0 & 0 & 0 & 0 \\ 0 & -1 & 1 & 0 & 0 & 0 \\ 0 & 0 & -1 & 0 & 0 & 0 \\ 0 & 0 & 0 & -1 & 1 & 0 \\ 0 & 0 & 0 & 0 & -1 & 1 \\ 0 & 0 & 0 & 0 & 0 & -1 \end{bmatrix} x + \begin{bmatrix} 0 & 0 \\ 0 & 0 \\ 1 & 0 \\ 0 & 0 \\ 0 & 0 \\ 0 & 1 \end{bmatrix} u,$$

$$y = \begin{bmatrix} 5 & 0 & 2 & 3 & 0 & 1 \\ 3 & 4 & 0 & 2 & 1 & 0 \end{bmatrix} x.$$

Example 2: Consider the transfer function matrix (1) given by

(1) This transfer function matrix was used by R.E. Kalman (ref. 2.9) among others.

$$s^4 Z(s) = M(s) = \begin{bmatrix} s^3-s^2+1 & 1 & -s^3+s^2-2 \\ 1.5s+1 & s+1 & -1.5s-2 \\ s^3-9s^2-s+1 & -s^2+1 & s^3-s-2 \end{bmatrix}$$

Following the procedure given above, we can write M(s) directly as a sum of matrices of rank 1 by using the column decomposition

$$M(s) = \begin{bmatrix} s^3-s^2+1 \\ 1.5s+1 \\ s^3-9s^2-s+1 \end{bmatrix} [1,\ 0,\ 0] + \begin{bmatrix} 1 \\ s+1 \\ -s^2+1 \end{bmatrix} [0,\ 1,\ 0] + \begin{bmatrix} -s^3+s^2-2 \\ -1.5s-2 \\ s^3-s-2 \end{bmatrix} [0,\ 0,\ 1]$$

$$= (Z_1 + Z_2 + Z_3)s^4.$$

To each of the elementary matrices Z_i, there corresponds a state representation, in which the A matrix appears as a Jordan block of order 4 associated with the eigenvalue zero, and the matrices B_i and C_i are determined in accordance with equations (2.30). We get

$$A = \begin{bmatrix} 0 & 1 & 0 & 0 \\ 0 & 0 & 1 & 0 \\ 0 & 0 & 0 & 0 \\ 0 & 0 & 0 & 0 \end{bmatrix}, \quad B_1 = \begin{bmatrix} 0 & 0 & 0 \\ 0 & 0 & 0 \\ 0 & 0 & 0 \\ 1 & 0 & 0 \end{bmatrix}, \quad C_1 = \begin{bmatrix} 1 & 0 & -1 & 1 \\ 1 & 1.5 & 0 & 0 \\ 1 & -1 & -9 & 1 \end{bmatrix},$$

$$B_2 = \begin{bmatrix} 0 & 0 & 0 \\ 0 & 0 & 0 \\ 0 & 0 & 0 \\ 0 & 1 & 0 \end{bmatrix}, \quad C_2 = \begin{bmatrix} 1 & 0 & 0 & 0 \\ 1 & 1 & 0 & 0 \\ 1 & 0 & -1 & 0 \end{bmatrix},$$

$$B_3 = \begin{bmatrix} 0 & 0 & 0 \\ 0 & 0 & 0 \\ 0 & 0 & 0 \\ 0 & 0 & 1 \end{bmatrix}, \quad C_3 = \begin{bmatrix} -2 & 0 & 1 & -1 \\ -2 & -1.5 & 0 & 0 \\ -2 & -1 & 0 & 1 \end{bmatrix},$$

and a state representation is given by

$$\dot{x} = \begin{bmatrix} A & 0 & 0 \\ 0 & A & 0 \\ 0 & 0 & A \end{bmatrix} x + \begin{bmatrix} B_1 \\ B_2 \\ B_3 \end{bmatrix} u, \qquad (2.39)$$

$$y = \begin{bmatrix} C_1 & C_2 & C_3 \end{bmatrix} x.$$

b) Reduction of the representation to minimal form. The state representation obtained above will be, in most cases, non-minimal (see, e.g. example 2). It is then necessary to reduce it by elimination of any uncontrollable or unobservable parts which it contains.

Let us briefly recall the criteria obtained in chapter 1 for the controllability and observability of a system in state-space form, where the A matrix contains a number of Jordan blocks with identical eigenvalues. The representation is controllable if the matrix B_C, composed of the last rows of the blocks B_i of B associated with the various Jordan blocks, is of full rank. Similarly, it is observable if the matrix C_0, constructed of the first columns of the corresponding blocks C_i of C, is of full rank.

For the form given by equations (2.37) and (2.38), the representation is obviously controllable since B_C is a unit matrix. It may, however, be unobservable, with C_0 having less than full rank. Let us assume that this is the case; the reduction procedure can then be carried out according to the following rules.

Rule 1: Take a linear combination of the blocks C_i such that a null first column appears in some block C_k (such a linear combination must exist if C_0 is rank-deficient). Since we do not want this procedure to alter the transfer function matrix, the corresponding combination must be applied to the blocks B_i.

Let us suppose, for instance, that the first column ${}^1C_1$ of C_1 is a linear combination of the first columns of the other C_j, i.e.

$${}^1C_1 = \alpha\, {}^1C_2 + \ldots + \nu\, {}^1C_m .$$

If the block C_1 is replaced by $C_1 - \alpha C_2 - \ldots - \nu C_m$, the first column of the new C_1 will be null. The transfer function matrix is

$$Z = C_1 \quad B_1 + \ldots + C_m \quad B_m,$$

where

$$= (sI-A)^{-1},$$

and it can be written as

$$Z = (C_1 - \alpha C_2 - \ldots - \nu C_m) \quad B_1 + C_2 \quad (B_2 + \alpha B_1) + \ldots + C_m \quad (B_m + \nu B_1)$$

since it is unaltered if, when the above modification is made to C_1, the blocks $B_2, \ldots B_m$ are replaced by $B_2 + \alpha B_1, \ldots B_m + \nu B_1$, respectively.

Rule 2: If the first column of a block C_i is null, the transfer function matrix will be unaltered by shifting all the columns of this block to the left, if the rows of the corresponding block B_i are simultaneously shifted upwards and a null last row introduced (the last column of C_i can be chosen arbitrarily).

Thus, consider an elementary matrix corresponding to the triple (C_i, A, B_i), where, for convenience, the subscript i will now be dropped:

$$Z = C \quad B, \text{ with } C = [0, {}^2C, \ldots {}^rC], \quad B = \begin{bmatrix} {}_rB \\ \vdots \\ {}_1B \end{bmatrix}.$$

We have

$$Z = [0, {}^2C, \ldots {}^rC] \begin{bmatrix} \frac{1}{s+\lambda} & \cdots & \frac{1}{(s+\lambda)^r} \\ 0 & \ddots & \vdots \\ 0 \cdots 0 & & \frac{1}{s+\lambda} \end{bmatrix} \begin{bmatrix} {}_rB \\ \vdots \\ {}_1B \end{bmatrix}$$

$$= [{}^2C, \ldots {}^rC, \times] \begin{bmatrix} \frac{1}{s+\lambda} & \cdots & \frac{1}{(s+\lambda)^r} \\ 0 & \ddots & \vdots \\ 0 \cdots 0 & & \frac{1}{s+\lambda} \end{bmatrix} \begin{bmatrix} {}_{r-1}B \\ \vdots \\ {}_1B \\ 0 \end{bmatrix}.$$

Rule 3: If the first ρ columns of a block C_i are null, ρ states can be eliminated from the corresponding block of A.

Indeed,

$$[0,\dots 0,\ {}^1C,\dots {}^{r-\rho}C] \begin{bmatrix} \frac{1}{s+\lambda} & \cdots & \frac{1}{(s+\lambda)^r} \\ 0 & \ddots & \vdots \\ 0 \cdots 0 & & \frac{1}{s+\lambda} \end{bmatrix} \begin{bmatrix} {}_rB \\ \vdots \\ {}_1B \end{bmatrix}$$

$$= [{}^1C,\dots {}^{r-\rho}C] \begin{bmatrix} \frac{1}{s+\lambda} & \cdots & \frac{1}{(s+\lambda)^{r-\rho}} \\ 0 & \ddots & \vdots \\ 0 \cdots 0 & & \frac{1}{s+\lambda} \end{bmatrix} \begin{bmatrix} {}_{r-\rho}B \\ \vdots \\ {}_1B \end{bmatrix}.$$

The successive application of the above rules will enable a full reduction of the original state representation to be made. Note that rule 3 should be used only when it is clear that the reduction process cannot otherwise be carried further. The purpose of rule 2 is to allow the repeated use of rule 1 after a first null column has appeared, in case more than one state can be eliminated in a single block.

Example: Let us return to example 2 of paragraph a) above. It is clear that the state representation (2.39) is not minimal. It is, in fact, unobservable, since the first columns of the blocks C_i are linearly dependent.

In order to reduce the representation according to the rules given above, we have first to combine the blocks C_i in such a way that a null column appears in first position (rule 1), and then shift the columns of C_i and rows of B_i (rule 2).

The sequence of steps is illustrated below (1).

1)

$$C = \left[\begin{array}{cccc:cccc:cccc} 1 & 0 & -1 & 1 & 1 & 0 & 0 & 0 & -2 & 0 & 1 & -1 \\ 1 & 1{,}5 & 0 & 0 & 1 & 1 & 0 & 0 & -2 & -1{,}5 & 0 & 0 \\ 1 & -1 & -9 & 1 & 1 & 0 & -1 & 0 & -2 & -1 & 0 & 1 \end{array}\right]$$

$$B^T = \left[\begin{array}{cccc:cccc:cccc} 0 & 0 & 0 & 1 & 0 & 0 & 0 & 0 & 0 & 0 & 0 & 0 \\ 0 & 0 & 0 & 0 & 0 & 0 & 0 & 1 & 0 & 0 & 0 & 0 \\ 0 & 0 & 0 & 0 & 0 & 0 & 0 & 0 & 0 & 0 & 0 & 1 \end{array}\right]$$

2a) $C_1 + C_2 + C_3 \rightarrow C_2$, $B_1 - B_2 \rightarrow B_1$, $B_3 - B_2 \rightarrow B_3$.

$$C = \left[\begin{array}{cccc:cccc:cccc} & & & & \boxed{0} & 0 & 0 & 0 & & & & \\ & & & & \boxed{0} & 1 & 0 & 0 & & & & \\ & & & & \boxed{0} & -2 & -10 & 2 & & & & \end{array}\right]$$

$$B^T = \left[\begin{array}{cccc:cccc:cccc} 0 & 0 & 0 & 1 & & & & & 0 & 0 & 0 & 0 \\ 0 & 0 & 0 & -1 & & & & & 0 & 0 & 0 & -1 \\ 0 & 0 & 0 & 0 & & & & & 0 & 0 & 0 & 1 \end{array}\right]$$

2b) Shift columns of C_2 to left and rows of B_2 upwards.

$$C = \left[\begin{array}{cccc:cccc:cccc} & & & & 0 & 0 & 0 & \boxed{\times} & & & & \\ & & & & 1 & 0 & 0 & \boxed{\times} & & & & \\ & & & & -2 & -10 & 2 & \boxed{\times} & & & & \end{array}\right]$$

$$B^T = \left[\begin{array}{cccc:cccc:cccc} & & & & 0 & 0 & 0 & 0 & & & & \\ & & & & 0 & 0 & 1 & 0 & & & & \\ & & & & 0 & 0 & 0 & 0 & & & & \end{array}\right]$$

(1) Note that the blocks are rewritten only when altered.

The following steps are similar and hence will not be given in detail.

3a) $2C_1 + C_3 \rightarrow C_3$, $B_1 - 2B_3 \rightarrow B_1$, in order to have a null first column in C_3. Then shift columns of C_3 and rows of B_3, giving:

$$C = \left[\begin{array}{cccc:c:cccc} & & & & \boxed{\times} & 0 & -1 & 1 & \boxed{\times} \\ & & & & \boxed{\times} & 1.5 & 0 & 0 & \boxed{\times} \\ & & & & \boxed{\times} & -3 & -18 & 3 & \boxed{\times} \end{array}\right]$$

$$B^T = \left[\begin{array}{cccc:c:cccc} 0 & 0 & 0 & 1 & \boxed{0} & 0 & 0 & 0 & \boxed{0} \\ 0 & 0 & 0 & 1 & \boxed{0} & 0 & 0 & -1 & \boxed{0} \\ 0 & 0 & 0 & -2 & \boxed{0} & 0 & 0 & 1 & \boxed{0} \end{array}\right]$$

4) $-1.5C_2 + C_3 \rightarrow C_3$, $B_2 + 1.5B_3 \rightarrow B_2$, followed by shifting columns of C_3 and rows of B_3.

$$C = \left[\begin{array}{c:cccc:cccc} & & & & \boxed{\times} & -1 & 1 & \boxed{\times} & \boxed{\times} \\ & & & & \boxed{\times} & 0 & 0 & \boxed{\times} & \boxed{\times} \\ & & & & \boxed{\times} & -3 & 0 & \boxed{\times} & \boxed{\times} \end{array}\right]$$

$$B^T = \left[\begin{array}{c:cccc:cccc} & 0 & 0 & 0 & \boxed{0} & 0 & 0 & \boxed{0} & \boxed{0} \\ & 0 & 0 & -0.5 & \boxed{0} & 0 & -1 & \boxed{0} & \boxed{0} \\ & 0 & 0 & 1.5 & \boxed{0} & 0 & 1 & \boxed{0} & \boxed{0} \end{array}\right]$$

5) First column of C_3 is still dependent on first columns of C_1 and C_2. Hence, $C_1 - C_2 + C_3 \rightarrow C_3$, $B_1 - B_3 \rightarrow B_1$, $B_2 + B_3 \rightarrow B_2$, followed by shifting columns of C_3 and rows of B_3.

$$C = \left[\begin{array}{cccc:cccc:cccc} & & & & & & & \boxed{\times} & 1 & \boxed{\times} & \boxed{\times} & \boxed{\times} \\ & & & & & & & \boxed{\times} & 1.5 & \boxed{\times} & \boxed{\times} & \boxed{\times} \\ & & & & & & & \boxed{\times} & 9 & \boxed{\times} & \boxed{\times} & \boxed{\times} \end{array}\right]$$

$$B^T = \left[\begin{array}{cccc:cccc:cccc} 0 & 0 & 0 & 1 & 0 & 0 & 0 & \boxed{0} & 0 & \boxed{0} & \boxed{0} & \boxed{0} \\ 0 & 1 & 0 & 1 & 0 & -1 & -0.5 & \boxed{0} & -1 & \boxed{0} & \boxed{0} & \boxed{0} \\ 0 & -1 & 0 & -2 & 0 & 1 & 1.5 & \boxed{0} & 1 & \boxed{0} & \boxed{0} & \boxed{0} \end{array}\right]$$

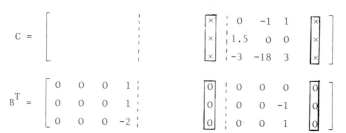

The matrix formed from the first columns of the C_i is now of full rank, and so the reduction procedure is completed. The minimal representation is of order 8 and the completed A matrix is composed of three Jordan blocks, of dimensions 4, 3, 1:

$$\dot{x} = \begin{bmatrix} 0 & 1 & 0 & 0 & 0 & 0 & 0 & 0 \\ 0 & 0 & 1 & 0 & 0 & 0 & 0 & 0 \\ 0 & 0 & 0 & 1 & 0 & 0 & 0 & 0 \\ 0 & 0 & 0 & 0 & 0 & 0 & 0 & 0 \\ 0 & 0 & 0 & 0 & 0 & 1 & 0 & 0 \\ 0 & 0 & 0 & 0 & 0 & 0 & 1 & 0 \\ 0 & 0 & 0 & 0 & 0 & 0 & 0 & 0 \\ 0 & 0 & 0 & 0 & 0 & 0 & 0 & 0 \end{bmatrix} x + \begin{bmatrix} 0 & 0 & 0 \\ 0 & 1 & -1 \\ 0 & 0 & 0 \\ 1 & 1 & -2 \\ 0 & 0 & 0 \\ 0 & -1 & 1 \\ 0 & -0.5 & 1.5 \\ 0 & -1 & 1 \end{bmatrix} u,$$

$$y = \begin{bmatrix} 1 & 0 & -1 & 1 & 0 & 0 & 0 & 1 \\ 1 & 1.5 & 0 & 0 & 1 & 0 & 0 & 1.5 \\ 1 & -1 & -9 & 1 & -2 & -10 & 2 & 9 \end{bmatrix} x.$$

<u>Note</u>: The same kind of argument can be used in the general case. However, while it is still easy to find a non-minimal representation, it will be more difficult to reduce it, since the criteria for controllability and observability are more complicated to use when the A matrix is of block diagonal form with companion blocks.

Consider, for instance, the example of paragraph 2.1.3.A.b), for which a minimal state representation was found to be of order 5:

$$Z(s) = \frac{1}{s^3+2s^2+2s+1} \begin{bmatrix} s+2 & s^2+4s+3 \\ 2s+3 & 2s^2+7s+5 \end{bmatrix}$$

The straightfoward decomposition into matrices of rank 1 gives

$$Z(s) = \frac{1}{s^3+2s^2+2s+1} \begin{bmatrix} s+2 \\ 2s+3 \end{bmatrix} [1, 0] + \frac{1}{s^3+2s^2+2s+1} \begin{bmatrix} s^2+4s+3 \\ 2s^2+7s+5 \end{bmatrix} [0, 1],$$

leading to the representation

$$\dot{x} = \begin{bmatrix} 0 & 1 & 0 & 0 & 0 & 0 \\ 0 & 0 & 1 & 0 & 0 & 0 \\ -2 & -2 & -1 & 0 & 0 & 0 \\ 0 & 0 & 0 & 0 & 1 & 0 \\ 0 & 0 & 0 & 0 & 0 & 1 \\ 0 & 0 & 0 & -2 & -2 & -1 \end{bmatrix} x + \begin{bmatrix} 0 & 0 \\ 0 & 0 \\ 1 & 0 \\ 0 & 0 \\ 0 & 0 \\ 0 & 1 \end{bmatrix} u,$$

$$y = \begin{bmatrix} 2 & 1 & 0 & 3 & 4 & 1 \\ 3 & 2 & 0 & 5 & 7 & 2 \end{bmatrix} x,$$

which is controllable but unobservable.

D - Construction of a state representation by flow graph reduction

a) Case of real poles. A minimal state representation of a transfer function matrix can also be advantageously obtained by flow graph reduction techniques. To begin with, we shall develop the method for the case where all the poles of the transfer function matrix are real, and afterwards proceed with the generalisation to complex modes. Thus, let

$$Z(s) = \frac{M(s)}{(s+\lambda)^r} \tag{2.40}$$

be the transfer function matrix, of dimensions (p, m), and let its expansion in descending powers of $(s+\lambda)^{-1}$ be

$$Z(s) = \frac{M_r}{(s+\lambda)^r} + \dots + \frac{M_1}{s+\lambda} . \tag{2.41}$$

i) Initial structure of the graph:- The graph possesses at most p independent rows, corresponding to the p outputs $y_1, \dots y_p$. In the case that no reduction occurs, the system has p invariants and each row will be constructed by cascading p blocks of transmittance $(s+\lambda)^{-1}$. However, care must be taken that:

- among all the possible graphs, only those for which independence of the rows is guaranteed will be considered; in no case will the output of a block be connected to the input of a block belonging to another row, since this would modify the structure of the A matrix, which we require to be in block-diagonal Jordan form;

- the connections between the graph and the inputs and outputs of the system are made in accordance with the following rules, which simply express the structure of the state equations:

1) the inputs of the system can only be connected to the inputs of integrators, the transmittances of these links fixing the rows of the B matrix;

2) the outputs of the integrators, apart from being connected to the integrator next in cascade, can only be connected to the outputs of the system, these links determining the columns of the C matrix.

ii) Graph construction and reduction:- We begin by establishing the first row. Since the first invariant is $(s+\lambda)^{-r}$, the first row will possess r blocks, and we can complete the specification of all the links ending on this row. If ${}_1M_i$ is the first row of the matrix M_i, then, numbering the blocks 1 to r from right to left, it follows from the equation

$$y_1 = \left[\frac{{}_1M_r}{(s+\lambda)^r} + \ldots + \frac{{}_1M_1}{s+\lambda}\right] u \tag{2.42}$$

that

$${}_1B_1 = {}_1M_1, \quad {}_2B_1 = {}_1M_2, \ldots \quad {}_rB_1 = {}_1M_r , \tag{2.43}$$

where ${}_iB_1$ denotes the ith row of the block B_1 of B:

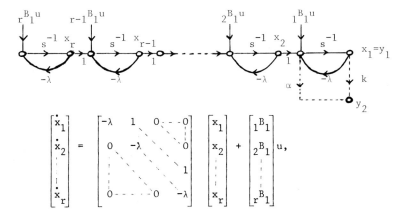

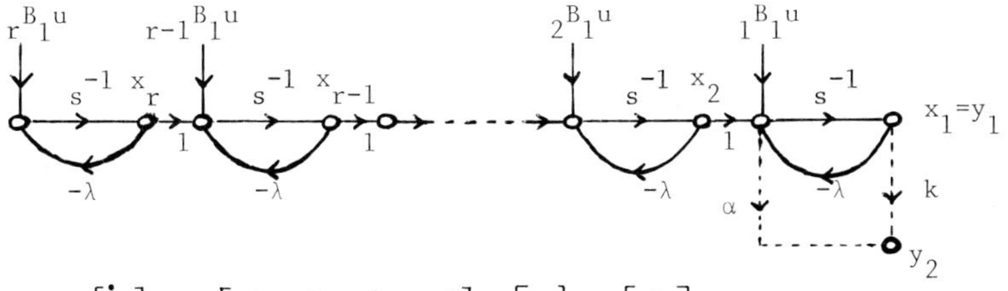

$$\begin{bmatrix} \dot{x}_1 \\ \dot{x}_2 \\ \vdots \\ \dot{x}_r \end{bmatrix} = \begin{bmatrix} -\lambda & 1 & 0 & \cdots & 0 \\ 0 & -\lambda & \ddots & & 0 \\ \vdots & & \ddots & \ddots & 1 \\ 0 & \cdots & & 0 & -\lambda \end{bmatrix} \begin{bmatrix} x_1 \\ x_2 \\ \vdots \\ x_r \end{bmatrix} + \begin{bmatrix} {}_1B_1 \\ {}_2B_1 \\ \vdots \\ {}_rB_1 \end{bmatrix} u,$$

$$y_1 = [1, 0, \dots 0]x.$$

We next establish the second row, corresponding to the output y_2. In building it up, we must take care to use the previous r blocks whenever possible, in order that the realisation obtained should be minimal. To do this, it is necessary to begin with the highest-order term in y_2, i.e. the term ${}_2M_r(s+\lambda)^{-r}u$. This term can be realised from the previous blocks only if ${}_2M_r$ is proportional to ${}_1M_r$. Otherwise, the second row will contain r blocks, like the first one, and the same procedure applies. If, however, ${}_2M_r = k_1M_r$, the highest-order term in y_2 can be generated by a direct link from x_1 to y_2, and so the second row will contain at most $r-1$ blocks. Next we have to realise the term ${}_2M_{r-1}(s+\lambda)^{r-1}u$ in y_2. The link k, which already exists, and a link α from x_2 to y_2, as in the diagram above, enable us to obtain a transmittance

$$(k\ {}_{r-1}B_1 + \alpha\ {}_rB_1)(s + \lambda)^{r-1}$$

and so, if there is a value of α such that

$$k\ {}_{r-1}B_1 + \alpha\ {}_rB_1 = {}_2M_{r-1},$$

this term can be obtained from the blocks of the first row, and then one more block can be deleted from the second row, so that it contains $r - 2$ blocks at most. Otherwise, it will contain $r - 1$ blocks. The rows ${}_iB_2$ of the block B_2 of B will, in any case, be determined as in the first step, except that the links between y_2 and the first row must now be taken into account.

We then continue the process, row after row. When the last row is completed, the graph will give directly:

- the number of invariants;
- their respective degrees;
- the matrices (A, B, C) of a minimal state representation.

Remark: We note that this procedure leads to a C matrix of the form

$$C = \begin{bmatrix} 1 & 0 \cdots 0 & 0 & 0 \cdots 0 & & 0 & 0 \cdots 0 \\ \times & \vdots \quad \vdots & 1 & \vdots \quad \vdots & \cdots & \vdots & \vdots \quad \vdots \\ \vdots & & \times & & \cdots & & \\ \vdots & & \vdots & & & 0 & \\ \times & 0 \cdots 0 & \times & 0 \cdots 0 & & 1 & 0 \cdots 0 \end{bmatrix}$$

which makes clear the observability of the representation.

Example: Consider the example of paragraph C.b) above, written in the form

$$y_1 = \frac{[1,\ 1,\ -2]}{s^4}u + \frac{[-1,\ 0,\ 1]}{s^2}u + \frac{[1,\ 0,\ -1]}{s}u$$

$$y_2 = \frac{[1,\ 1,\ -2]}{s^4}u + \frac{[1.5,\ 1,\ -1.5]}{s^3}u,$$

$$y_3 = \frac{[1,\ 1,\ -2]}{s^4}u + \frac{[-1,\ 0,\ -1]}{s^3}u + \frac{[-9,\ -1,\ 0]}{s^2}u + \frac{[1,\ 0,\ 1]}{s}u.$$

The first invariant is s^{-4}. The first row of the graph, corresponding to the realisation of y_1, can be drawn directly, with four integrators in cascade:

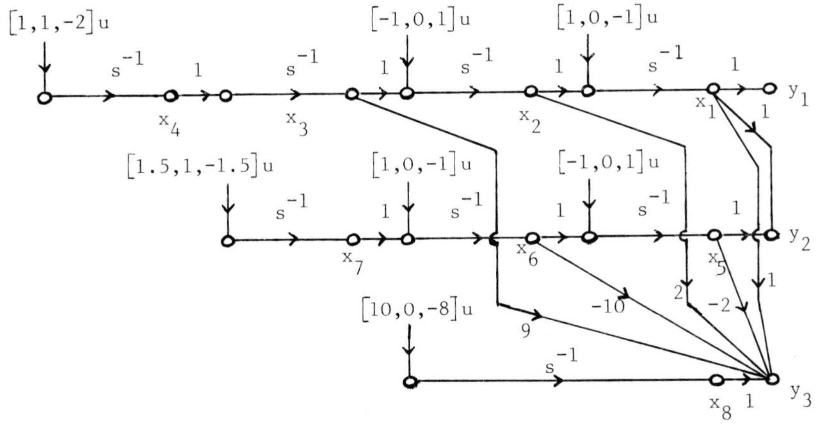

To this row corresponds a Jordan block A_1 of order 4; the corresponding block B_1 is

$$B_1 = \begin{bmatrix} 1 & 0 & -1 \\ -1 & 0 & 1 \\ 0 & 0 & 0 \\ 1 & 1 & -2 \end{bmatrix}$$

and the first row of the block C_1 is $[1, 0, 0, 0]$.

Next, consider the realisation of y_2. The first term can be generated by a direct link from x_1 to y_2. Since the term in s^{-3} cannot be generated from the first row, the second row will necessarily have three integrators, and consequently the block A_2 associated with it is of order 3. The block B_2 can be written in the form

$$B_2 = \begin{bmatrix} -1 & 0 & 1 \\ 1 & 0 & -1 \\ 1.5 & 1 & -1.5 \end{bmatrix}$$

where the first two rows are chosen so as to compensate the terms in s^{-1} and s^{-2} contributed by the link x_1 y_2. The first two rows of C_1 and C_2 are given by

$$\begin{bmatrix} 1 & 0 & 0 & 0 & 0 & 0 & 0 \\ 1 & 0 & 0 & 0 & 1 & 0 & 0 \end{bmatrix}.$$

Now we have to generate the last output y_3. The first term can be obtained from a direct link between x_1 and y_3. Also, the second term can be obtained from the previous blocks if the equation

$$\alpha[1, 1, -2] + \beta[1.5, 1, -1.5] = [-1, 0, -1]$$

can be satisfied, where the coefficients α and β correspond to links x_2 y_3 and x_5 y_3, respectively. Since this equation is satisfied by $\alpha = 2$, $\beta = -2$, the third row of the graph will include at most two integrators. We note, however, that the links will generate in y_3 a term of order s^{-2} with the coefficient

$$[-1, 0, 1]u - 2[1, 0, -1]u = [-3, 0, 3]u.$$

Hence, without introducing new integrators, direct links $x_6\, y_3$ and $x_3\, y_3$, with coefficients γ and δ, respectively, will generate the desired transmittance from the previous blocks if it is possible to satisfy

$$[-3,\ 0,\ 3] + \gamma[1.5,\ 1,\ -1.5] + \delta[1,\ 1,\ -2] = [-9,\ -1,\ 0].$$

This is indeed satisfied for $\gamma = -10$, $\delta = 9$, and so the third row includes only one integrator, at most. The links established up to now will contribute to y_3 a term in $s^{-1}u$, with coefficient

$$[1,\ 0,\ -1] + 2[-1,\ 0,\ 1] - 2[-1,\ 0,\ 1] - 10[1,\ 0,\ -1] = [-9,\ 0,\ 9],$$

to which we could add contributions from links $x_4\, y_3$ and $x_7\, y_3$, with coefficients ε and ϕ, respectively. However it is easy to see that there is no solution to the equation

$$[-9,\ 0,\ 9] + \varepsilon[1,\ 1,\ -2] + \phi[1.5,\ 1,\ -1.5] = [1,\ 0,\ 1],$$

and hence one integrator must be included in the last row. For instance, we can choose $\varepsilon = \phi = 0$ and $B_3 = [10,\ 0,\ -8]$.

The minimal state representation finally obtained is:

$$\dot{x} = \begin{bmatrix} 0 & 1 & 0 & 0 & 0 & 0 & 0 & 0 \\ 0 & 0 & 1 & 0 & 0 & 0 & 0 & 0 \\ 0 & 0 & 0 & 1 & 0 & 0 & 0 & 0 \\ 0 & 0 & 0 & 0 & 0 & 0 & 0 & 0 \\ 0 & 0 & 0 & 0 & 0 & 1 & 0 & 0 \\ 0 & 0 & 0 & 0 & 0 & 0 & 1 & 0 \\ 0 & 0 & 0 & 0 & 0 & 0 & 0 & 0 \\ 0 & 0 & 0 & 0 & 0 & 0 & 0 & 0 \end{bmatrix} x + \begin{bmatrix} 1 & 0 & -1 \\ -1 & 0 & 1 \\ 0 & 0 & 0 \\ 1 & 1 & -2 \\ -1 & 0 & 1 \\ 1 & 0 & -1 \\ 1.5 & 1 & -1.5 \\ 10 & 0 & -8 \end{bmatrix} u,$$

$$y = \begin{bmatrix} 1 & 0 & 0 & 0 & 0 & 0 & 0 & 0 \\ 1 & 0 & 0 & 0 & 1 & 0 & 0 & 0 \\ 1 & 2 & 9 & 0 & -2 & -10 & 0 & 1 \end{bmatrix} x .$$

b) Case of complex poles. The method can be used in the same way for the case of multiple complex poles. The only difference from the previous procedure is that we must now consider blocks of order 2 instead of 1. Let us take, for instance, the transfer function matrix

$$Z(s) = \frac{M(s)}{(s^2+\alpha^2)^r} = \sum_{i=1}^{r} \frac{M_i + s\dot{M}_i}{(s^2+\alpha^2)^i} \, .$$

The A matrix of the state representation will be determined in block diagonal form; the block A_μ, corresponding to an invariant $(s^2+\alpha^2)^\mu$, represented in the graph by μ blocks $(s^2+\alpha^2)^{-1}$ in cascade, has the form

$$A_\mu = \begin{bmatrix} 0 & 1 & & & & & \\ -\alpha^2 & 0 & 1 & & & & \\ & & 0 & 1 & & & \\ & & -\alpha^2 & 0 & \ddots & & \\ & & & & \ddots & 1 & \\ & & & & & 0 & 1 \\ & & & & & -\alpha^2 & 0 \end{bmatrix}$$

which is an extension of Jordan form:

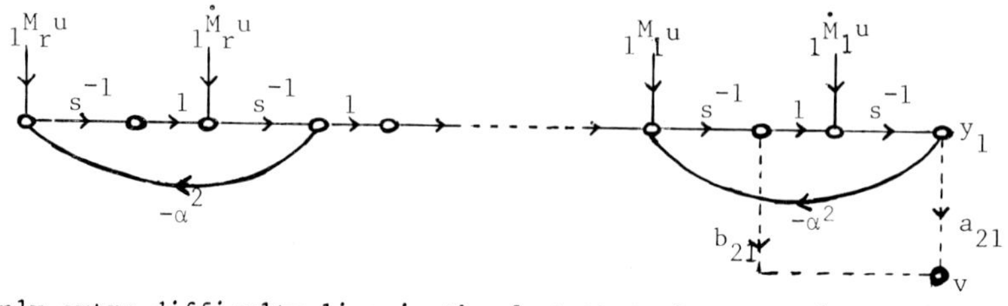

The only extra difficulty lies in the fact that the transmittances are a little harder to compute. For the case shown in the diagram, let us suppose that the first row includes r blocks and that r is generated through the links a_{21} and b_{21}. Then,

$$y_1 = \sum_{i=1}^{r} \frac{({}_1M_i + s\,{}_1\dot{M}_i)u}{(s^2+\alpha^2)^i}$$

and

$$v = a_{21}\, y_1 + \frac{b_{21}}{s^2+\alpha^2}\left[s\,{}_1M_1 - \alpha^2\,{}_1\dot{M}_1 + s\sum_{i=2}^{r} \frac{({}_1M_i + s\,{}_1\dot{M}_i)}{(s^2+\alpha^2)^{i-1}} \right] u.$$

More specifically, the highest order terms in v are

$$[(a_{21}\,{}_1M_r - b_{21}\,\alpha^2\,{}_1\dot{M}_r) + s(a_{21}\,{}_1\dot{M}_r + b_{21}\,{}_1M_r)]\;(s^2+\alpha^2)^{-r},$$

using the fact that

$$\frac{b_{21}\,s^2\,{}_1\dot{M}_r}{(s^2+\alpha^2)^r} = \frac{b_{21}\,{}_1\dot{M}_r}{(s^2+\alpha^2)^{r-1}} - \frac{b_{21}\,\alpha^2\,{}_1\dot{M}_r}{(s^2+\alpha^2)^r}\,.$$

Example: Let

$$Z(s) = \frac{1}{(s^2+2)^2}\begin{bmatrix} 2s^3+5s+1 & 2 & 4s+1 \\ 3s^3-10s^2+11s-27 & 6s^3+s^2+20s+4 & 3s^3+17s^2+14s+3 \\ 18s^3-14s^2+38s-26 & 7s^3-84s^2+14s-164 & 7s^3-42s^2+22s-82 \end{bmatrix}$$

$$= \frac{1}{(s^2+2)^2}\left(\begin{bmatrix} 1 & 2 & 1 \\ -7 & 2 & -31 \\ 2 & 4 & 2 \end{bmatrix} + s\begin{bmatrix} 1 & 0 & 4 \\ 5 & 8 & 8 \\ 2 & 0 & 8 \end{bmatrix}\right)$$

$$+ \frac{1}{s^2+2}\left(\begin{bmatrix} 0 & 0 & 0 \\ -10 & 1 & 17 \\ -14 & -84 & -42 \end{bmatrix} + s\begin{bmatrix} 2 & 0 & 0 \\ 3 & 6 & 3 \\ 18 & 7 & 7 \end{bmatrix}\right).$$

The minimal graph of this system is found to be:

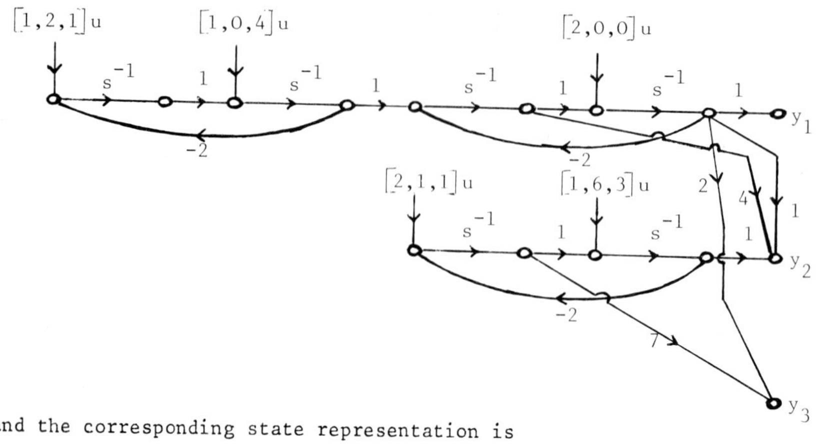

and the corresponding state representation is

$$\dot{x} = \begin{bmatrix} 0 & 1 & 0 & 0 & 0 & 0 \\ -2 & 0 & 1 & 0 & 0 & 0 \\ 0 & 0 & 0 & 1 & 0 & 0 \\ 0 & 0 & -2 & 0 & 0 & 0 \\ 0 & 0 & 0 & 0 & 0 & 1 \\ 0 & 0 & 0 & 0 & -2 & 0 \end{bmatrix} x + \begin{bmatrix} 2 & 0 & 0 \\ 0 & 0 & 0 \\ 1 & 0 & 4 \\ 1 & 2 & 1 \\ 1 & 6 & 3 \\ 2 & 1 & 1 \end{bmatrix} u,$$

$$y = \begin{bmatrix} 1 & 0 & 0 & 0 & 0 & 0 \\ 1 & 4 & 0 & 0 & 1 & 0 \\ 2 & 0 & 0 & 0 & 0 & 7 \end{bmatrix} x.$$

Remarks: The problem of obtaining a minimal state representation of a transfer function matrix with multiple poles, in generalised Jordan form, has been considered by many authors (cf. refs. 2.1, 2.10, 2.13, 2.16). The methods given here are those which appear to be the most efficient from a practical point of view. However, the method proposed by Alessandro (ref. 2.1) should also be mentioned, since it may have computational advantages in some cases. This method makes use of the matrices B_h^k, formed of the kth rows of the blocks of B corresponding to the Jordan blocks of order h in A, and the matrices C_h^k formed in the same way from the columns of the blocks of C. These matrices can be computed from $M_1, \ldots M_r$.

2.1.4 The state representation and the transfer function matrix

It is often necessary to consider the inverse problem to the one studied above, that is to say, the construction of the transfer function matrix corresponding to a given state-space form. This is a much simpler problem, since it presents no theoretical difficulties but merely involves some computation.

Of course, given the state equations

$$\dot{x} = Ax + Bu, \quad y = Cx,$$

the transfer function matrix can be obtained, in accordance with its definition, by computing the expression

$$C(sI-A)^{-1}B.$$

Such a direct approach, which involves the inversion of a polynomial matrix, is, however, very cumbersome in general.

A - Evolution matrix in companion form

In the case where the A matrix is in companion form, or, more generally in the multivariable case, where A is block-diagonal with companion blocks, the properties exhibited in paragraph 2.1.3.A.b) lead to a very efficient way of computing Z(s).

Suppose the matrices of the state representation are in the forms

$$A = \begin{bmatrix} A(1) & & 0 \\ & \ddots & \\ 0 & & A(r) \end{bmatrix} \qquad B = \begin{bmatrix} B(1) \\ \vdots \\ B(r) \end{bmatrix}, \quad C = \left[C(1) \dots C(r)\right].$$

Then,

$$Z = \sum_{k=1}^{r} C(k)\,\left[sI-A(k)\right]^{-1} B(k) = \sum_{k=1}^{r} Z(k).$$

If A(k) is a companion block, with the coefficients of the characteristic polynomial in the last row, for instance, then, from equations (2.15) to (2.20), with n_k denoting the degree of ψ_k, we have:

$$Z(k) = \frac{C(k)}{\psi_k} \begin{bmatrix} 1 \\ s \\ \vdots \\ s^{n_k-1} \end{bmatrix} \begin{bmatrix} s^{n_k-1}, \dots s, & 1 \end{bmatrix} \mathcal{A}(k)\ B(k) \text{ modulo } \psi_k,$$

or, more concisely,

$$Z(k) = \frac{\mathcal{C}(k)\,\mathcal{B}(k)}{\psi_k} \text{ modulo } \psi_k,$$

where

$$\mathcal{C}(k) = C(k) \begin{bmatrix} 1 \\ \vdots \\ s^{n_{k-1}} \end{bmatrix}, \mathcal{B}(k) = \begin{bmatrix} s^{n_k-1}, \dots & 1 \end{bmatrix} \mathcal{A}(k)\ B(k).$$

Example: Consider the system

$$\dot{x} = \begin{bmatrix} 0 & 1 & 0 \\ 0 & 0 & 1 \\ 0 & -2 & -3 \end{bmatrix} x + \begin{bmatrix} 0 & 0 \\ 1 & 0 \\ -1 & 1 \end{bmatrix} u$$

$$y = \begin{vmatrix} -1 & 1 & 0 \\ -1 & -1 & 1 \end{vmatrix} x$$

Since A is in companion form, the characteristic polynomial can be written directly

$$\psi(s) = s^3 + 3s^2 + 2s.$$

We obtain

$$\mathcal{C} = \begin{bmatrix} -1 & 1 & 0 \\ -1 & -1 & 1 \end{bmatrix} \begin{bmatrix} 1 \\ s \\ s^2 \end{bmatrix} = \begin{bmatrix} s-1 \\ s^2-s-1 \end{bmatrix},$$

$$\mathcal{B} = \begin{bmatrix} s^2, & s, & 1 \end{bmatrix} \begin{bmatrix} 1 & 0 & 0 \\ 3 & 1 & 0 \\ 2 & 3 & 1 \end{bmatrix} \begin{bmatrix} 0 & 0 \\ 1 & 0 \\ -1 & 1 \end{bmatrix} = \begin{bmatrix} s+2, & 1 \end{bmatrix},$$

$$\frac{\mathcal{C}\mathcal{B}}{\psi} = \frac{1}{s^3+3s^2+2s} \begin{bmatrix} s^2+s-2 & s-1 \\ s^3+s^2-3s-2 & s^2-s-1 \end{bmatrix}$$

whence, taking the modulo operation into account,

$$Z = \frac{\mathcal{C}\mathcal{B}}{\psi} - \begin{bmatrix} 0 & 0 \\ 1 & 0 \end{bmatrix}$$

$$= \frac{1}{s^3+3s^2+2s} \begin{bmatrix} s^2+s-2 & s-1 \\ -2s^2-5s-2 & s^2-s-1 \end{bmatrix} .$$

B - Leverrier's algorithm

If the A matrix is not in the previous form, we can still obtain an analytical expression for $(sI-A)^{-1}$ by using Leverrier's algorithm, which leads to the following procedure:

$$(sI-A)^{-1} = \psi^{-1}(s)R(s)$$

with

$$\psi(s) = s^n + a_{n-1}\, s^{n-1} + \ldots + a_1 s + a_0,$$

$$R(s) = s^{n-1}I + s^{n-2}R_1 + \ldots + R_{n-1},$$

where the a_i and R_i are computed from the relations:

Then,

$$Z(s) = \frac{1}{\psi(s)} \sum_{i=0}^{n-1} CR_iB\; s^{n-1-i} .$$

Note: This algorithm is interesting, not only on computational grounds but also from a theoretical viewpoint. It will be used, for instance, in chapter 5.

Example: Consider

$$\dot{x} = \begin{vmatrix} 0 & -1 & 0 \\ -1 & -2 & -2 \\ 1 & 0 & 0 \end{vmatrix} x + \begin{vmatrix} 4 & 2 \\ 1 & 3 \\ 2 & 1 \end{vmatrix} u, \quad y = \begin{vmatrix} 1 & 2 & 1 \\ 3 & 1 & 2 \end{vmatrix} x$$

We get successively

$$A_1 = A \qquad a_2 = 2 \qquad R_1 = \begin{bmatrix} 2 & -1 & 0 \\ -1 & 0 & -2 \\ 1 & 0 & 2 \end{bmatrix}$$

$$A_2 = AR_1 = \begin{bmatrix} 1 & 0 & 2 \\ -2 & 1 & 0 \\ 2 & -1 & 0 \end{bmatrix} \qquad a_1 = -1 \qquad R_2 = \begin{bmatrix} 0 & 0 & 2 \\ -2 & 0 & 0 \\ 2 & -1 & -1 \end{bmatrix}$$

$$A_3 = AR_2 = \begin{bmatrix} 2 & 0 & 0 \\ 0 & 2 & 0 \\ 0 & 0 & 2 \end{bmatrix} \qquad a_0 = -2 \qquad R_3 = 0$$

whence

$$Z(s) = \frac{1}{s^3+2s^2-s-2}\left(\begin{bmatrix} 8 & 9 \\ 17 & 11 \end{bmatrix} s^2 + \begin{bmatrix} -1 & -3 \\ 29 & 7 \end{bmatrix} s + \begin{bmatrix} -7 & -6 \\ 14 & 2 \end{bmatrix}\right)$$

2.2 DIFFERENTIAL-OPERATOR REPRESENTATIONS AND STATE-SPACE FORMS

2.2.1 Properties

At the beginning of this chapter, we remarked that the laws of Mechanics, when applicable, usually lead to a representation of a system in the form of a set of linear differential equations connecting the inputs u_i and the outputs y_i. This set of equations can be written in the more compact matrix form

$$L(D)y = M(D)u \tag{2.44}$$

where L(D) and M(D) are polynomial matrices, of respective dimensions (p, p) and (p, m), in the differential operator D. As before, L(D) will be assumed to be invertible, so that a transfer function matrix

$$Z(s) = L^{-1}(s)M(s) \tag{2.45}$$

can be associated with the system.

A - Order of the system

Given the system (2.44), the first question to arise is: what is the order of the system, i.e. how many variables will an equivalent state representation require? This question is easily answered by recalling some properties of polynomial matrices. It is well-known that, given a square polynomial matrix L(D), it is always possible to find a polynomial matrix V(D), with constant determinant, such that:

i) the product V(D)L(D) is an upper triangular matrix T(D);

ii) the matrix T(D) has the property that, in each of its columns, the highest-degree term lies on the diagonal.

The matrix V(D) results from elementary operations on the rows of L(D), i.e. premultiplication by elementary matrices, as will be explained in paragraph 2.2.2.C.

Thus, premultiplying both sides of the equation (2.44) by V(D), we get the completely equivalent system

$$T(D)y = V(D)M(D)u = M^*(D)u$$

or, written out in full,

$$\begin{bmatrix} t_{11}(D) & \cdots & t_{1p}(D) \\ 0 & \ddots & \vdots \\ 0 \cdots 0 & & t_{pp}(D) \end{bmatrix} \begin{bmatrix} y_1 \\ \vdots \\ y_p \end{bmatrix} = \begin{bmatrix} {}_1M^*(D) \\ \vdots \\ {}_pM^*(D) \end{bmatrix} u \tag{2.46}$$

In this form, the output variables are separated and a state representation can be found using classical single-variable methods. The last equation

$$t_{pp}(D)y_p = {}_pM^*(D)u$$

requires a number of state variables equal to

$$n_p = \text{degree } (t_{pp})$$

and its state representation will have the form:

$$\dot{x}_p = A_p x_p + B_p u,$$

$$y_p = C_p x_p.$$

The penultimate equation of the set (2.46) is

$$t_{p-1,p-1}(D)\ y_{p-1} + t_{p-1,p}(D)\ y_p = {}_{p-1}M^*(D)u, \qquad (2.47)$$

and, in view of property ii) above, we have

$$\text{degree } (t_{p-1,p}) < \text{degree } (t_{p,p}),$$

so that, in equation (2.47), we can replace the term $t_{p-1,p}(D)\ y_p$ by its expression in terms of x_p and derivatives of u, giving an equation

$$t_{p-1,p-1}(D)\ y_{p-1} = Kx_p + M^{**}(D)u, \qquad (2.48)$$

a state representation of which will require a number of variables equal to

$$n_{p-1} = \text{degree } (t_{p-1,p-1}).$$

Proceeding in this way up to the first of equations (2.46), the number of variables required is found to be

$$n = \sum_{i=1}^{p} n_i = \sum_{i=1}^{p} \text{degree } (t_{ii}).$$

Further, since T is triangular and V has constant determinant, we have, within a multiplicative constant,

$$\det L = \det T = \prod_{i=1}^{p} t_{ii}$$

so that

$$n = \text{degree (det L)}. \qquad (2.49)$$

Thus, the number of state variables associated with the system

$$L(D)y = M(D)u$$

is the degree of the determinant of L(D).

Note 1: As in paragraph 2.1.3.A, L(D) could also be written in the form

$$L = V^* \Gamma W^*$$

where V^* and W^* are matrices with constant determinants and Γ is diagonal. Since V^* and W^* are invertible, we can define new output variables

$$z = W^* y$$

and hence rewrite the system (2.44) as

$$\Gamma z = V^{*-1} Mu \qquad (2.50)$$

where

$$y = W^{*-1} z.$$

The original system has thus been transformed into an equivalent system (2.50) where all the outputs are decoupled. A state representation can then be obtained by combining the partial representations associated with the separate single-variable systems, and the total order will again be

$$n = \sum_{i=1}^{p} \text{degree } (\gamma_{ii}) = \text{degree (det L)},$$

where

$$\Gamma = \text{diag } (\gamma_{ii}).$$

Note 2: These properties will be used in paragraph 2.2.2.C to find a state representation of the system (2.44).

B - State variables and initial conditions on the outputs

In the system (2.44), let the matrices L(D) and M(D) be, explicitly,

$$L(D) = L_\nu D^\nu + \ldots L_1 D + L_0,$$

$$M(D) = M_r D^r + \dots + M_0,$$

where L_i and M_i are scalar coefficient matrices, and ν and r are the degrees of the highest-degree terms in L and M, respectively. Equation (2.44) thus becomes

$$L_\nu \overset{(\nu)}{y} + \dots + L_1 \dot{y} + L_0 y = M_r \overset{(r)}{u} + \dots + M_0 u.$$

Then, taking Laplace transforms with nonzero initial conditions, we get

$$L_\nu \left[s^\nu Y(s) - s^{\nu-1} y_0 - \dots - \overset{(\nu-1)}{y_0} \right] + \dots + L_1 \left[sY(s) - y_0 \right] + L_0 Y(s)$$

$$= M_r \left[s^r U(s) - s^{r-1} u_0 - \dots - \overset{(r-1)}{u_0} \right] + \dots + M_0 U(s)$$

i.e.

$$\left[L_0 + L_1 s + \dots + L_\nu s^\nu \right] Y(s) - s^{\nu-1} L_\nu y_0 - \dots - \left[L_\nu \overset{(\nu-1)}{y_0} + \dots + L_1 y_0 \right]$$

$$= \left[M_0 + \dots + M_r s^r \right] U(s) - s^{r-1} M_r u_0 - \dots - \left[M_r \overset{(r-1)}{u_0} + \dots + M_1 u_0 \right]$$

which can be written in matrix form:

$$L(s)Y(s) - \left[s^{\nu-1} I, \dots, sI, I \right] \begin{bmatrix} L_\nu & 0 & \cdots & 0 \\ \vdots & \ddots & \ddots & \vdots \\ \vdots & & \ddots & 0 \\ L_1 & \cdots & \cdots & L_\nu \end{bmatrix} \begin{bmatrix} y_0 \\ \vdots \\ \overset{(\nu-1)}{y_0} \end{bmatrix}$$

$$= M(s)U(s) - \left[s^{\nu-1}, \dots I \right] \begin{bmatrix} 0 & \cdots & 0 \\ 0 & \ddots & \vdots \\ M_r & \ddots & 0 \\ \vdots & \ddots & \\ M_1 & \cdots & M_r \end{bmatrix} \begin{bmatrix} u_0 \\ \vdots \\ \overset{(r-1)}{u_0} \end{bmatrix} \tag{2.51}$$

or, in more compact notation,

$$L(s)Y(s) - [s^{\nu-1}I,\ldots I]\mathcal{L}\mathcal{Y}_0 = M(s)U(s) - [s^{\nu-1}I,\ldots I]\mathcal{M}\mathcal{U}_0 \qquad (2.52)$$

where $\mathcal{L}$ and $\mathcal{M}$ are matrices of dimensions $(p\nu,\ p\nu)$ and $(p\nu,\ mr)$, respectively. Even more compactly, equation (2.52) can be rewritten

$$L(s)Y(s) = M(s)U(s) + [s^{\nu-1}I,\ldots I][-\mathcal{L},\mathcal{M}]\begin{bmatrix} 0 \\ 0 \end{bmatrix}$$

and, since $[-\mathcal{L},\mathcal{M}]$ is a matrix of dimensions $(p\nu,\ p\nu + mr)$, its rank is limited by the number of rows, i.e. by $p\nu$, the number of initial conditions on the outputs.

Equation (2.51) will be used later, to find, very simply, a state representation for the system (2.44).

C - Controllability and Observability

We saw in the introductory example of this chapter that, if a system of known structure has unobservable modes, they will be absent in the representation by differential operators [cf. equation (2.5) and figure 2.2]. This is, indeed, a general property; the representation of a system by differential equations exhibits only the observable modes. This representation does, however, reveal more about the system than does the transfer function matrix, since the uncontrollable modes are retained. The situation is summarised in figure 2.4, where a cross indicates that the modes appear in the representation while a circle means that they do not.

Type of representation \ Modes	Controllable unobservable	Controllable observable	Uncontrollable unobservable	Uncontrollable observable
State $\dot{x} = Ax + Bu$, $y = Cx$	×	×	×	×
Differential operator $Ly = Mu$	0	×	0	×
Transfer function matrix $y = Zu$	0	×	0	0

Fig. 2.4 Appearance of different types of modes in various representations

It can be shown that the system

$$L(D)y = M(D)u$$

is controllable if and only if

$$\text{rank}\,[L(s),\ M(s)] = p$$

for every s such that

$$\det L(s) = 0.$$

Example 1: Let the system be defined by

$$\begin{bmatrix} D^2 & -1 \\ 0 & D^2 \end{bmatrix} y = \begin{bmatrix} 1 & D \\ D & m(D) \end{bmatrix} u$$

so that

$$\det L = D^4,\quad [L,\ M] = \begin{bmatrix} D^2 & -1 & 1 & D \\ 0 & D^2 & D & m(D) \end{bmatrix}$$

and, for D = 0,

$$\text{rank}\,[L,\ M] = \text{rank}\begin{bmatrix} 0 & -1 & 1 & 0 \\ 0 & 0 & 0 & m(0) \end{bmatrix}.$$

Hence: if $m(0) = 0$, rank $[L,\ M] = 1$ and the system is uncontrollable; if $m(0) \neq 0$, rank $[L,\ M] = 2$ and the system is controllable.

Example 2: Consider the linearised equations of a satellite with an inertial wheel on each axis and with no external torques:

$$A_x \dot{\omega}_x + J_x \dot{\nu}_x - A_z \Omega \omega_y + A_y \Omega \omega_y + J_y \Omega \nu_y = 0$$

$$A_y \dot{\omega}_y + J_y \dot{\nu}_y - A_x \Omega \omega_x - J_x \Omega \nu_x + A_z \Omega \omega_x = 0$$

$$A_z \dot{\omega}_z + J_z \dot{\nu}_z = 0$$

$$J_x(\dot{\omega}_x + \dot{\nu}_x) = u_x$$

$$J_y(\dot{\omega}_y + \dot{\nu}_y) = u_y$$

$$J_z(\dot{\omega}_z + \dot{\nu}_z) = u_z$$

where ν_x, ν_y, ν_z are the angular velocities of the wheels, and ω_x, ω_y, ω_z are the components of the instantaneous rotation vector.

These equations can be put in the form

$$Ly = Mu$$

with the output vector defined by

$$y = [\omega_x, \omega_y, \omega_z, \nu_x, \nu_y, \nu_z]^T .$$

It is easy to see that the determinant of L is of the form D^4 $(D^2$ + constant) and that, for D = 0,

$$[L, M] = \left[\begin{array}{cccccc|ccc} 0 & (A_y-A_z)\Omega & 0 & 0 & J_y\Omega & 0 & 0 & 0 & 0 \\ (A_z-A_x)\Omega & 0 & 0 & -J_x\Omega & 0 & 0 & 0 & 0 & 0 \\ 0 & 0 & 0 & 0 & 0 & 0 & 0 & 0 & 0 \\ 0 & 0 & 0 & 0 & 0 & 0 & 1 & 0 & 0 \\ 0 & 0 & 0 & 0 & 0 & 0 & 0 & 1 & 0 \\ 0 & 0 & 0 & 0 & 0 & 0 & 0 & 0 & 1 \end{array}\right]$$

which is of rank 5. Thus, the system is uncontrollable (it can be shown that the uncontrollability appears in the pitch motion).

2.2.2 State representations associated with a differential-operator form

Given a system in the form $L(D)y = M(D)u$, let us try to construct an equivalent state representation. Using the properties discussed in the preceding paragraphs, various methods can be devised. We shall see that this problem can very often be solved quite easily if systematic methods are used.

A - The case where L_ν is invertible

We consider first the case where the matrix L_ν defined in paragraph 2.2.1.B (i.e. the matrix of coefficients of the highest power of D in L) is invertible. If L_ν is invertible, the determinant of L(D) is of degree $p\nu$.

The state vector then has $p\nu$ components, and these can easily be related to the outputs and their derivatives.

Thus, consider equation (2.51) and, since L_ν is invertible, take $L_\nu = I$ (this can be achieved by multiplying both sides of the equation $Ly = Mu$ by L_ν^{-1}, which evidently does not change the system). Equation (2.51) then becomes

$$L(s)Y(s) - \left[s^{\nu-1}I, \ldots I\right] \begin{bmatrix} I & 0 & \cdots & 0 \\ L_{\nu-1} & \ddots & \ddots & \vdots \\ \vdots & \ddots & \ddots & 0 \\ L_1 & \cdots & L_{\nu-1} & I \end{bmatrix} \begin{bmatrix} y_0 \\ \vdots \\ \vdots \\ \overset{(\nu-1)}{y_0} \end{bmatrix}$$

$$= M(s)U(s) - \left[s^{r-1}I, \ldots I\right] \begin{bmatrix} M_r & 0 & \cdots & 0 \\ \vdots & \ddots & \ddots & 0 \\ M_1 & \cdots & \cdots & M_r \end{bmatrix} \begin{bmatrix} u_0 \\ \vdots \\ \overset{(r-1)}{u_0} \end{bmatrix}$$

Let us now take as state variables:

$$\begin{aligned} x_1 &= y, \\ x_2 &= L_{\nu-1}y + \dot{y}, \\ &\vdots \\ x_\nu &= L_1 y + \ldots + \overset{(\nu-1)}{y} - M_1 u - \ldots - M_r \overset{(r-1)}{u}, \end{aligned}$$

giving the correct total of $p\nu$ components since y has dimension p. We then have

$$\begin{aligned} \dot{x}_1 &= \dot{y} = x_2 - L_{\nu-1}x_1, \\ \dot{x}_2 &= L_{\nu-1}\dot{y} + \ddot{y} = x_3 - L_{\nu-2}x_1, \\ &\vdots \\ \dot{x}_{\nu-1} &= L_2\dot{y} + \ldots + \overset{(\nu-1)}{y} - M_2\dot{u} - \ldots - M_r \overset{(r-1)}{u} = x_\nu - L_1 x_1, \\ \dot{x}_\nu &= L_1\dot{y} + \ldots + \overset{(\nu)}{y} - M_1\dot{u} - \ldots - M_r \overset{(r)}{u} = -L_0 x_1 + M_0 u, \end{aligned}$$

or, in matrix form:

$$\begin{bmatrix} \dot{x}_1 \\ \vdots \\ \dot{x}_\nu \end{bmatrix} = \begin{bmatrix} -L_{\nu-1} & I & 0 & \cdots & 0 \\ \vdots & 0 & \ddots & \ddots & 0 \\ \vdots & \vdots & & \ddots & I \\ -L_0 & 0 & \cdots & \cdots & 0 \end{bmatrix} \begin{bmatrix} x_1 \\ \vdots \\ x_\nu \end{bmatrix} + \begin{bmatrix} 0 \\ \vdots \\ 0 \\ M_r \\ \vdots \\ M_0 \end{bmatrix} u, \tag{2.53}$$

$$y = [I, 0, \ldots 0] \begin{bmatrix} x_1 \\ \vdots \\ x_\nu \end{bmatrix}$$

Note 1: The state representation can be written directly, using the matrices L_i and M_i, in blocks of dimensions (p, p) and (p,m). The A matrix is a block companion matrix.

Note 2: In the multivariable case, we assumed that L_ν was invertible and put it in the form of a unit matrix by premultiplication. However, in the single-variable case, L_ν is a nonzero scalar and the method is always applicable.

Example 1: Consider the two-input, two-output system defined by

$$\ddot{y}_1 + 2\dot{y}_1 + 3y_1 + \dot{y}_2 + 2y_2 = \dot{u}_1 + u_1 + 2u_2,$$

$$\dot{y}_1 + y_1 + \ddot{y}_2 + y_2 = u_1 + 2\dot{u}_2 + u_2.$$

In matrix form, we have

$$\begin{bmatrix} D^2+2D+3 & D+2 \\ D+1 & D^2+1 \end{bmatrix} y = \begin{bmatrix} D+1 & 2 \\ 1 & 2D+1 \end{bmatrix} u$$

so that

$$\text{degree } (\det L) = 4$$

and hence four variables are necessary. Also,

$$L_\nu = L_2 = \begin{bmatrix} 1 & 0 \\ 0 & 1 \end{bmatrix}, \quad L_1 = \begin{bmatrix} 2 & 1 \\ 1 & 0 \end{bmatrix}, \quad L_0 = \begin{bmatrix} 3 & 2 \\ 1 & 1 \end{bmatrix}, \quad M_1 = \begin{bmatrix} 1 & 0 \\ 0 & 2 \end{bmatrix}, \quad M_0 = \begin{bmatrix} 1 & 2 \\ 1 & 1 \end{bmatrix},$$

giving directly the state-space form

$$\begin{bmatrix} \dot{x}_1 \\ \dot{x}_2 \end{bmatrix} = \begin{bmatrix} -2 & -1 & 1 & 0 \\ -1 & 0 & 0 & 1 \\ -3 & -2 & 0 & 0 \\ -1 & -1 & 0 & 0 \end{bmatrix} \begin{bmatrix} x_1 \\ x_2 \end{bmatrix} + \begin{bmatrix} 1 & 0 \\ 0 & 2 \\ 1 & 2 \\ 1 & 1 \end{bmatrix} u,$$

$$y = \begin{bmatrix} 1 & 0 & 0 & 0 \\ 0 & 1 & 0 & 0 \end{bmatrix} \begin{bmatrix} x_1 \\ x_2 \end{bmatrix},$$

with

$$x_1 = y = \begin{bmatrix} y_1 \\ y_2 \end{bmatrix}, \quad x_2 = L_1 y + L_2 \dot{y} - M_1 u = \begin{bmatrix} 2y_1 + y_2 + \dot{y}_1 - u_1 \\ y_1 + \dot{y}_2 - 2u_2 \end{bmatrix}$$

<u>Example 2</u>: Consider now the system

$$\begin{bmatrix} D^2+D & D^2+1 \\ D+3 & 2D^2+D+1 \end{bmatrix} y = \begin{bmatrix} D+1 & 1 & 2 \\ 3 & D+2 & D+1 \end{bmatrix} u.$$

In this case,

$$L_2 = \begin{bmatrix} 1 & 1 \\ 0 & 2 \end{bmatrix},$$

which is invertible but not a unit matrix, so the first step is to multiply both sides of the equation by L_2^{-1}, giving

$$\begin{bmatrix} D^2+0.5D-1.5 & -0.5D+0.5 \\ 0.5D+1.5 & D^2+0.5D+0.5 \end{bmatrix} y = \begin{bmatrix} D-0.5 & -0.5D & -0.5D+1.5 \\ 1.5 & 0.5D+1 & 0.5D+0.5 \end{bmatrix} u,$$

whence the state representation follows directly:

$$\begin{bmatrix} \dot{x}_1 \\ \dot{x}_2 \end{bmatrix} = \begin{bmatrix} -0.5 & 0.5 & 1 & 0 \\ -0.5 & -0.5 & 0 & 1 \\ 1.5 & -0.5 & 0 & 0 \\ -1.5 & -0.5 & 0 & 0 \end{bmatrix} \begin{bmatrix} x_1 \\ x_2 \end{bmatrix} + \begin{bmatrix} 1 & -0.5 & -0.5 \\ 0 & 0.5 & 0.5 \\ -0.5 & 0 & 1.5 \\ 1.5 & 1 & 0.5 \end{bmatrix} u,$$

$$y = \begin{bmatrix} 1 & 0 & 0 & 0 \\ 0 & 1 & 0 & 0 \end{bmatrix} \begin{bmatrix} x_1 \\ x_2 \end{bmatrix}.$$

B - Extension to the case where L_ν is not invertible

If L_ν is not invertible, the previous method (1) cannot be used directly. However, it can be generalised by a minor modification which consists of splitting the system into subsystems to each of which the method does apply.

Since L_ν is a square matrix of dimension p, it can only fail to be invertible if its rank is less than p. Moreover, it is non-null by definition, so we may take rank $(L_\nu) = \rho$ with $0 < \rho < p$.

It is then always possible to find a non-singular matrix A_0, defining new output variables $z = A_0 y$, and an elementary transformation T, such that the system Ly = Mu can be written in the equivalent form

$$L'(D)z = M'(D)u$$

where the matrix L'_ν corresponding to L' has the form

$$L'_\nu = \begin{bmatrix} I_\rho & 0 \\ 0 & 0 \end{bmatrix} . \qquad (2.54)$$

For example, the following procedure may be used:-

i) First, by elementary operations on the rows of L, make the last $p - \rho$ rows of L_ν null. The system then becomes TLy = TMu, where the matrix of coefficients of the highest-degree term in T(D)L(D) has the form

$$\begin{bmatrix} \ell'_{11} & \ell'_{12} \\ 0 & 0 \end{bmatrix}$$

where the matrices ℓ'_{11}, ℓ'_{12} are of respective dimensions (ρ,ρ), $(\rho, p-\rho)$.

(1) Another method has been proposed by F.M. Brown (cf. ref. 2.4).

ii) Then, make a change of variables $z = A_0 y$, transforming the system into

$$TLA_0^{-1}z = TMu.$$

Writing A_0 in the form

$$A_0 = \begin{bmatrix} A_{11} & A_{12} \\ A_{21} & A_{22} \end{bmatrix},$$

we require

$$\begin{bmatrix} \ell'_{11} & \ell'_{12} \\ 0 & 0 \end{bmatrix} A_0^{-1} = \begin{bmatrix} I_\rho & 0 \\ 0 & 0 \end{bmatrix},$$

giving

$$\begin{bmatrix} \ell'_{11}, & \ell'_{12} \end{bmatrix} = \begin{bmatrix} A_{11}, & A_{12} \end{bmatrix},$$

which determines A_{11} and A_{12}. For the other blocks, we can take, e.g., $A_{21} = 0$, $A_{22} = I_{p-\rho}$.

Example: Consider the three-input, three-output system defined by

$$\begin{bmatrix} (D+1)^2 & 2D^2+3D+1 & D+1 \\ D^2+D+1 & 2D^2+5 & 2D+3 \\ 2D^2+D+1 & D^2+D & D^2+D+3 \end{bmatrix} \begin{bmatrix} y_1 \\ y_2 \\ y_3 \end{bmatrix} = \begin{bmatrix} u_1 \\ u_2 \\ u_3 \end{bmatrix}$$

so that

$$L_\nu = L_2 = \begin{vmatrix} 1 & 2 & 0 \\ 1 & 2 & 0 \\ 2 & 1 & 1 \end{vmatrix}$$

and

$$\text{rank } (L_2) = 2 < 3.$$

Premultiplying by the matrix

$$T = \begin{bmatrix} 1 & 0 & 0 \\ 0 & 0 & 1 \\ -1 & 1 & 0 \end{bmatrix},$$

we get

$$TL_2 = \begin{bmatrix} 1 & 2 & 0 \\ 2 & 1 & 1 \\ 0 & 0 & 0 \end{bmatrix}.$$

Then, defining, as above,

$$A_0 = \begin{bmatrix} 1 & 2 & 0 \\ 2 & 1 & 1 \\ 0 & 0 & 1 \end{bmatrix},$$

we have

$$\frac{1}{3} \begin{bmatrix} 3D^2+4D+1 & D+1 & 2(D+1) \\ D-1 & 3D^2+D+2 & 2D+7 \\ -5D+8 & D-4 & 2D+10 \end{bmatrix} z = \begin{bmatrix} 1 & 0 & 0 \\ 0 & 0 & 1 \\ -1 & 1 & 0 \end{bmatrix} u,$$

where L'_ν has the required form

$$L'_2 = \begin{bmatrix} 1 & 0 & 0 \\ 0 & 1 & 0 \\ 0 & 0 & 0 \end{bmatrix}.$$

Let us now decompose the system $L'z = M'u$ into

$$\begin{bmatrix} L_{11} & L_{12} \\ L_{21} & L_{22} \end{bmatrix} \begin{bmatrix} z_1 \\ z_2 \end{bmatrix} = \begin{bmatrix} M_1 \\ M_2 \end{bmatrix} u, \tag{2.55}$$

where z_1, z_2 are vectors of dimensions ρ, $p-\rho$, respectively, and the matrices L_{11}, L_{12}, L_{21}, L_{22} are of respective dimensions (ρ,ρ), $(\rho, p-\rho)$, $(p-\rho,\rho)$, $(p-\rho, p-\rho)$.

The first of equations (2.55) is

$$L_{11}z_1 + L_{12}z_2 = M_1u,$$

where, in consequence of the manipulations previously carried out, the matrix of coefficients $(L_{11})_\nu$ is the unit matrix I_ρ. Hence, if, to begin with, we regard z_2 as a vector of independent variables, like u, the method already used for the case of invertible L_ν can be applied to find a state-space form for the system

$$L_{11} z_1 = M_1 u - L_{12} z_2. \tag{2.56}$$

[We note that degree (M_1) < degree (L_{11}) and degree (L_{12}) < degree (L_{11})]. A state representation of (2.56) is thus obtained in the form:

$$\begin{aligned} \dot{x}_1 &= A_1 x_1 + B_1 u - K_1 z_2, \\ z_1 &= C_1 x_1, \end{aligned} \tag{2.57}$$

where x_1 is a vector of dimension $\rho\nu$.

The second of equations (2.55) gives

$$L_{22} z_2 = M_2 u - L_{21} z_1, \tag{2.58}$$

where the last term comprises a sum of terms in z_1, $\dot{z}_1, \ldots$, corresponding to the various powers of D contained in $L_{21}(D)$. According to equations (2.57), these quantities can be expressed in terms of x, u, $\dot{u}, \ldots, z_2, \dot{z}_2, \ldots$, since

$$\begin{aligned} z_1 &= C_1 x_1, \\ \dot{z}_1 &= C_1 (A_1 x_1 + B_1 u - K_1 z_2), \\ &\vdots \end{aligned}$$

or, in matrix form,

$$\begin{bmatrix} z_1 \\ \dot{z}_1 \\ \vdots \\ z_1^{(k)} \end{bmatrix} = \begin{bmatrix} C_1 & 0 & \cdots & 0 \\ C_1 A_1 & \ddots & & \vdots \\ \vdots & & \ddots & 0 \\ C_1 A_1^k & \cdots & \cdots & C_1 \end{bmatrix} \begin{bmatrix} x_1 \\ B_1 u - K_1 z_2 \\ \vdots \\ B_1 u^{(k-1)} - K_1 z_2^{(k-1)} \end{bmatrix}.$$

Substituting these expressions into (2.58), we get an equation of the form

$$L'_{22} z_2 = M'_2 u + Q_1 x_1 \tag{2.59}$$

by grouping together all the terms in z_2, $\dot{z}_2, \ldots$, and also all those in u and its derivatives.

Now, if $(L'_{22})_\nu$ is invertible, the previous method can be used to put (2.59) into a state-space form:

$$\dot{x}_2 = A_2 x_2 + B_2 u + K_2 x_1,$$

$$z_2 = C_2 x_2,$$

and a state representation of the complete system is then given by

$$\begin{bmatrix} \dot{x}_1 \\ \dot{x}_2 \end{bmatrix} = \begin{bmatrix} A_1 & -K_1 C_2 \\ K_2 & A_2 \end{bmatrix} \begin{bmatrix} x_1 \\ x_2 \end{bmatrix} + \begin{bmatrix} B_1 \\ B_2 \end{bmatrix} u,$$

$$y = A_0^{-1} z = A_0^{-1} \begin{bmatrix} C_1 & 0 \\ 0 & C_2 \end{bmatrix} \begin{bmatrix} x_1 \\ x_2 \end{bmatrix}. \tag{2.59a}$$

If $(L'_{22})_\nu$ is not invertible, we again decompose the second subsystem into smaller subsystems, and repeat the procedure given above.

Example: Consider the system

$$\begin{bmatrix} D^2+D+1 & 2D+3 & 1 \\ D+1 & D^2+2D+1 & D+3 \\ D+2 & D+5 & D+2 \end{bmatrix} \begin{bmatrix} z_1 \\ z_2 \\ z_3 \end{bmatrix} = \begin{bmatrix} 1 & 0 & 1 \\ 1 & 1 & D+1 \\ 1 & 1 & 0 \end{bmatrix} \begin{bmatrix} u_1 \\ u_2 \\ u_3 \end{bmatrix}$$

where the determinant of L(D) is of degree 5, so that five state variables will be required. We have

$$L_2 = \begin{bmatrix} 1 & 0 & 0 \\ 0 & 1 & 0 \\ 0 & 0 & 0 \end{bmatrix},$$

and the first subsystem is

$$\begin{bmatrix} D^2+D+1 & 2D+3 \\ D+1 & D^2+2D+1 \end{bmatrix} \begin{bmatrix} z_1 \\ z_2 \end{bmatrix} = \begin{bmatrix} 1 & 0 & 1 \\ 1 & 1 & D+1 \end{bmatrix} \begin{bmatrix} u_1 \\ u_2 \\ u_3 \end{bmatrix} - \begin{bmatrix} 1 \\ D+3 \end{bmatrix} z_3, \tag{2.60}$$

regarding z_3 as an independent variable.

Using equations (2.53), we obtain immediately a state representation of the system (2.60), using four variables:

$$\dot{x}_1 = \begin{bmatrix} -1 & -2 & 1 & 0 \\ -1 & -2 & 0 & 1 \\ -1 & -3 & 0 & 0 \\ -1 & -1 & 0 & 0 \end{bmatrix} x_1 + \begin{bmatrix} 0 & 0 & 0 \\ 0 & 0 & 1 \\ 1 & 0 & 1 \\ 1 & 1 & 1 \end{bmatrix} u - \begin{bmatrix} 0 \\ 1 \\ 1 \\ 3 \end{bmatrix} z_3$$

$$= A_1x_1 + B_1u - K_1z_3 , \tag{2.61}$$

$$\begin{bmatrix} z_1 \\ z_2 \end{bmatrix} = \begin{bmatrix} 1 & 0 & 0 & 0 \\ 0 & 1 & 0 & 0 \end{bmatrix} x_1 = C_1x_1 .$$

The second subsystem is

$$(D+2)z_3 = [1, 1, 0]u - [D+2, D+5] \begin{bmatrix} z_1 \\ z_2 \end{bmatrix}$$

$$= [1, 1, 0]u - [1, 1] \begin{bmatrix} \dot{z}_1 \\ \dot{z}_2 \end{bmatrix} - [2, 5] \begin{bmatrix} z_1 \\ z_2 \end{bmatrix} . \tag{2.62}$$

From (2.61), we have

$$[2 , 5] \begin{bmatrix} z_1 \\ z_2 \end{bmatrix} = [2, 5, 0, 0]x_1$$

and

$$\begin{bmatrix}\dot z_1\\ \dot z_2\end{bmatrix} = \begin{bmatrix}-1 & -2 & 1 & 0\\ -1 & -2 & 0 & 1\end{bmatrix} x_1 + \begin{bmatrix}0 & 0 & 0\\ 0 & 0 & 1\end{bmatrix} u - \begin{bmatrix}0\\ 1\end{bmatrix} z_3$$

so that

$$\begin{vmatrix}1, & 1\end{vmatrix} \begin{bmatrix}\dot z_1\\ \dot z_2\end{bmatrix} = [-2, -4, 1, 1]x_1 + [0, 0, 1]u - z_3$$

whence, by substituting in (2.62) and collecting like terms,

$$(D+1)z_3 = [0, -1, -1, -1]x_1 + [1, 1, -1]u, \qquad (2.63)$$

which has the state representation

$$\begin{aligned}\dot x_2 &= [-1]x_2 + [0, -1, -1, -1]x_1 + [1, 1, -1]u,\\ z_3 &= [1]x_2\end{aligned} \qquad (2.64)$$

Hence, putting together equations (2.61) and (2.64), and writing

$$x = \begin{bmatrix}x_1\\ x_2\end{bmatrix},$$

we have a state representation for the complete system:

$$\dot x = \begin{bmatrix}-1 & -2 & 1 & 0 & 0\\ -1 & -2 & 0 & 1 & -1\\ -1 & -3 & 0 & 0 & -1\\ -1 & -1 & 0 & 0 & -3\\ 0 & -1 & -1 & -1 & -1\end{bmatrix} x + \begin{bmatrix}0 & 0 & 0\\ 0 & 0 & 1\\ 1 & 0 & 1\\ 1 & 1 & 1\\ 1 & 1 & -1\end{bmatrix} u,$$

$$z = \begin{bmatrix}1 & 0 & 0 & 0 & 0\\ 0 & 1 & 0 & 0 & 0\\ 0 & 0 & 0 & 0 & 1\end{bmatrix} x .$$

Note: The first step, when applying the above method, should be to compute the degree of the determinant of $L(D)$, since this fixes the order of the state representation. It is worth noticing, however, that, even if this step is omitted, the method is perfectly safe. In fact, it will eventually reveal any non-minimality which may have arisen in the earlier stages. The following example will help to make this clear.

Example: Consider the case of a two-input, three-output system defined by the differential equations

$$\begin{bmatrix} D^3 & D^2 & D^2 \\ D+1 & D^3 & D^2+D+1 \\ D^2 & D+1 & D \end{bmatrix} \begin{bmatrix} y_1 \\ y_2 \\ y_3 \end{bmatrix} = \begin{bmatrix} D+1 & D \\ D+2 & 1 \\ 1 & 1 \end{bmatrix} \begin{bmatrix} u_1 \\ u_2 \end{bmatrix},$$

where det L(D) is of degree 5. Since, however,

$$L_3 = \begin{bmatrix} 1 & 0 & 0 \\ 0 & 1 & 0 \\ 0 & 0 & 0 \end{bmatrix},$$

we may be led to take, as the first subsystem,

$$\begin{bmatrix} D^3 & D^2 \\ D+1 & D^3 \end{bmatrix} \begin{bmatrix} y_1 \\ y_2 \end{bmatrix} = \begin{bmatrix} D+1 & D \\ D+2 & 1 \end{bmatrix} u - \begin{bmatrix} D^2 \\ D^2+D+1 \end{bmatrix} y_3,$$

which leads to a state representation of order 6:

$$\dot{x}_1 = \begin{bmatrix} 0 & -1 & 1 & 0 & 0 & 0 \\ 0 & 0 & 0 & 1 & 0 & 0 \\ 0 & 0 & 0 & 0 & 1 & 0 \\ -1 & 0 & 0 & 0 & 0 & 1 \\ 0 & 0 & 0 & 0 & 0 & 0 \\ -1 & 0 & 0 & 0 & 0 & 0 \end{bmatrix} x_1 + \begin{bmatrix} 0 & 0 \\ 0 & 0 \\ 1 & 1 \\ 1 & 0 \\ 1 & 0 \\ 2 & 1 \end{bmatrix} u - \begin{bmatrix} 1 \\ 1 \\ 0 \\ 1 \\ 0 \\ 1 \end{bmatrix} y_3,$$

$$\begin{bmatrix} y_1 \\ y_2 \end{bmatrix} = \begin{bmatrix} 1 & 0 & 0 & 0 & 0 & 0 \\ 0 & 1 & 0 & 0 & 0 & 0 \end{bmatrix} x_1 .$$

The non-minimality of this representation will show up when we write the equations of the second subsystem:

$$[D] y_3 = [1, 1] u - [D^2, D+1] \begin{bmatrix} y_1 \\ y_2 \end{bmatrix},$$

which is to say

$$\dot{y}_3 = [1,\ 1]u - [1,\ 0]\begin{bmatrix}\ddot{y}_1\\ \ddot{y}_2\end{bmatrix} - [0,1]\begin{bmatrix}\dot{y}_1\\ \dot{y}_2\end{bmatrix} - [0,\ 1]\begin{bmatrix}y_1\\ y_2\end{bmatrix},$$

whence, by substitution from the previous equations,

$$0 = [0,\ -1,\ 0,\ 0,\ -1,\ 0]x_1.$$

Thus, the second and fifth components of x_1 are not independent. Eliminating the latter, we find a minimal state representation in the form:

$$\dot{x} = \begin{bmatrix} 0 & -1 & 1 & -1 & 0\\ 0 & 0 & 0 & 0 & 0\\ 0 & -1 & 0 & 0 & 0\\ -1 & 0 & 0 & -1 & 1\\ -1 & 0 & 0 & -1 & 0 \end{bmatrix} x + \begin{bmatrix} -1 & 0\\ -1 & 0\\ 1 & 1\\ 0 & 0\\ 1 & 1 \end{bmatrix} u,$$

$$y = \begin{bmatrix} 1 & 0 & 0 & 0 & 0\\ 0 & 1 & 0 & 0 & 0\\ 0 & 0 & 0 & 1 & 0 \end{bmatrix} x + \begin{bmatrix} 0 & 0\\ 0 & 0\\ 1 & 0 \end{bmatrix} u.$$

<u>Note</u>: We observe that the above system is not proper. In fact, the transfer function matrix is

$$Z(s) = L^{-1}(s)M(s) = \frac{1}{s^5+s^4-s^2}\begin{bmatrix} -s^4+s^3+s^2+3s+1 & s^3+s\\ -s^4-s^3+s & 0\\ s^5+s^4+2s^3-2s^2-3s-1 & s^3-s^2-s \end{bmatrix}.$$

<u>C - Construction of a state representation by triangularisation or diagonalisation of the L matrix</u>

a) <u>The methods</u>. In the case where L_ν is invertible, the method presented in paragraph A above is so easy to apply that we can recommend its automatic use. If L_ν is not invertible, however, the necessity of performing successive decompositions into subsystems makes the method, though still very valuable, less superior. Other methods of a more theoretical nature, based on elementary matrix operations as considered in paragraph 2.2.1A, may then be used.

Suppose the system

$$L(D)y = M(D)u$$

has been rewritten in the form (2.46), i.e.

$$T(D)y = M^*(D)u$$

with $T(D) = V(D)L(D)$ being a triangular matrix such that the element of highest degree in each column lies on the diagonal (the method of obtaining the matrix V will be explained at the end of this paragraph):

$$|L(D)| = \prod_{i=i}^{p} t_{ii}(D) \qquad (2.46)$$

On account of the triangular form of T, the system has been decomposed into p subsystems, partially decoupled.

The subsystem defined by the last of equations (2.46),

$$t_{pp}(D)y_p = {}_pM^*(D)u,$$

is completely decoupled since only the last component y_p of y appears in it. A state representation of this system can therefore be obtained directly, using, for instance, the method of paragraph A (which always applies in the single-variable case) and the formula (2.53). Thus, we have:

$$\begin{aligned} \dot{x}_p &= A_p x_p + B_p u, \\ y_p &= C_p x_p, \end{aligned} \qquad (2.65)$$

where x_p is a vector whose dimension n_p is the degree of $t_{pp}(D)$.

The next equation from the set (2.46) involves y_p and y_{p-1}:

$$t_{p-1,p-1}(D)y_{p-1} = {}_{p-1}M^*(D)u - t_{p-1,p}(D)y_p, \qquad (2.66)$$

where the last term contains y_p and its derivatives up to the order of degree $(t_{p-1,p})$, which is less than n_p. These can all be replaced by the expressions for them derived from equations (2.65), and equation (2.66) will then take the form

$$t_{p-1,p-1}(D)y_{p-1} = {}_{p-1}M^{*'}(D)u + Nx_p, \tag{2.67}$$

which is a differential equation for y_{p-1} only, since x_p can be regarded as an independent input. A state representation of (2.67) will require n_{p-1} variables, with n_{p-1} = degree $(t_{p-1,p-1})$:

$$\begin{aligned} \dot{x}_{p-1} &= A_{p-1}x_{p-1} + B_{p-1}u + K_{p-1,p}x_p, \\ y_{p-1} &= C_{p-1}x_{p-1}. \end{aligned} \tag{2.68}$$

Proceeding in the same way for the other equations, obtained by reading (2.46) row-by-row upwards, we obtain, for, e.g., the first equation,

$$\begin{aligned} \dot{x}_1 &= A_1x_1 + B_1u + K_{1p}x_p + \dots K_{12}x_2, \\ y_1 &= C_1x_1 \end{aligned} \tag{2.69}$$

and then, collecting all these representations together, a complete state representation is given by:

$$\begin{bmatrix} \dot{x}_p \\ \dot{x}_{p-1} \\ \vdots \\ \dot{x}_1 \end{bmatrix} = \begin{bmatrix} A_p & 0 & \cdots & 0 \\ K_{p-1,p} & \ddots & \ddots & 0 \\ \vdots & \ddots & \ddots & \\ K_{1p} & \cdots & K_{12} & A_1 \end{bmatrix} \begin{bmatrix} x_p \\ x_{p-1} \\ \vdots \\ x_1 \end{bmatrix} + \begin{bmatrix} B_p \\ B_{p-1} \\ \vdots \\ B_1 \end{bmatrix} u,$$

$$\begin{bmatrix} y_1 \\ \vdots \\ \vdots \\ y_p \end{bmatrix} = \begin{bmatrix} 0 & \cdots & 0 & C_1 \\ \vdots & & \iddots & 0 \\ 0 & \iddots & & \vdots \\ C_p & 0 & \cdots & 0 \end{bmatrix} \begin{bmatrix} x_p \\ \vdots \\ \vdots \\ x_1 \end{bmatrix} \tag{2.70}$$

Example: Consider the system

$$\begin{bmatrix} D(D+1) & D+2 & D+1 \\ 0 & (D+1)(D+2) & D \\ 0 & 0 & (D+2)(D+1)^2 \end{bmatrix} \begin{bmatrix} y_1 \\ y_2 \\ y_3 \end{bmatrix} = \begin{bmatrix} 1 & 0 & 0 \\ D & 0 & 1 \\ D+3 & 1 & D \end{bmatrix} \begin{bmatrix} u_1 \\ u_2 \\ u_3 \end{bmatrix} \qquad (2.71)$$

which is already in the form (2.46).

The last equation is

$$\dddot{y}_3 + 4\ddot{y}_3 + 5\dot{y}_3 + 2y_3 = \dot{u}_1 + \dot{u}_3 + 3u_1 + u_2,$$

and formula (2.53) gives

$$\dot{x}_3 = \begin{bmatrix} -4 & 1 & 0 \\ -5 & 0 & 1 \\ -2 & 0 & 0 \end{bmatrix} x_3 + \begin{bmatrix} 0 & 0 & 0 \\ 1 & 0 & 1 \\ 3 & 1 & 0 \end{bmatrix} u = A_3x_3 + B_3u,$$

$$y_3 = [1, \ 0, \ 0]x_3 = C_3x_3.$$

Next, the second row of (2.71) gives

$$\ddot{y}_2 + 3\dot{y}_2 + 2y_2 = -\dot{y}_3 + \dot{u}_1 + u_3,$$

and we have

$$\dot{y}_3 = C_3\dot{x}_3 = C_3(A_3x_3 + B_3u) = [-4, \ 1, \ 0]x_3,$$

so that

$$\ddot{y}_2 + 3\dot{y}_2 + 2y_2 = [4, \ -1, \ 0]x_3 + \dot{u}_1 + u_3,$$

with the state representation, of order 2:

$$\dot{x}_2 = \begin{bmatrix} -3 & 1 \\ -2 & 0 \end{bmatrix} x_2 + \begin{bmatrix} 0 & 0 & 0 \\ 4 & -1 & 0 \end{bmatrix} x_3 + \begin{bmatrix} 1 & 0 & 0 \\ 0 & 0 & 1 \end{bmatrix} u = A_2x_2 + K_{23}x_3 + B_2u,$$

$$y_2 = [1, \ 0]x_2 = C_2x_2.$$

Finally, the first row of (2.71) is

$$\ddot{y}_1 + \dot{y}_1 = [1,\ 0,\ 0]u - (\dot{y}_2 + 2y_2) - (\dot{y}_3 + y_3),$$

and, substituting for y_2, $\dot{y}_2$, y_3, $\dot{y}_3$, we get

$$\ddot{y}_1 + \dot{y}_1 = [1,\ -1]x_2 + [3,\ -1,\ 0]x_3,$$

which has the state representation

$$\dot{x}_1 = \begin{bmatrix} -1 & 1 \\ 0 & 0 \end{bmatrix} x_1 + \begin{bmatrix} 0 & 0 \\ 1 & -1 \end{bmatrix} x_2 + \begin{bmatrix} 0 & 0 & 0 \\ 3 & -1 & 0 \end{bmatrix} x_3,$$

$$y_1 = [1,\ 0]x_1.$$

A state representation of the whole system (2.71) is thus, according to (2.70):

$$\dot{x} = \begin{bmatrix} -4 & 1 & 0 & 0 & 0 & 0 & 0 \\ -5 & 0 & 1 & 0 & 0 & 0 & 0 \\ -2 & 0 & 0 & 0 & 0 & 0 & 0 \\ 0 & 0 & 0 & -3 & 1 & 0 & 0 \\ 4 & -1 & 0 & -2 & 0 & 0 & 0 \\ 0 & 0 & 0 & 0 & 0 & -1 & 1 \\ 3 & -1 & 0 & 1 & -1 & 0 & 0 \end{bmatrix} x + \begin{bmatrix} 0 & 0 & 0 \\ 1 & 0 & 1 \\ 3 & 1 & 0 \\ 1 & 0 & 0 \\ 0 & 0 & 1 \\ 0 & 0 & 0 \\ 0 & 0 & 0 \end{bmatrix} u,$$

$$y = \begin{bmatrix} 0 & 0 & 0 & 0 & 0 & 1 & 0 \\ 0 & 0 & 0 & 1 & 0 & 0 & 0 \\ 1 & 0 & 0 & 0 & 0 & 0 & 0 \end{bmatrix} x$$

Note: It should be noted that the required substitutions for the components of y and their derivatives are very easy to implement, because of the special simple forms of the C_i.

b) Polak's algorithm for triangularisation of a polynomial matrix . Given a polynomial matrix M(s), consider the problem of finding a polynomial matrix V(s) such that

$$\det V(s) = \text{constant} \neq 0,$$
$$V(s)M(s) = T(s) = \text{triangular matrix.}$$

We shall require also (since, although not necessary here, it will be useful later) that T(s) is such that, in each column, the element which has the highest degree is on the diagonal, i.e.

$$\text{degree } (t_{ii}) > \text{degree } (t_{ji}) \text{ for } j \neq i.$$

The matrix V(s) will be obtained as a product $V_k(s)\ldots V_1(s)$, where each $V_i(s)$ is a matrix with constant determinant, corresponding to an elementary operation on the rows. By an elementary operation, we mean one of the following:

- interchange two rows;
- multiply a row by a constant nonzero factor;
- add to one row a multiple, by a polynomial factor, of another row.

The procedure consists of the following steps:-

1) Among the nonzero elements of the first column of M(s), find the one of lowest degree, say m_{i1}. Put it on the first row by interchanging rows 1 and i. This corresponds to multiplying M(s) by the elementary matrix V_1:

$$V_1 = \begin{bmatrix} 0 & 0 & \cdots & 0 & 1 & 0 & \cdots & 0 \\ 0 & 1 & 0 & \cdots & 0 & & & \\ \vdots & & \ddots & \ddots & \vdots & & & \vdots \\ 0 & \cdots & 0 & 1 & 0 & & & \\ 1 & 0 & \cdots & \cdots & 0 & 0 & & \\ 0 & \cdots & \cdots & \cdots & 0 & 1 & \ddots & \\ \vdots & & & & & \ddots & \ddots & 0 \\ 0 & \cdots & \cdots & \cdots & \cdots & \cdots & 0 & 1 \end{bmatrix} \leftarrow \text{row } i$$

column i ↑

2) Then, divide each element of the first column of the matrix V_1M (now called M) by the first element m_{11}, giving

$$m_{j1} = m_{11}\, q_{j1} + r_{j1} \quad (j = 2, \ldots p),$$

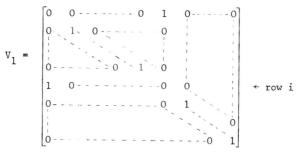

where the q_{j1} and r_{j1} are polynomials. Now, multiply the first row by q_{j1} and subtract it from the jth row, so that m_{j1} is replaced by r_{j1}. This amounts to pre-multiplying M(s) by a matrix V_2 of the form

$$V_2 = \begin{bmatrix} 1 & 0 & \cdots & 0 \\ -q_{21} & 1 & \ddots & \vdots \\ \vdots & 0 & \ddots & 0 \\ -q_{p1} & 0 \cdots 0 & & 1 \end{bmatrix} .$$

After this operation, the first column of V_2M (now referred to as M) will be $\left[m_{11}, r_{21}, \ldots r_{p1}\right]^T$. If all the remainders r_{j1} are zero, we can proceed to the second column. Usually this will not be the case, but, since degree (r_{j1}) < degree (m_{11}), the repetition of steps 1) and 2) on the first column will eventually achieve this result.

3) Repeat the above procedure on the last p-1 elements of the second column, until it reaches the form $\left[m_{12}, m_{22}, 0, \ldots 0\right]^T$. Then, proceed to the third column, and so on. Eventually, M(s) will reach the form

$$M(s) = \begin{bmatrix} m_{11} & m_{12} & \cdots \\ 0 & m_{22} & \cdots \\ \vdots & \ddots & \ddots \\ 0 \cdots & 0 & m_{pp} \end{bmatrix} .$$

4) It may happen that the conditions on the degrees of the elements of each column are not satisfied. Suppose, for instance, that, in the kth column, there is an element m_{jk} whose degree is greater than or equal to that of m_{kk}. Then, with

$$m_{jk} = m_{kk}\, q_{jk} + r_{jk},$$

premultiplication by

$$\begin{bmatrix} 1 & 0 & \cdots & 0 & \cdots & 0 \\ 0 & \ddots & & 0 & & \vdots \\ \vdots & & \ddots & -q_{jk} & & \vdots \\ & & & 0 & & \\ & & & 0 & \ddots & \\ & & & 1 & & \\ \vdots & & & & \ddots & 0 \\ 0 & \cdots & & \cdots & 0 & 1 \end{bmatrix}$$

will put the kth column in the form

$$\left[m_{1k}, \ldots\ m_{j-1,k},\ r_{jk},\ m_{j+1,k}, \ldots m_{kk},\ 0, \ldots 0\right]^T.$$

Repetition of this procedure will thus ultimately achieve the desired form.

<u>Example</u>: Let

$$M(s) = \begin{bmatrix} s^2+4s+4 & s+3 \\ s+1 & s^2+1 \end{bmatrix}.$$

The successive steps are as follows:

$$V_1M = \begin{bmatrix} 0 & 1 \\ 1 & 0 \end{bmatrix} M = \begin{bmatrix} s+1 & s^2+1 \\ s^2+4s+4 & s+3 \end{bmatrix};$$

$$V_2V_1M = \begin{bmatrix} 1 & 0 \\ s+1 & -1 \end{bmatrix} V_1M = \begin{bmatrix} s+1 & s^2+1 \\ -1 & s^3+3s^2 \end{bmatrix};$$

$$V_3V_2V_1M = \begin{bmatrix} 0 & 1 \\ 1 & 0 \end{bmatrix} V_2V_1M = \begin{bmatrix} -1 & s^3+3s^2 \\ s+1 & s^2+1 \end{bmatrix};$$

$$V_4V_3V_2V_1M = \begin{bmatrix} 1 & 0 \\ s+1 & 1 \end{bmatrix} V_3V_2V_1M = \begin{bmatrix} -1 & s^3+3s^2 \\ 0 & s^4+4s^3+4s^2+1 \end{bmatrix}.$$

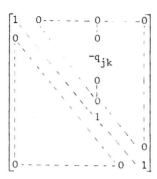

c) Diagonalisation. When a triangular form has been obtained, the analogous procedure, acting on columns instead of rows, can be followed. This corresponds to postmultiplication by elementary matrices.

For instance, in the case of the last example, we have:

$$\Gamma = V_4V_3V_2V_1MW_1 = V_4V_3V_2V_1M \begin{bmatrix} -1 & s^3+3s^2 \\ 0 & 1 \end{bmatrix} = \begin{bmatrix} 1 & 0 \\ 0 & s^4+4s^3+4s^2+1 \end{bmatrix}$$

and it follows that

$$M = (V_4V_3V_2V_1)^{-1}\Gamma W_1^{-1} = V\Gamma W$$

$$= \begin{bmatrix} -(s^2+4s+4) & s+3 \\ -(s+1) & 1 \end{bmatrix} \begin{bmatrix} 1 & 0 \\ 0 & s^4+4s^3+4s^2+1 \end{bmatrix} \begin{bmatrix} -1 & s^3+3s^2 \\ 0 & 1 \end{bmatrix}$$

Note: We remark that

$$\det V = -1 = \text{constant},$$
$$\det W = -1 = \text{constant},$$

and that the invariants of M are 1 and $s^4+4s^3+4s^2+1$.

2.2.3 Differential representations associated with a given transfer function matrix or state representation

We shall briefly consider here the problem of finding a differential representation equivalent to a given transfer function matrix or state-space form. It should be noted at the outset that, although an equivalent (L, M) representation can always be found for any transfer function matrix Z(s), the corresponding statement for the case of a system initially in state-space form is true only if the system is observable.

A - Derivation of a differential representation from a transfer function matrix.

Suppose a system is defined in terms of its transfer function matrix Z(s), of dimensions (p, m), by

$$Y(s) = Z(s)U(s) \tag{2.72}$$

and we are looking for an equivalent representation in the form

$$L(D)y = M(D)u. \tag{2.73}$$

Clearly, the matrices L and M must satisfy the following conditions:

i) dimensions of L, M are (p, p), (p, m) respectively;

ii) L is invertible;

iii) $L^{-1}M = Z$.

However, these conditions, though necessary, are not sufficient, for, if the system (2.73) is to be equivalent to (2.72), it must be controllable, and so we have the extra condition:

iv) degree (det L) = order of minimal state representation for Z(s).

Example: Suppose, for instance,

$$Z(s) = \frac{1}{(s+2)^2}\begin{bmatrix} s+3 & 3s+8 \\ 1 & s+4 \end{bmatrix} \tag{2.74}$$

It may easily be verified that a minimal state representation for Z(s) is of order 2 only.

If we consider, as a candidate for an equivalent representation , the system

$$\begin{bmatrix} (D+2)^2 & 0 \\ 0 & (D+2)^2 \end{bmatrix} y = \begin{bmatrix} D+3 & 3D+8 \\ 1 & D+4 \end{bmatrix} u, \tag{2.75}$$

we see that the first three conditions are satisfied. Since, however, det $L(s) = (s+2)^4$, the system (2.75) cannot be considered as equivalent to (2.74); it has, in fact, two uncontrollable modes.

We recall from paragraph 2.1.3.A that it is possible to cast a transfer function matrix Z(s) into Smith-McMillan form:

$$Z = V\Gamma W. \tag{2.76}$$

If this expression is compared with the transfer function matrix arising from (2.73), i.e.

$$Z = L^{-1}M,$$

it is clear that, provided Γ (and, equivalently, Z) is invertible, conditions i) to iv) can be satisfied by taking

$$L^{-1} = V\Gamma, \ M = W. \tag{2.77}$$

In the example above, we get

$$Z(s) = \begin{bmatrix} s+3 & 1 \\ 1 & 0 \end{bmatrix} \begin{bmatrix} \frac{1}{(s+2)^2} & 0 \\ 0 & 1 \end{bmatrix} \begin{bmatrix} 1 & s+4 \\ 0 & -1 \end{bmatrix} = V(s)\Gamma(s)W(s),$$

and so, making the identifications according to (2.77), we obtain the differential representation:

$$\begin{bmatrix} 0 & (D+2)^2 \\ 1 & -(D+3) \end{bmatrix} y = \begin{bmatrix} 1 & D+4 \\ 0 & -1 \end{bmatrix} u.$$

B - Differential representations associated with a state-space form

Let us suppose that we know an observable state representation of a given system and that we are looking for an equivalent differential representation. We could try to solve this problem by elimination of various variables from the state representation. However, if the order of the system is large, such a procedure becomes very cumbersome and is likely to introduce many errors. Consequently, a more systematic approach is highly desirable.

Suppose the state representation is in companion form:

$$\dot{x} = Ax + Bu,$$

$$y = Cx.$$

As in paragraph 2.1.3.A.b), we can write the transfer function matrix in the form

$$\frac{\mathcal{C}(s)\mathcal{B}(s) \text{ modulo } \psi(s)}{\psi(s)} = \frac{P(s)}{\psi(s)}$$

where

$$\mathcal{B}(s) = [s^{n-1}, \ldots s, 1]B, \quad \mathcal{C}(s) = C\mathcal{A}\begin{bmatrix} 1 \\ s \\ \vdots \\ s^{n-1} \end{bmatrix},$$

if A is in vertical companion form. Then, writing P(s) in Smith form

$$P(s) = V^*(s)\Gamma^*(s)W^*(s),$$

we have to satisfy the equation

$$L^{-1}M = \frac{V^*\Gamma^*W^*}{\psi}.$$

Thus, again provided that Γ^* is invertible, the matrices L and M of the differential representation can be taken as

$$L = \psi\Gamma^{*-1}V^{*-1}, \quad M = W^* \tag{2.78}$$

Example 1: Consider the system

$$\dot{x} = \begin{bmatrix} -2 & 1 \\ -1 & 0 \end{bmatrix} x + \begin{bmatrix} 0 & 1 \\ 1 & 1 \end{bmatrix} u \qquad y = \begin{bmatrix} 0 & 1 \\ 1 & 0 \end{bmatrix} x$$

for which

$$\mathcal{C}(s) = [s, 1]\begin{bmatrix} 0 & 1 \\ 1 & 1 \end{bmatrix} = [1, s+1],$$

$$\mathcal{B}(s) = \begin{bmatrix} 0 & 1 \\ 1 & 0 \end{bmatrix}\begin{bmatrix} 1 & 0 \\ 2 & 1 \end{bmatrix}\begin{bmatrix} 1 \\ s \end{bmatrix} = \begin{bmatrix} s+2 \\ 1 \end{bmatrix},$$

$$\psi(s) = s^2+2s+1 = (s+1)^2,$$

so that

$$\frac{\mathcal{C}(s)\mathcal{B}(s)}{\psi(s)} = \frac{1}{(s+1)^2}\begin{bmatrix} s+2 & s^2+3s+2 \\ 1 & s+1 \end{bmatrix}$$

whence

$$\frac{P(s)}{\psi(s)} = \frac{1}{(s+1)^2}\begin{bmatrix} s+2 & s+1 \\ 1 & s+1 \end{bmatrix}$$

and so

$$P(s) = \begin{bmatrix} s+2 & 1 \\ 1 & 0 \end{bmatrix}\begin{bmatrix} 1 & 0 \\ 0 & (s+1)^2 \end{bmatrix}\begin{bmatrix} 1 & s+1 \\ 0 & -1 \end{bmatrix} = V^* \Gamma^* W^* .$$

Hence, according to (2.78), the differential representation will be

$$\begin{bmatrix} 0 & (D+1)^2 \\ 1 & -(D+2) \end{bmatrix} y = \begin{bmatrix} 1 & D+1 \\ 0 & -1 \end{bmatrix} u.$$

Example 2: Let us now consider the uncontrollable single-variable system

$$\dot{x} = \begin{bmatrix} -1 & 0 & 0 \\ 0 & -2 & 0 \\ 0 & 0 & -3 \end{bmatrix} x + \begin{bmatrix} 1 \\ 2 \\ 0 \end{bmatrix} u, \quad y = [1, 1, 1]x.$$

With the transformation $x = M\tilde{x}$, where

$$M = \begin{bmatrix} 6 & 5 & 1 \\ 3 & 4 & 1 \\ 2 & 3 & 1 \end{bmatrix},$$

the system is put in companion form:

$$\dot{\tilde{x}} = \begin{bmatrix} 0 & 1 & 0 \\ 0 & 0 & 1 \\ -6 & -11 & -6 \end{bmatrix} \tilde{x} + \frac{1}{2}\begin{bmatrix} 3 \\ 7 \\ -15 \end{bmatrix} u, \quad y = [11, 12, 3]\tilde{x}.$$

Note: We have chosen here a horizontal companion form, and so

$$\mathcal{B}(s) = [s^{n-1}, \ldots 1]\mathcal{A}B, \quad \mathcal{C}(s) = C\begin{bmatrix} 1 \\ \vdots \\ s^{n-1} \end{bmatrix}.$$

Thus, in the example,

$$\mathcal{B}(s) = [s^2, s, 1]\begin{bmatrix} 1 & 0 & 0 \\ 6 & 1 & 0 \\ 11 & 6 & 1 \end{bmatrix}\begin{bmatrix} -1.5 \\ 3.5 \\ -7.5 \end{bmatrix} = -\frac{1}{2}(3s^2+11s+6),$$

$$\mathcal{C}(s) = [11, 12, 3]\begin{bmatrix} 1 \\ s \\ s^2 \end{bmatrix} = 3s^2+12s+11,$$

$$P(s) = -\frac{1}{2}(9s^4+69s^3+183s^2+193s+66) \text{ modulo } (s^3+6s^2+11s+6)$$

$$= 3s^2+13s+12,$$

$$\psi(s) = s^3+6s^2+11s+6,$$

and so the differential equation obtained is

$$\dddot{y} + 6\ddot{y} + 11\dot{y} + 6y = 3\ddot{u} + 13\dot{u} + 12u.$$

We note, however, that the differential equation associated with the transfer function is

$$\ddot{y} + 3\dot{y} + 2y = 3\dot{u} + 4u.$$

Thus, the mode s = 3, because of its uncontrollability, would be lost if the transfer function were used as an intermediate stage.

Conclusions: In so far as the internal structure of a dynamical system is precisely known, it appears that the most adequate representation is a state-space model like (2.1), since this preserves all the modes, whether or not they are controllable or observable.

In many cases, however, a theoretical study of the system leads naturally to a differential representation of the form (2.3). We have seen that such a representation is necessarily observable; consequently, this kind of model loses all information about any unobservable parts which the system may have.

Furthermore, in some cases, knowledge of the system is limited to what can be obtained from input-output relations. This leads us to use a transfer function matrix representation, which exhibits only the controllable and observable part of the system.

It follows that there is a hierarchy among these representations, which are, in general, not equivalent, and that care must be taken in choosing which form to use for representing a system. When a model is available in one form, and a representation in a different one is required, it is important to make a distinction between the following two cases.

a) The representation sought is less general than the original one, i.e. we are transforming in the direction $x \rightarrow (L,M) \rightarrow Z$. In this case, modes may be lost.

b) A representation more general than the initial one is sought, i.e. the transformation is in the direction $Z \rightarrow (L,M) \rightarrow x$. Then, care should be taken that no extra modes, absent from the original model, are introduced.

Chapter 3

Structures

It is already several years since Professors Gille, Pélegrin and Decaulne wrote: "In contrast to what is sometimes thought, the casting of a problem into the form of a set of equations is more important than the solution of the equations themselves. Indeed, experience shows that the majority of errors in research projects arise from incorrect equations and not from inaccurate solutions" (ref.3.15).This warning, which incidentally has not always been taken at its full value in the past, finds a new immediacy in the developments currently being made in the analysis, design and simulation of multivariable systems, and this is why, in a cohesive study devoted to the control of multivariable systems, we felt it unavoidable to start from the fundamental bases which form the subjects of chapters 1 and 2. The aim of this third chapter is to establish the connection between the first two and the following ones which deal with control in the true sense and its applications, setting out the practical implications which the properties of various representations can have for the construction of models, their reduction and their stability. It finishes moreover with a teleological approach to state-space representations, which is directly adapted to control problems and whose results will be extensively used in chapters 4 and 5.

3.1 STRUCTURES AND REPRESENTATIONS

In the first rank of the problems which we encounter in the study of multivariable systems, there appear those of structure. They show themselves to be among the most important because they are the ones which, in the end, most often influence those of synthesis, of compensation and especially of optimisation. The problem moreover cannot but become more acute in proportion as the systems under consideration become more complex, and this will be the case in the great majority of industrial systems. If the structure of such a system is ignored, and if we proceed to a global identification of the system starting from input-output relations, the dimension of the mathematical model will usually be restricted in such a way that all synthesis problems can be solved by well-tried methods. The problem will clearly be more complicated, however, if the mathematical model is obtained by a detailed analysis of the system, of which each element is studied separately and the complete model is established by taking account of the various connections existing between the subsystems.

In fact, when we talk about the structure of a system, it is important to make clear at the beginning what we mean by this and on what level the problem is to be envisaged; the concepts of system, representation and structure are indeed closely linked. It will be convenient to use the word system for an arrangement or organised combination of parts into a whole, thus implying that the combination is made according to a rational principle and represents a methodical arrangement of parts. The term "subsystem" is, on the contrary, reserved for an entity whose physical structure is unknown and is not accessible except via its inputs and outputs.

3.1.1 Review of the properties of multivariable systems

Input-output interaction

A non-degenerate multivariable system is characterised by the phenomenon of interaction or coupling, which is to say that one input in general affects several outputs.

Interaction of outputs

Whether the above property exists or not, a perturbation (or an initial condition) acting on one output can, by propagation through the system, affect other outputs. This would be, for example, the effect on the outputs of an initial condition on one of them, $y_i(0)$, all the inputs being zero. If this condition affects only the output y_i, we shall say that the system is not interactive in respect of this output. In the case that this holds whichever output is considered, the system is described, in the American literature, as the possessing the property of "independent output restoration".

Independence of outputs

The purpose of a multivariable control system is, ultimately, to generate desired outputs. The question which thus arises is whether or not, by using suitable input functions, the outputs can be made to evolve independently of each other and in a prescribed manner.

It is evident that a necessary condition for this to happen is that the number of outputs (p) should be at most equal to the number of inputs (m), since, if $p > m$, $p - m$ outputs will necessarily be dependent on the others. More exactly, if the matrix Z(s) is of rank p, the p outputs can be changed independently of one another if Z(s) is of rank $p - \rho$, then ρ outputs can be expressed as functions of the $p - \rho$ independent outputs.

Nevertheless, it is clear that these conditions express only the possibility of making the outputs vary independently and certainly not of making them follow arbitrary paths. We know that the latter property is directly connected with the concept of reproducibility (cf. chapter 1, section 1.4.).

3.1.2 Structures at the subsystem level

A - Mesarovic canonical structures (1)

Let us consider a multivariable system accessible only through its inputs and outputs, and let us suppose that a certain number of tests made

(1) For the material of this paragraph, one may have recourse to the book of M.D. Mesarovic, ref. (3.26). The names, which are preserved, of P-, V- and H- canonical structures are in particular due to this author.

on the system have enabled us to determine its response behaviour, that is to say, the relations connecting inputs to outputs. These relations can be obtained in various ways, depending on the nature of the experiments performed.

a) We have p relations of the form:

$$y_j(t) = f_j(u_1, u_2, \ldots, u_m ; y_1, y_2, \ldots, y_{j-1}, y_{j+1}, \ldots, y_p) \qquad (3.1)$$

Although these relations permit us to predict the responses of the system under given conditions, they do not enable us to determine which, among the p + m variables, influence a particular output. Indeed, these equations can be manipulated in various ways. We can, for example, eliminate all the outputs and deduce a set of equations of the form

$$y_j(t) = f_j(u_1, u_2, \ldots, u_m) . \qquad (3.2)$$

In this case, each output is influenced by all the inputs. This form of representation is such as we use when, for example, we represent a system by its transfer function matrix.

$$Y(s) = Z(s)\, U(s) \qquad (3.3)$$

It is nevertheless important to notice that when we pass from equations (3.1) to (3.2), we are making a supplementary hypothesis, which at this stage has not been physically justified, about the structure of the system. Indeed, in the latter form, the system can be represented by the functional diagram of figure 3.1 which shows that all the subsystems are non-interactive and, in particular, that there is no interaction between the outputs. (Here, the term subsystems is to be understood in the sense of elements of the transfer function matrix).

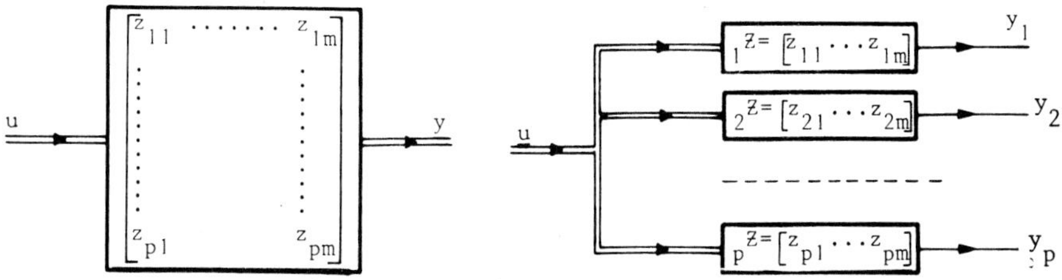

Fig. 3.1 P-canonical structure

By carrying out the elimination process differently, we could just as well, starting from equations (3.1), eliminate from the right-hand side not all the output variables but only a certain number, and obtain equations of the form

$$y_j(t) = f_j^{**}(u_1, u_2, \ldots, u_{m-p} ; y_1, \ldots, y_{j-1}, y_{j+1}, \ldots, y_p) \qquad (3.4)$$

As in the previous case, by doing this, we are arbitrarily imposing a certain structure on the system. There will now be interaction between a certain number of the outputs, that is to say, every internal perturbation of a subsystem will react on the others.

In particular, if p = m, the preceding equations take the form:

$$y_j(t) = f_j(u_j, y_1, \ldots, y_{j-1}, y_{j+1}, \ldots, y_p)$$

or again, in matrix form (cf. figure 3.2)

$$Y = F[U + VY] \qquad (3.5)$$

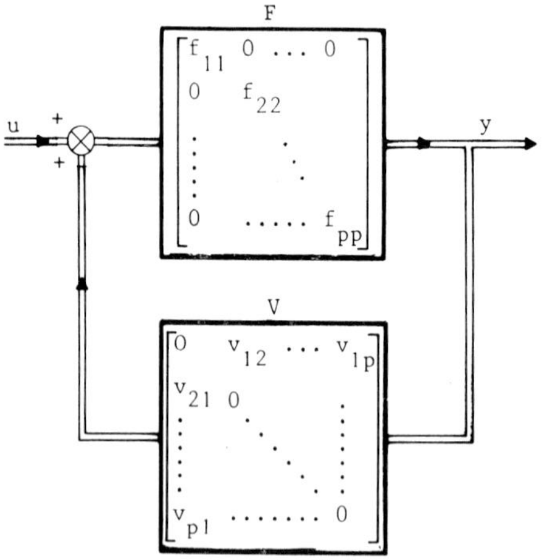

Fig. 3.2 V-canonical structure

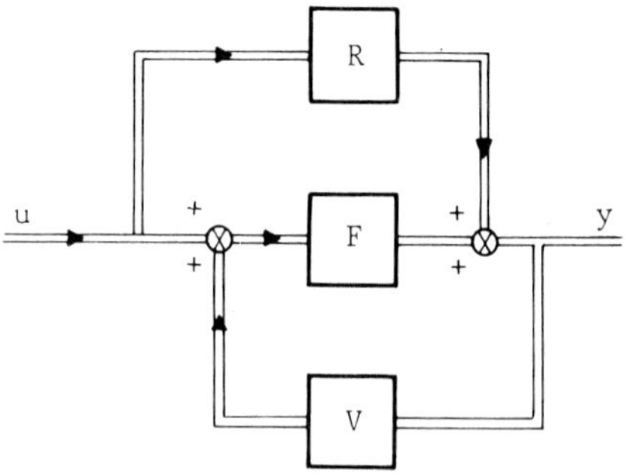

Fig. 3.3 H-canonical structure

If m ≠ p, the situation is slightly different in the sense that the whole system cannot be put in the above form, part of it remaining in the original form. The system can be represented in the matrix form (cf. figure 3.3)

$$Y = RU + F\left[U + VY\right] \qquad (3.6)$$

The three types of structures which have just been mentioned only represent typical cases and in no way characterise all the possible situations. They were, to our knowledge, defined by Mesarovic (cf. ref. 3.26), and are frequently found mentioned in the literature under the names of P-canonical, V-canonical and H-canonical structures.

b) The responses characterising a multivariable system can alternatively be obtained by starting from a set of differential equations of the form:

$$L(D)\ Y = M(D)\ U \qquad (3.7)$$

as we saw in chapter 2. These equations are no different in spirit from equations (3.1), except that, at least in the case where L is invertible, there is a tendency to write them in the form:

$$Y(s) = L^{-1}(s)\ M(s)\ U(s) \qquad (3.8)$$

by taking Laplace transforms with zero initial conditions. The matrix Z(s) thus obtained is not of general form, since all its elements have the same poles, which are the zeroes of the determinant of L. By doing this, we are again implicitly assuming for the system a certain structure, as we shall see in the following paragraph.

c) The system responses can be obtained by methods giving directly the transmission between an input e_i and an output y_j. This will be the case, for example, if we look at correlation functions between these variables. We then again obtain a matrix Z, but one whose elements no longer have the property of the preceding section, in the sense that the poles can now be arbitrary.

The cases (b) and (c) seem to impose particular structures on the system. In case (b), the fact that the elements of Z have the same poles suggests the presence of a closed-loop structure, the common denominator corresponding to the dynamics of the loop, as we see in the graph associated with a V-canonical structure. Case (c) leads us to assume a P-canonical structure. There evidently remains a certain arbitrariness in these conclusions since the reduction of the matrix Z to a form with a common denominator always allows us to pass from a P-canonical to a V-canonical structure. These considerations suggest that, from an external point of view, the structural composition of the system can hardly be determined at all. This is what Mesarovic calls the principle of structural uncertainty, which is to say that the manner in which the subsystems are coupled together cannot be determined from outside the system (1).

The ideas of structural uncertainty at the subsystem level will be the justification for the teleological approach which we shall make in paragraph 3.3. If, at the subsystem level, no further investigation of internal structure can be made, we can always choose the most convenient structure within a given framework, without concerning ourselves about its physical justification.

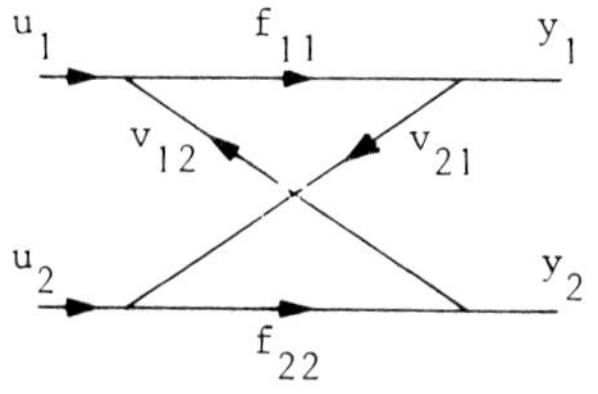

Fig. 3.4 Signal-flow graph of a two-dimensional system having a V-canonical structure

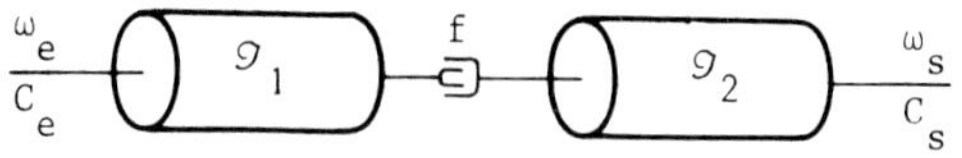

Fig. 3.5 Example of linear clutch-coupling

(1) This principle of structural uncertainty is valid only in a linear framework. We shall see in paragraph D, in a particular example, that the introduction of nonlinearities can allow us, starting from input-output observations, to discover part of the internal structure.

B - Manifestation of interactions in a transfer function matrix representation (2)

We saw in the preceding paragraph that a strictly external knowledge of the system does not allow us to represent it, and in particular the possible interactions between the outputs, correctly, in consequence of the structural properties which can be more or less arbitrarily imposed. This is the consideration which leads us, in general, to adopt, in the transfer function matrix representation, a P-canonical structure in the sense defined above.

It is important to notice that this deficiency is not necessarily attached to the notion of a transfer function matrix representation, but can arise from an incomplete description of the system: other things being equal, moreover, the same problem arises in single-variable theory, and we know that it can often be solved by the introduction of the idea of a "quadripole", replacing that of a transfer function matrix.

With each of the two terminals of a single-variable system ("input" and "output") we know that we can associate two variables, one independent, the other dependent. (Angular velocity and torque, pressure and flow rate, voltage and current, temperature and heat flux...). Under these conditions, a correct description of a single-variable system capable of being coupled to another system necessitates the determination of a transfer function quadripole.

Let us consider, for example, the case of a linear clutch-coupling as illustrated in figure 3.5. With each terminal we can associate two variables, speed and torque.

We have the equations:

$$C_e = \mathcal{J}_1 \, d\omega_e/dt + f(\omega_e - \omega_s),$$

$$f(\omega_e - \omega_s) = \mathcal{J}_2 \, d\omega_s/dt + C_s ,$$

(2) cf. ref. (3.24) of L.E. McBride and K.S. Narendra.

and hence

$$\begin{bmatrix} \omega_s \\ C_e \end{bmatrix} = \frac{1}{\mathcal{J}_2 s+f} \begin{bmatrix} f & -1 \\ \mathcal{J}_2\mathcal{J}_1 s^2 + f(\mathcal{J}_1+\mathcal{J}_2)s & f \end{bmatrix} \begin{bmatrix} \omega_e \\ C_s \end{bmatrix}$$

We thus see the influence of the output torque C_s on the output speed ω_s as a result of the coupling term $-1/(\mathcal{J}_2 s + f)$.

Other things being equal, the problem is the same in the multivariable case. If such a system has m + p terminals (m inputs and p outputs) we can associate with it 2(m + p) terminal variables, and as one of the two variables associated with each terminal will be dependent, we can regard the system as comprising (cf. ref. 3.24):

m independent terminal variables corresponding to real inputs, u;
p dependent terminal variables corresponding to real outputs, y;
p independent terminal variables corresponding to variables associated with the p real outputs, w;
m dependent terminal variables corresponding to variables associated with the m real inputs, v.

If we take as input vector the vector $[u,w]^T$ (primary inputs u, secondary inputs w) and as output vector the vector $[y,v]^T$, a representation of the system can be written in the form (1):

$$\begin{bmatrix} Y \\ V \end{bmatrix} = \begin{bmatrix} Z_a & Z_b \\ Z_c & Z_d \end{bmatrix} \begin{bmatrix} U \\ W \end{bmatrix} = Z^* \begin{bmatrix} U \\ W \end{bmatrix} \qquad (3.9)$$

where Z_a, Z_b, Z_c, Z_d are matrices of dimensions respectively (p,m), (p,p), (m,m), (m,p) (cf. figure 3.6). In the case where we can suppose that w, which is to say the vector of independent variables conjugate to the outputs y, is identically zero, the equation $Y = Z_a U + Z_b W$ reduces to the

(1) We are designating by Y, V,... the Laplace transforms of y, v,...

form $Y = Z_a U$ corresponding to a classical transfer function matrix. Furthermore, we can ignore the second equation

$$V = Z_c U + Z_d W$$

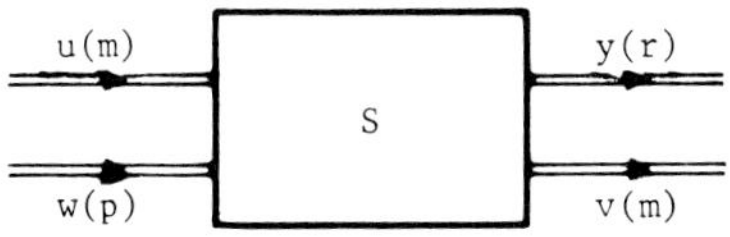

Fig. 3.6 Terminal quantities associated with a single-variable system

if the matrix Z^* represents the whole system being studied; consideration of the secondary outputs is then in fact superfluous. On the contrary, it is not so if the system represented by a matrix Z_1^* is to be connected to another system, represented by a matrix Z_2^*. It is not sufficient, in fact, to make $y_2 = u_1$ (cf. figure 3.7) in order to establish the connection of the two systems. It is necessary also to ensure that $v_1 = w_2$. (In the case, for example, where two electrical networks are connected, it is not enough to make the voltages equal but also the currents. The omission of the latter condition amounts to supposing that the systems are connected through an isolating amplifier stage).

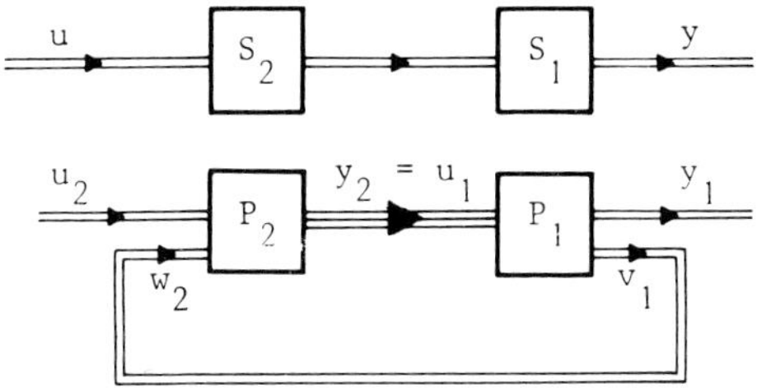

Fig. 3.7 Cascade connection of two systems. Coupling of secondary variables.

This type of representation of a system with m + p inputs and outputs by a matrix of dimension m + p was introduced, to our knowledge, by Narendra and McBride (ref. 3.24). Its advantage is to permit us to describe not only the one-sided properties of the system (which are fixed by Z_a) but also, thanks to the matrices Z_b, Z_c, Z_d, the two-sided properties associated with the action of the conjugate variables, in particular the phenomena of output interaction. Also, a P-canonical structure in the sense of Mesarovic (complete non-interaction among the outputs) will correspond to a diagonal matrix Z_b, a V-canonical structure to a matrix Z_d with no null element, etc.

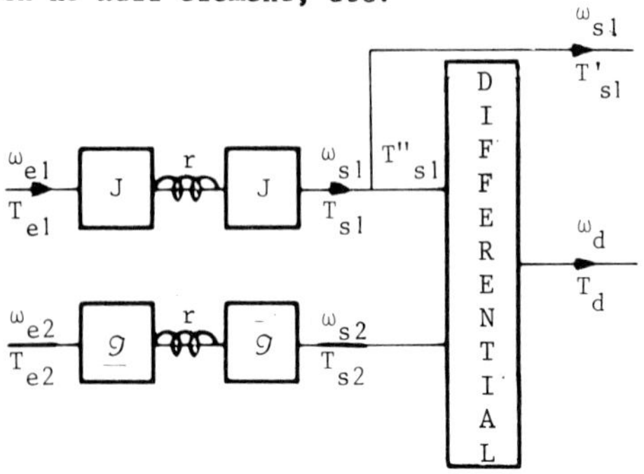

Fig. 3.8 Two-dimensional example

Example: by way of example, let us consider the mechanical system represented by figure 3.8 where ω represents an angular velocity, T a torque, $\mathcal{J}$ an inertia and r a spring constant. If we are only interested in the velocity variables, we may tend to represent the system by the matrix

$$\begin{bmatrix} \omega_{s1} \\ \omega_d \end{bmatrix} = Z \begin{bmatrix} \omega_{e1} \\ \omega_{e2} \end{bmatrix}$$

with

$$Z = \begin{bmatrix} L & -K \\ -n(L-K) & -n(M-K) \end{bmatrix}$$

where

$$L = EB + \mathcal{J} \qquad K = FB \qquad M = BD + C$$

$$E = (B + \mathcal{J}\, p)\, p \qquad C = \mathcal{J}\, p^2/(r+1) \qquad F = (B + \mathcal{J}\, p)\, p \qquad B = p/r$$

Such a representation fails to reveal the interaction between the outputs, which is, however, brought about physically via the torques. The matrix Z corresponds only to the part Z_a of the complete matrix written in the form:

$$Z^* = \left[\begin{array}{c|cc} Z_a & -BL & -nB(L-K) \\ & nB(L-K) & n^2B(L+M-2K) \\ \hline Z_c & \multicolumn{2}{c}{Z_d} \end{array}\right]$$

which itself represents well the properties of the physical system.

C - Structure and state-space representations

We have seen in the preceding paragraphs that, at the subsystem level, no external relation enables us to determine the internal structure of the system. This "principle of structural indeterminacy", and the consequences which one can draw from it with regard to transfer function matrix representations, evidently remain valid in the case of state-space representations. The consequences are fundamental, practically as much as theoretically and for the development of algorithms as much as for the representation problem itself.

Indeed:

(a) We know (cf. chapter 1) that a state-space representation of given dimension is defined only to within a linear transformation. Since this representation does not indicate the internal structure of the system, we thus have the possibility of choosing, among the infinity of possible representations, one or another which turns out to be particularly well adapted to the solution of the given problem. In view of the importance of this teleological conception of the structural problem, its examination will be made the subject of a separate section (cf. section 3.3).

(b) We also know (cf. chapter 2) that an external relation expressed by a transfer function matrix does not enable us to fix the order of the system, which can be arbitrarily augmented by incorporating any number of uncontrollable or unobservable modes.

Suppose, for example, that a process is defined by its transfer function matrix

$$Z(s) = \frac{1}{s(s+1)(s+2)(s+3)} \begin{bmatrix} s+3 & (s+1)(s+2) \\ (s+1)(s+3) & s+2 \end{bmatrix}$$

whose minimal state-space representation, of order 5, can be put in the form

$$\dot{x} = \begin{bmatrix} 0 & & & & \\ & -1 & & & \\ & & -1 & & \\ & & & -2 & \\ & & & & -3 \end{bmatrix} x + \begin{bmatrix} 3 & 2 \\ -1 & 0 \\ 0 & -0,5 \\ 1 & 0 \\ 0 & 1 \end{bmatrix} u$$

$$y = \begin{bmatrix} \frac{1}{6} & 1 & 0 & \frac{1}{2} & -\frac{1}{3} \\ \frac{1}{6} & 0 & 1 & -\frac{1}{2} & \frac{1}{6} \end{bmatrix} x$$

All the structures indicated in table 3.1 (among others), whose dimensions can vary from 5 to 10, correspond to the same transfer function matrix, and it is not possible, starting from the knowledge of inputs and outputs, to specify which one corresponds to the real physical system. This is the reason why, in the absence of other information, we agree to adopt a representation of minimal order even though this may not represent the actual system (and despite the disadvantages which this may have in certain applications).

D - Structural information brought by the presence of nonlinearities

Everything we have said so far assumes the process to be isolated and linear. Without going into details which do not belong in this investigation devoted to the study of multivariable linear systems, it is nevertheless important to indicate that the combination of such a system with nonlinear elements can allow a better understanding of the internal structure of the linear part to be obtained.

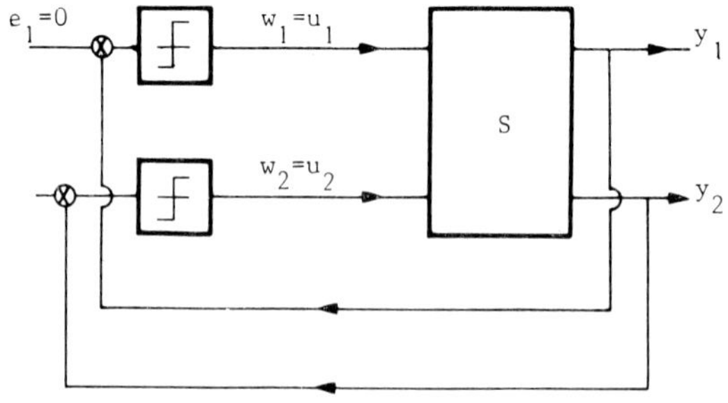

Fig. 3.9 Feedback coupling of a linear system through nonlinear elements

TABLE 3.1

Configuration	Structure	Degree of the Representation
1	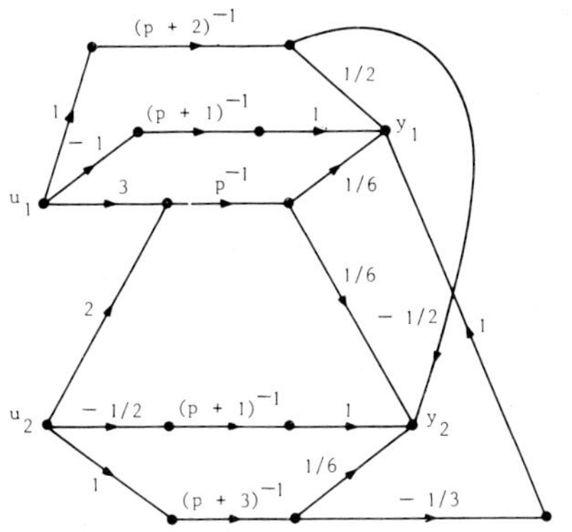	5
2	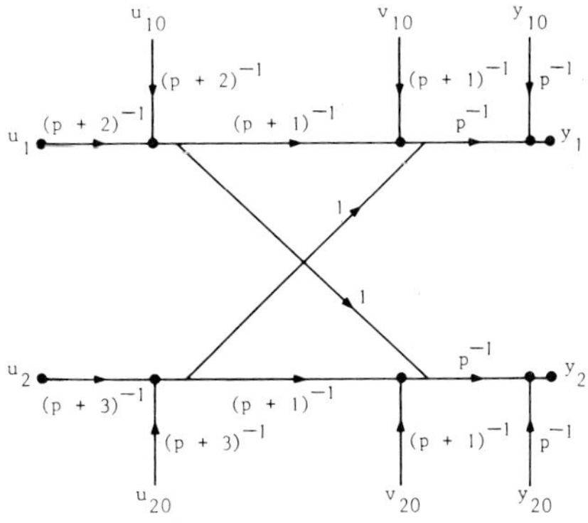	6
3	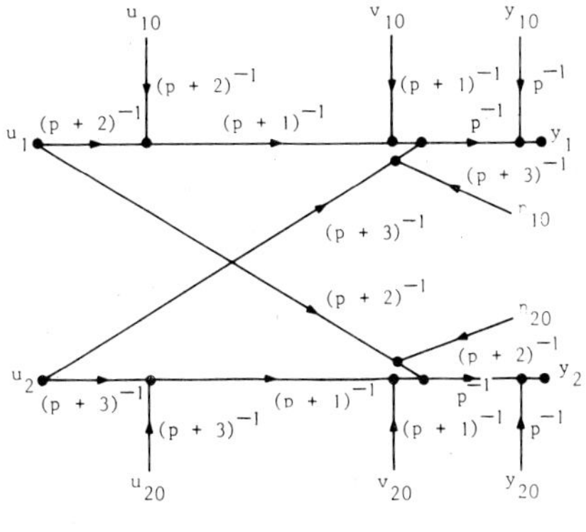	8

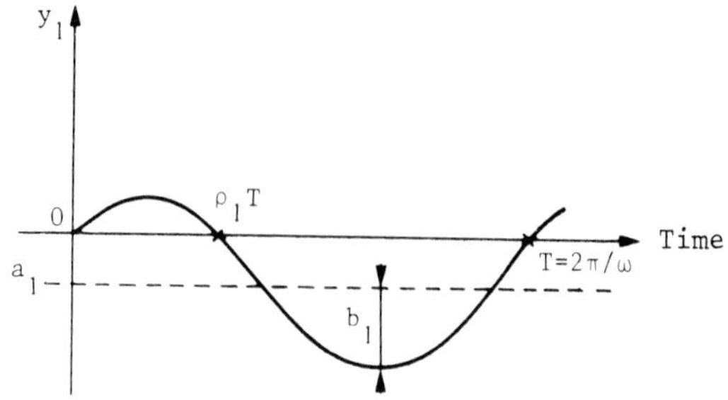

Fig. 3.10 Characteristics of the off-centre oscillation.

Let us suppose for example that the linear process defined in the previous paragraph is subjected to feedback through two nonlinear elements of relay type, as shown in figure 3.9. For the various structures indicated in table 3.1, it is possible to determine the conditions of oscillation of the nonlinear system thus constructed, and, in so far as they pertain to a given configuration, to deduce, from the observations made, the real structure of the process being analysed. In the present case, the existence of free stable oscillations of nonzero average value depends on the transmittances upstream of the integrators, which must be such that the continuous components of their input signals are null. Hence, if we actually observe stable off-centre oscillations, configurations 4 and 5 can be excluded from table 3.1. In the case of configurations 1 to 3, we may notice in addition that off-centre oscillations can arise when one relay only is in process of switching. Suppose for example that it is relay 1 and let (cf. figure 3.10):

a_1, a_2	be the average continuous values of y_1, y_2
b_1, b_2	be their amplitudes of oscillation,
ω	be the common frequency,
ϕ	be the phase-shift between y_1 and y_2.

Simple analytic calculations, for example a generalisation of Tsypkin's method to the multivariable case (cf. ref. 3.13), allow us easily to determine the parameters ω, ϕ, a_1, b_1, b_2. On the other hand, we note that this method, which utilises a matrix representation (the same thing would happen if we employed a generalised first-harmonic method) does not permit us to find the average value a_2 of the output y_2. This gap is explained by the facts that, in order to evaluate a_2, we need a knowledge of the system structure which Z(s) does not give, and that the value of this quantity depends on the controllability of the system. If the system is represented according to the minimal configuration 1 of table 3.1 and if we designate by 6α the average value of the output of the integrator, we find that

$$a_1 = \alpha - M(1 - 2\rho_1) + \frac{M}{4}(1 - 2\rho_1) - \frac{\varepsilon M}{9}$$

$$a_2 = \alpha - \frac{M}{4}(1 - 2\rho_1) + \frac{\varepsilon M}{18} - \frac{\varepsilon M}{2}$$

εM being the relay output level and ρ the oscillation parameter defined in figure 3.10. We deduce immediately that (1):

$$a_2 = a_1 - 2M/3$$

and that, a_1 being fixed, so is a_2.

If, on the contrary, the structure assumed for the linear system corresponds to configuration 2 or 3, the same calculations show that we have

$$a_2 - a_1 = -\frac{2M\varepsilon}{3} + (y_2(0) - y_1(0)) + (v_2(0) - v_1(0))$$

$y_i(0)$ and $v_i(0)$ being the initial conditions on the integrators corresponding to s^{-1} and $(s+1)^{-1}$, and that

$$a_2 - a_1 = -\frac{2M\varepsilon}{3} + (y_2(0) - y_1(0)) + (v_2(0) - v_1(0)) + \left(\frac{u_2(0)}{3} - \frac{u_1(0)}{2}\right) + \frac{\eta_2(0)}{2} - \frac{\eta_1(0)}{3}$$

$u_i(0)$ and $\eta_i(0)$ being the initial conditions on the integrators corresponding to $(s+2)^{-1}$ and $(s+3)^{-1}$.

The average value a_2 thus depends in these cases on the initial values, and, if we make several tests, we shall find different values for this average component. Hence if, in the course of a series of tests, we find identical values for a_2, we shall be able to conclude that the physical structure of the system corresponds to that of configuration 1

(1) We are taking account of the fact that the vanishing of the continuous component of the input to the integrator implies that
$$3M(1 - 2\rho_1) + 2\varepsilon M = 0$$

(a single integrator). If this is not the case, we should, on the contrary, exclude this configuration and adopt configuration 2. (Having in fact no means of deciding between configurations 2 and 3, since the initial conditions are inaccessible, we choose the one of smaller degree).

It is also interesting to note that in case 2, the controllability conditions may be written:

$$y_2(0) - y_1(0) + v_2(0) - v_1(0) = 0 \tag{3.10}$$

and that in case 3, it is convenient to adjoin to this equation the conditions

$$u_2(0) = \eta_1(0) \qquad u_1(0) = \eta_2(0) \tag{3.11}$$

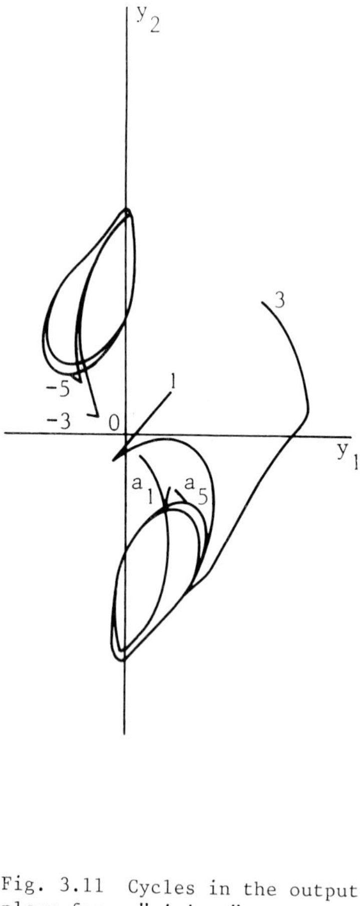

Fig. 3.11 Cycles in the output plane for a "minimum" system.

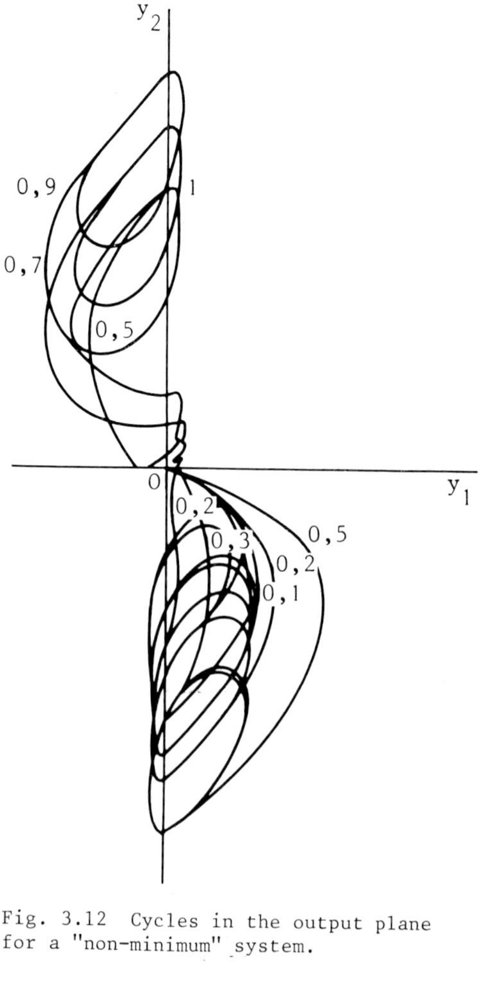

Fig. 3.12 Cycles in the output plane for a "non-minimum" system.

If the initial conditions are thus fixed in the controllable subspaces defined by equations (3.10) and (3.11), we find a definite constant value for a_2, regardless of the initial conditions satisfying these constraints. If an accidental, even fugitive, nonlinearity appears, equations (3.10) and (3.11) cease to be satisfied, the right-hand sides taking nonzero values which remain constant from the moment when the nonlinearity plays no further part; we then have a displacement of the limit cycle. Figures 3.11 and 3.12 display this phenomenon.

3.1.3 Structures and representations at the system level

A - Introductory example

In section 3.1, we defined a system as an agglomeration of subsystems, each having its own physical identity, connected to one another according to an organised arrangement. The fundamental objective will be to connect mathematically the characteristics of this agglomeration with those of its components, by taking account of their manner of combination. While the subsystem is a black box whose structure is unknown and for which any representation is equally valid, the model of the system should take account of the knowledge of internal structure given by its known organisation. In this sense, one of the first problems arising will be to establish the intrinsic properties of the simplest systems which one can consider, those constituted by the parallel and cascade connections of two elementary subsystems. It is in fact in this area that the practical implications of the ideas of controllability and observability generally emerge.

By way of introduction, let us consider for example a system which is really made up by the cascade of two subsystems S_1 and S_2, which we will suppose to be defined by the differential equations

$$S_1 : \begin{cases} \dot{v}_1 - 2 v_1 + \dot{v}_2 - 2 v_2 = 3 u_1 \\ -\dot{v}_1 - v_1 + 2 \dot{v}_2 + 2 v_2 = 6 u_2 \end{cases} \tag{3.12}$$

$$S_2 : \begin{cases} 3 \ddot{y}_1 + 15 \dot{y}_1 + 18 y_1 + 2 \ddot{y}_2 + 11 \dot{y}_2 + 12 y_2 = -2 \dot{w}_1 + 3 w_1 + \dot{w}_2 \\ 3 \ddot{y}_1 + 15 \dot{y}_1 + 18 y_1 + 2 \ddot{y}_2 + 14 \dot{y}_2 + 24 y_2 = -2 \dot{w}_1 + \dot{w}_2 + 6 w_2 \end{cases} \tag{3.13}$$

where u and w are respectively the inputs of S_1 and S_2, and v and y their outputs. It is important to note that, although the representations of S_1 and S_2 are given here in the form of differential operators, any other representation, in particular by the transfer function matrices $Z_1(s)$, $Z_2(s)$, would be equally valid, since we have supposed that only an external knowledge of the subsystems is available and consequently have assumed each to be controllable and observable.

The problem we pose is to give a valid representation for the system S resulting from the cascade (v = w), which reveals simultaneously the properties of each subsystem, defined by equations (3.12) and (3.13), and the existence of the cascade connection.

a) a first possibility would be to eliminate directly the intermediate variable v = w from equations (3.12) and (3.13). From these we deduce

$$\dot{y}_2 + 4\, y_2 = 2\, w_2 - w_1 \tag{3.14}$$

$$u_1 + 2\, u_2 = -\, w_1 + \dot{w}_2 \tag{3.15}$$

and from (3.14), taking account of (3.15) and (3.12),

$$\ddot{y}_2 + 4\, \dot{y}_2 = 2\, \dot{w}_2 - \dot{w}_1 = 6\, u_2 + w_1 - 2\, w_2 = 6\, u_2 - \dot{y}_2 - 4\, y_2$$

whence

$$\ddot{y}_2 + 5\, \dot{y}_2 + 4\, y_2 = 6\, u_2 \tag{3.16}$$

Further, we have

$$3\, \ddot{y}_1 + 15\, \dot{y}_1 + 18\, y_1 + 2\, \ddot{y}_2 + 8\, \dot{y}_2 = 6\, w_1 - 6\, \dot{w}_1 + \dot{w}_2 - 2\, \dot{w}_1$$
$$= 3(w_1 - 2\, w_2) - 3(u_1 + 2\, u_2) + 2\, \ddot{y}_2 + 8\, \dot{y}_2$$

so that

$$\ddot{y}_1 + 5\, \dot{y}_1 + 6\, y_1 + y_2 + 4\, y_2 = -\, u_1 - 2\, u_2 \tag{3.17}$$

From equations (3.16) and (3.17), we deduce a representation of the cascade in the form of differential operators

$$\begin{bmatrix} (s+2)(s+3) & s+4 \\ 0 & (s+1)(s+4) \end{bmatrix} Y = \begin{bmatrix} -1 & -2 \\ 0 & 6 \end{bmatrix} U \tag{3.18}$$

We note that a minimal representation corresponding to equation (3.18) is of order 4.

b) from the transfer function matrix point of view, we should have

$$Z_1 = \begin{bmatrix} \frac{2}{s-2} & \frac{-2}{s+1} \\ \frac{1}{s-2} & \frac{2}{s+1} \end{bmatrix}, \quad Z_2 = \begin{bmatrix} \frac{2}{(s+2)(s+3)} & \frac{-1}{s+3} \\ \frac{-1}{s+4} & \frac{2}{s+4} \end{bmatrix}$$

and

$$Z_2 Z_1 = \frac{1}{(s+1)(s+2)(s+3)(s+4)} \begin{bmatrix} -(s+1)(s+4) & -2(s+4)^2 \\ 0 & 6(s+2)(s+3) \end{bmatrix} \tag{3.19}$$

for which a minimal representation is also of order 4, the matrices associated with each pole being of rank 1 (one null row or column). We note also, in case a) as well as b), that the resulting system is stable (taking account of the given representations).

c) if, on the other hand, we utilise a (minimal) state-space representation of each of the systems, we have

$$S_1 : \dot{x}_1 = \begin{bmatrix} 1 & 2 \\ 1 & 0 \end{bmatrix} x_1 + \begin{bmatrix} 2 & -2 \\ 1 & 2 \end{bmatrix} u, \quad v = \begin{bmatrix} 1 & 0 \\ 0 & 1 \end{bmatrix} x_1$$

$$S_2 : \dot{x}_2 = \begin{bmatrix} -2 & 0 & 0 \\ 0 & -3 & 0 \\ 0 & 0 & -4 \end{bmatrix} x_2 + \begin{bmatrix} 1 & 0 \\ 2 & 1 \\ -1 & 2 \end{bmatrix} w, \quad y = \begin{bmatrix} 2 & -1 & 0 \\ 0 & 0 & 1 \end{bmatrix} x_2$$

whence the representation of the cascade is

$$\dot{x} = \begin{bmatrix} 1 & 2 & 0 & 0 & 0 \\ 1 & 0 & 0 & 0 & 0 \\ 1 & 0 & -2 & 0 & 0 \\ 2 & 1 & 0 & -3 & 0 \\ -1 & 2 & 0 & 0 & -4 \end{bmatrix} x + \begin{bmatrix} 2 & -2 \\ 1 & 2 \\ 0 & 0 \\ 0 & 0 \\ 0 & 0 \end{bmatrix} u, \quad y = \begin{bmatrix} 0 & 0 & 2 & -1 & 0 \\ 0 & 0 & 0 & 0 & 1 \end{bmatrix} x \tag{3.20}$$

We note that this representation requires five state variables. It is easy to see how the discrepancy between the results found in paragraphs a), b) and c) comes about. If we put equations (3.20) into the diagonal canonical form

$$\dot{z} = \begin{bmatrix} -1 & & & & \\ & 2 & & & \\ & & -2 & & \\ & & & -3 & \\ & & & & -4 \end{bmatrix} z + \begin{bmatrix} 0 & -2 \\ 0,5 & 0 \\ -0,5 & 2 \\ -1 & 1 \\ 0 & -2 \end{bmatrix} u$$

$$y = \begin{bmatrix} 1,5 & 0 & 2 & -1 & 0 \\ -1 & 0 & 0 & 0 & 1 \end{bmatrix} z$$

it becomes apparent that the representation (3.20) is not minimal and that the cascade connection has "made unobservable" the unstable mode $s = 2$ of subsystem S_1. The unobservability of this mode explains why it does not appear in the overall representation by a transfer function matrix (there is a cancellation of factors $s - 2$ which appear as pole and zero in the product $Z_2\ Z_1$) or through the differential operators: the first representation in fact reveals only the controllable and observable part of the resultant system, while the second, being necessarily observable, does not reveal the parts which happen to be unobservable (cf. chapter 2).

This simple example illustrates all the care which must be brought into the representation of a system, when its constitution in terms of subsystems is known. No representation of the form (3.18) or (3.19) will reveal the effect of the inherent instability which is physically present in the system. The fact that the unstable mode is unobservable, so far from saving the situation, serves to make it more obscure. Indeed, if the system, seen from its outputs, is stable, one can see the disadvantage of the internal divergence of the $s = 2$ mode which can bring about, if not the destruction of the system, at least its operation in such a manner that the hypothesis of linearity will not necessarily be satisfied.

Furthermore, in relation to this example, we can make two other remarks whose practical character seems to us fundamental.

a) in the simulation context, the recognition of the unobservable nature of the system defined by equation (3.20) can avoid much waste of time. Experience shows, indeed, that, in analogous situations, experimenters, deceived by the stability of the outputs, have tended, in view of the inevitable saturations produced in certain integrators, to overcalibrate the corresponding variables at first. With one over-calibration following another (and still obviously with few results) they then come to doubt the functioning of these integrators ... then they try replacing them by others ... which of course give the same result. Even in the fortunate case where one refrains from doubting the whole machine and sending for the manufacturer, the time wasted soon becomes important.

b) still more serious can be the case where the object of the study is to envisage a control system which will allow the resulting process to have suitable dynamics. If one uses a representation of the form (3.18) or (3.19) (or a state-space representation deduced from one of them), there is a great risk of having, despite the excellent results on paper, a severe disillusionment when the control system is applied to the actual process.

Such remarks lead us to consider urgently the properties of controllability and observability for composite systems, and, to begin with, the simplest among them: those formed by the parallel or cascade connections of two elementary subsystems. This will be the aim of the following paragraphs.

B - System formed by the parallel connection of two subsystems represented by their state equations

Let two systems S_1 and S_2 be defined by the equations:

$$S_1 : \begin{cases} \dot{x}_1 = A_1 x_1 + B_1 u \\ y_1 = C_1 x_1 + D_1 u \end{cases} \quad (3.21) \qquad \begin{cases} \dot{x}_2 = A_2 x_2 + B_2 u \\ y_2 = C_2 x_2 + D_2 u \end{cases} \quad (3.22)$$

The states x_1 of S_1 generate the space Σ_1, those of S_2 the space Σ_2. The equations of the system S, resulting from the parallel connection of S_1 and S_2, may be written directly, setting $x = [x_1 \; x_2]^T$:

$$\begin{bmatrix} \dot{x}_1 \\ x_2 \end{bmatrix} = \begin{bmatrix} A_1 & \\ & A_2 \end{bmatrix} \begin{bmatrix} x_1 \\ x_2 \end{bmatrix} + \begin{bmatrix} B_1 \\ B_2 \end{bmatrix} u$$

$$y = \begin{bmatrix} C_1 & C_2 \end{bmatrix} \begin{bmatrix} x_1 \\ x_2 \end{bmatrix} + \begin{bmatrix} D_1 + D_2 \end{bmatrix} u \qquad (3.23)$$

It is clear that a necessary condition, for the resulting system to be controllable and observable, is that the constituent systems S_1 and S_2 should be so individually. This condition is sufficient if the sets of eigenvalues associated with S_1 and S_2 are disjoint. Otherwise, this condition may not be sufficient, as is easily seen.

Let us suppose, in fact, without loss of generality, that A_1 and A_2 are in Jordan form, and let λ be a common eigenvalue of S_1 and S_2, associated with ρ_1 elementary Jordan blocks in S_1 and with ρ_2 blocks in S_2.

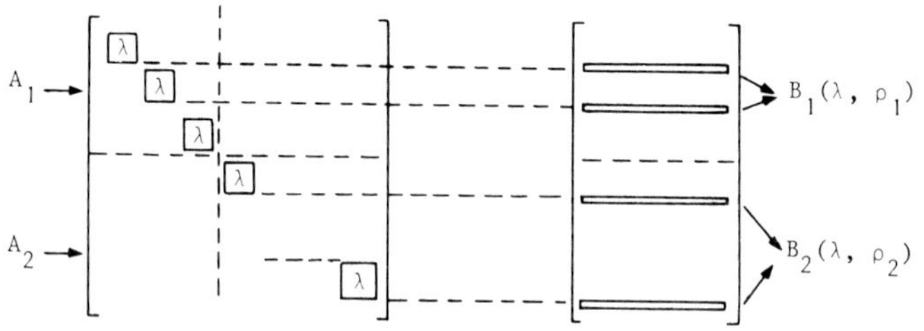

The assumption of controllability for S_1 requires (1) that the ρ_1 rows of B_1 associated with the last rows of the blocks of $A_1(\lambda)$ should be independent, and for S_2 entails the same property for the corresponding rows of B_2. The controllability of the combined system requires the independence of the $\rho_1 + \rho_2$ rows just defined, that is to say, if we designate by $B_1(\lambda, \rho_1)$ and $B_2(\lambda, \rho_2)$ the sets of ρ_1 and ρ_2 rows:

$$\text{rank} \begin{vmatrix} B_1(\lambda, \rho_2) \\ B_2(\lambda, \rho_1) \end{vmatrix} = \rho_1 + \rho_2 \qquad (3.24)$$

It follows, in particular, that, if the input of the system is an m-vector, the parallel connection is necessarily uncontrollable if the number of elementary Jordan blocks associated with an eigenvalue in the combination of the two systems is greater than m.

(1) cf. section 1.2.3

<u>Remark</u>: the observability properties can be deduced in the same way by applying the argument to the first columns of the blocks of C associated with elementary Jordan blocks related to a common eigenvalue.

<u>Example</u>: Let the two systems S_1 and S_2 be defined by:

$$\dot{x}_1 = \begin{bmatrix} -2 & 1 & & \\ 0 & -2 & & \\ & & -3 & 0 \\ & & 0 & -4 \end{bmatrix} x_1 + \begin{bmatrix} 2 & 1 \\ 2 & 4 \\ 4 & 1 \\ 1 & 3 \end{bmatrix} u$$

$$\dot{x}_2 = \begin{bmatrix} -1 & & \\ & -1 & \\ & & -2 \end{bmatrix} x_2 + \begin{bmatrix} 2 & 1 \\ 1 & 3 \\ 1 & 2 \end{bmatrix} u$$

with the above notation we have:

$$B_1(-2, \rho_1) = [2, 4] \qquad \text{rank} \begin{bmatrix} B_1(-2, \rho_1) \\ B_2(-2, \rho_1) \end{bmatrix} = 1$$

$$B_2(-2, \rho_1) = [1, 2]$$

and the controllable subspace is defined by $2x_{23} = x_{12}$.

The same results can obviously be obtained in other ways which do not assume that the systems have been put in Jordan form. We could for example apply:

- the criteria using truncated functions (cf. chapter 1): in this case we can restrict ourselves to the scheme corresponding simply to the common modes. In the case of the above example, the scheme of figure 3.13 gives immediately:

$$x_{12}(0) + 2\, u_1(-2) + 4\, u_2(-2) = 0$$

$$x_{23}(0) + u_1(-2) + 2\, u_2(-2) = 0$$

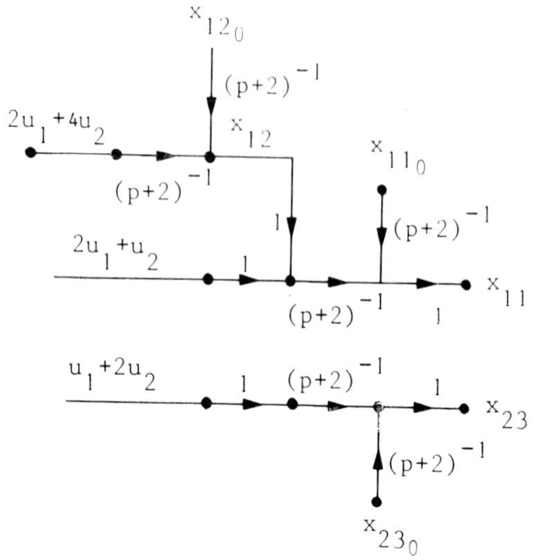

Fig. 3.13 Signal-flow graph for the determination of controllability (modes s + 2).

equations which are compatible only if:

$$2\, x_{23}(0) = x_{12}(0) \ ;$$

- the criteria based on the decomposition of the system into rank 1 matrices, discussed in chapter 1.

Remark: if the system resulting from the parallel connection of two individually controllable systems is not controllable, we can always find an input to control the most interesting states from the point of view of the output, whether they belong to S_1 or S_2. We shall see that the same thing does not happen in the case of a cascade connection.

C - Cascade connection of two subsystems

Let two systems S_1 and S_2 (defined as before) be connected in cascade (we suppose, of course, that their dimensions are compatible: $m_2 = p_1$). The resulting system is defined by the equations:

$$\begin{bmatrix} \dot{x}_2 \\ \dot{x}_1 \end{bmatrix} = \begin{bmatrix} A_2 & B_2\, C_1 \\ 0 & A_1 \end{bmatrix} \begin{bmatrix} x_2 \\ x_1 \end{bmatrix} + \begin{bmatrix} B_2\, D_1 \\ B_1 \end{bmatrix} u$$

$$[y] = \begin{bmatrix} C_2 & D_2 C_1 \end{bmatrix} \begin{bmatrix} x_2 \\ x_1 \end{bmatrix} + \begin{bmatrix} D_2 & D_1 \end{bmatrix} u \qquad (3.25)$$

taking the vector $[x_2 \ x_1]^T$ as state-vector. This choice has the advantage of giving the evolution matrix a triangular form, which is more convenient, whenever the matrices A_1 and A_2 are in Jordan form. We propose to examine the fundamental properties of the system thus constructed, and, in particular, the structure of this system in regard to the subsystems of which it is composed.

1) Problem of series-parallel equivalence:

Without detracting from the generality of the problem, we can suppose each of the subsystems S_1 and S_2 to be put in Jordan form, that is to say, themselves decomposed into the form of the maximum possible number of elementary systems (the simplest possible from the point of view of dimension) in parallel and not coupled among themselves. The cascade connection of S_1 and S_2 leads, under these conditions, to the scheme of figure 3.14 where each elementary subsystem S_{1i} or S_{2j} is associated with a single elementary divisor of S_1 or S_2. At the expense of a change of basis in the state-space of the cascaded system, it is possible to represent it by an equivalent system (in the input-output sense) composed simply of elementary subsystems in parallel, since it is sufficient to put the evolution matrix in Jordan form. We then obtain a scheme of just

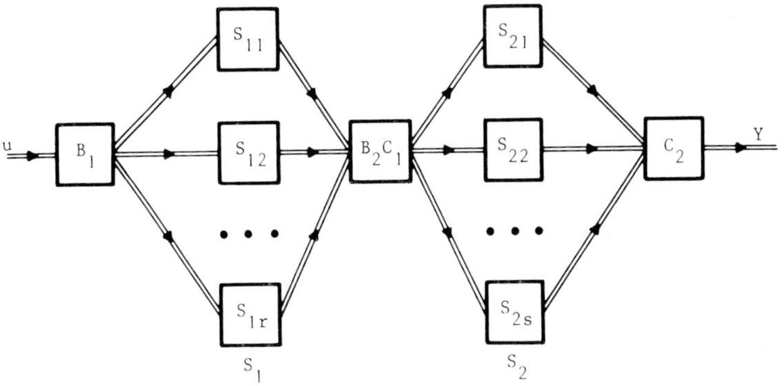

Fig. 3.14 Cascade connection of two subsystems

the form required, illustrated in figure 3.15. Thanks to this transformation, the study of the controllability and observability of the cascaded system is reduced to that of a parallel connection, a problem which has already been treated.

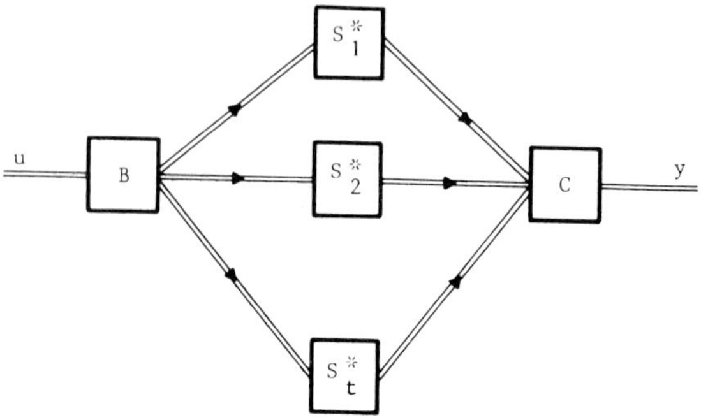

Fig.3.15 Series-parallel equivalence

It is appropriate to point out that, although the set of eigenvalues of the resultant system S includes all the eigenvalues of the systems S_1 and S_2 and only these (to convince oneself, it is sufficient to notice that the characteristic polynomial of the cascaded system is equal to the product of those of S_1 and S_2, which is obvious given the triangular form of the evolution matrix), on the other hand there is no reason for the elementary divisors to be the same, which is to say that, in general, the cascade does not preserve the elementary divisors of the systems S_1 and S_2.

This destruction and recombination of elementary divisors is brought about by the coupling matrix $B_2\ C_1$ (1).

Example: Consider the two subsystems S_1 and S_2 given by the equations:

$$S_1 : A_1 = \begin{bmatrix} \lambda & 1 & 0 & 0 \\ 0 & \lambda & 1 & 0 \\ 0 & 0 & \lambda & 1 \\ 0 & 0 & 0 & \lambda \end{bmatrix}, B_1 = \begin{bmatrix} 0 & 0 \\ 0 & 2 \\ -1 & 0 \\ 0 & 1 \end{bmatrix}, C_1 = \begin{bmatrix} 0 & 1 & 1 & 0 \\ 1 & 0 & 2 & -1 \end{bmatrix}$$

(1) Mathematical developments relevant to this important question will be found in ref. 3.13.

$$S_2 : A_2 = \begin{bmatrix} \lambda & 1 & 0 \\ 0 & \lambda & 1 \\ 0 & 0 & \lambda \end{bmatrix}, \quad B_2 = \begin{bmatrix} 0 & 1 \\ 2 & -1 \\ 1 & 0 \end{bmatrix}, \quad C_2 = \begin{bmatrix} -1 & 2 & -2 \\ 0 & 1 & 3 \end{bmatrix}$$

These two systems have elementary divisors $(s - \lambda)^4$ and $(s - \lambda)^3$ respectively, and the system constituted by the cascade $S_1\ S_2$ has state equations:

$$S_{12} : \begin{bmatrix} \dot{x}_2 \\ \dot{x}_1 \end{bmatrix} = \begin{bmatrix} A_2 & B_2\,C_1 \\ & A_1 \end{bmatrix} \begin{bmatrix} x_2 \\ x_1 \end{bmatrix} + \begin{bmatrix} 0 \\ B_1 \end{bmatrix} u, \qquad y = \begin{bmatrix} C_2 & 0 \end{bmatrix} \begin{bmatrix} x_2 \\ x_1 \end{bmatrix}$$

whose elementary divisors are $(s - \lambda)^5$ and $(s - \lambda)^2$.

2) Controllability and observability of a system composed of two subsystems (Jordan form):

As in the case of parallel connection, it is clear that a necessary condition for controllability (or observability) of the system resulting from the cascade connection of two subsystems S_1 and S_2 is the controllability (observability) of each taken separately. This condition, however, is not sufficient, as one can easily convince oneself (1).

We gave, in chapter 1, criteria for controllability based on the representation of the evolution matrix in Jordan form. These criteria evidently remain usable if we can easily put the system resulting from a cascade into this form. However, although one can easily deduce the elementary divisors of the cascade, starting from the examination of $B_2\ C_1$ (that is, if one knows the Jordan form of the evolution matrix), one cannot

(1) See for example the case proposed by C.T. Chen and C.A. Desoer: Controllability and Observability of Composite Systems, IEEE Trans. Auto. Control Vol. AC-12, no. 4, August 1967.

so easily study its controllability and observability, since the criteria depend on the new B and C matrices formed when the cascade is put in Jordan form. The knowledge of these matrices requires the identification of the transformation matrix T which implements the conversion of the evolution matrix A into its Jordan form J. The discovery of such a matrix T is often quite a delicate matter. However, if we restrict our attention to the case where the matrices A_1 and A_2 of the systems constituting the cascade are given in Jordan form, there are precise rules which allow us to identify simply and quickly a matrix T adequate for the task (cf. reference 3.13 , chapter III, appendix 3).

3.2 TELEOLOGY OF STATE-SPACE REPRESENTATIONS; CANONICAL STRUCTURAL FORMS

We have seen that, at the level of the subsystem (in the sense in which this term has been defined), that is to say, a part of a system which is accessible only via its inputs and outputs, and for which no physical consideration allows us to discover the existence or nature of its internal interactions, the principle of structural indeterminacy justifies a teleological approach to the problem. Although, indeed, at the subsystem level, all structures are equivalent in the input-output sense, we can always choose the structure best adapted to a particular application, without the necessity, at the calculational level, of giving it a physical interpretation. This amounts to saying that, since the control, evolution and observation matrices B, A, C, are only defined to within a linear transformation, any change of basis which allows us to put these matrices in a form which facilitites the calculations associated with the solution of such and such a problem should be utilised. (This is, moreover, the same approach as was followed in chapter 2, where the passage from a representation by Z(s) to a state-space representation was facilitated by the choice of a particular structure for A).

Three fundamental decompositions will be studied here:

- Decomposition into r single-input controllable subsystems.
- Decomposition into m single-input controllable subsystems.
- Decomposition into p single-output observable subsystems.

The first two decompositions have the purpose of putting the A and B matrices in particularly simple forms, and we shall see their importance in the study of compensation problems (chapter 4). The third decomposition, whose purpose is to simplify the forms of A and C, will be used in problems of observation (cf. chapter 4 also). All of them derive from the celebrated works of D.G. Luenberger (refs. 3.22-3.23).

We shall see also, in chapter 5, that the solution of non-interactive control problems is facilitated by putting the A, B and C matrices into appropriate forms. This latter decomposition has, however, been separated from the preceding ones, since it involves, in fact, two transformations, of which one, being nonlinear, falls outside the scope of this section.

3.2.1 Decomposition of a system into r subsystems, each controllable from a single input (1)

A - Structure of the decomposition

We are attempting to decompose the total system, of dimension n and possessing m inputs, into a number r of subsystems ($r \leq m$) such that:

- each of them is controllable from a single input
- they are hierarchically arranged in the sense that, if $S_1, S_2, \ldots S_r$ are the r subsystems, S_r acts on $S_{r-1}, \ldots S_1$, and so on, S_3 on S_2 and S_1, while S_2 acts only on S_1. Schematically, the decomposition of the system is as indicated in figure 3.16 which shows the one-sided nature of the interactions. S_i is controllable from the single input u_i. If r is less than m, then m - r components of u (denoted by $u^* = [u_{r+1} \ldots u_m]^T$) may possibly act on all the subsystems. Finally, we denote by n_i the respective dimensions of the subsystems S_i and by $\tilde{x}_i$ their state-vectors.

(1) cf. ref. 3.22 of Luenberger

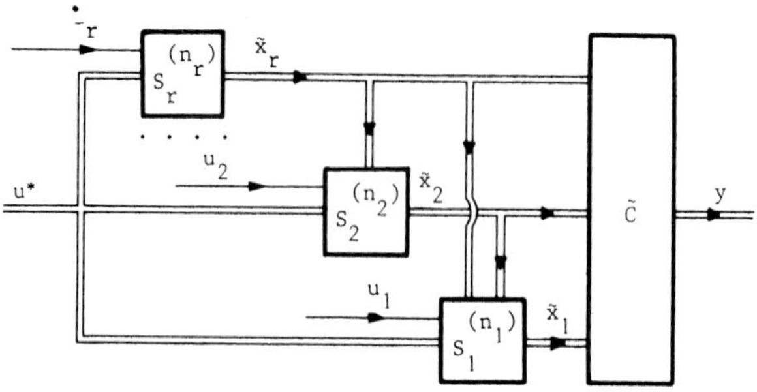

Fig. 3.16 The principle of the decomposition of a multivariable system into r subsystems controllable from single inputs.

The state equation of subsystem S_i is thus of the form:

$$\dot{\tilde{x}}_i = A_{ii}\,\tilde{x}_i + \left[A_{i,i+1}\,\tilde{x}_{i+1} \dots + A_{ir}\,\tilde{x}_r\right] + {}^{i}Bu_i + \beta(i)\,u^*$$

the quantity in brackets representing the action on the subsystem S_i of the systems S_{i+1} to S_r. From the assumption of controllability of S_i by u_i, we have also

$$\text{rank}\left[{}^{i}B \quad A_{ii}\,{}^{i}B \dots (A_{ii})^{n_i-1}\,{}^{i}B\right] = n_i$$

The evolution matrix $\tilde{A}$ and the input matrix $\tilde{B}$ will then appear, after the linear transformation M which we are seeking and with $\tilde{x}$ defined by $\tilde{x}^T = [\tilde{x}_r \dots \tilde{x}_1]$, in the block-triangular forms:

$$\tilde{A} = M^{-1}AM = \left[\begin{array}{c|c|c|c} A_{r,r} & 0 & & \\ \hline A_{r-1,r} & A_{r-1,r-1} & & \\ & \ddots & & \\ \hline & & A_{2,2} & 0 \\ \hline A_{1,r} & & A_{1,2} & A_{11} \end{array}\right] \qquad (3.26)$$

$$\tilde{B} = M^{-1} B = \left[\begin{array}{c|c|c|c|c} & & & {}^{r}B & \beta(r) \\ \hline & & \vdots & & \beta(r-1) \\ & & & & \\ \hline & {}^{2}B & & & \beta(2) \\ \hline {}^{1}B & & & & \beta(1) \end{array}\right] \qquad (3.27)$$

We are also trying to put the matrices A_{ii} (of dimensions n_i, n_i) and the column vectors ${}^{i}B$ (n_i, 1) in the forms:

$$A_{ii} = \begin{bmatrix} 0 & 1 & 0 & & & \\ 0 & 0 & 1 & 0 & \cdots & \\ & & & & & 0 \\ & & & & 0 & 1 \\ -a_0^i & -a_1^i & -a_2^i & & \cdots & -a_{n_i-1}^i \end{bmatrix} \qquad (3.28)$$

$${}^{i}B = \begin{bmatrix} 0 \\ \vdots \\ 0 \\ 0 \\ 1 \end{bmatrix} \qquad (3.29)$$

with $a_{n_i}^i = 1$, the upper index i in a_x^i referring to the number of the subsystem under consideration. We note that the block-triangular form of A, combined with the corresponding form of A_{ii}, allows us to write the overall characteristic polynomial directly in the form:

$$\phi(s) = \prod_{i=1}^{r} \phi_i(s), \quad \phi_i(s) = s^n + a_{n-1}^i s^{n-1} + \ldots + a_0^i .$$

<u>B - Determination of the transformation matrix</u>

If, in the defining equation $AM = M\tilde{A}$, we partition the matrix M in the form

$$M = [M(r) \ \ldots \ M(2) \ \ M(1)]$$

with r column blocks, of dimensions $n_r, \ldots n_2, n_1$, corresponding to the dimensions of the subsystems $S_r, \ldots S_2, S_1$, the nth order matrix equation

$$AM = M\tilde{A} \ ,$$

which can also be written

$$A\begin{bmatrix} M(r) & \ldots & M(2) & M(1)\end{bmatrix} = \begin{bmatrix} M(r) & \ldots & M(2) & M(1)\end{bmatrix}\begin{bmatrix} A_{rr} & 0 & 0 \\ A_{jr} & \ddots & 0 \\ A_{1r} & \ldots & A_{11}\end{bmatrix}$$

decomposes into r matrix equations of n_i^{th} order

$$AM(i) = \sum_{k=1}^{i} M(k)\, A_{ki} \qquad (3.30)$$

Let us consider the first of these equations:

$$AM(1) = M(1)\, A_{11} \qquad (3.31)$$

in which we display the n_1 columns of the block M(1)

$$A\begin{bmatrix} {}^1M(1) & \ldots & {}^{n_1}M(1)\end{bmatrix} = \begin{bmatrix} {}^1M(1) & \ldots & {}^{n_1}M(1)\end{bmatrix}\begin{bmatrix} 0 & 1 & 0 & & 0 \\ 0 & 0 & 1 & & 0 \\ & & & \ddots & \\ & & & & 1 \\ -a_0^1 & & & & -a_{n_1-1}^1\end{bmatrix}$$

The above n_1 vector equations can be written, by identifying the columns:

$$\begin{aligned}
{}^{n_1-1}M(1) &= \left[A + Ia_{n_1-1}^1\right]\, {}^{n_1}M(1) \\
{}^{n_1-2}M(1) &= \left[A^2 + Aa_{n_1-1}^1 + Ia_{n_1-2}^1\right]\, {}^{n_1}M(1) \\
&\ \vdots \\
{}^1M(1) &= \left[A^{n_2-1} + a_{n_1-1}^1\, A^{n_1-2} + \ldots + Ia_1^1\right]\, {}^{n_1}M(1) \qquad (3.32) \\
A^{n_1}\, {}^{n_1}M(1) &= -\left[A^{n_1-1}\, a_{n_1-1}^1 + \ldots + Aa_1^1 + Ia_0^1\right]\, {}^{n_1}M(1) \ .
\end{aligned}$$

As to the equation relating to the control matrix $B = M\tilde{B}$, it can be written

$$\begin{bmatrix} M(r) & \ldots & M(2) & M(1)\end{bmatrix}\begin{bmatrix} & & {}^rB & \beta(r) \\ & & \cdot^{\cdot^{\cdot}} & \vdots \\ & {}^2B & & \vdots \\ {}^1B & & & \beta(1)\end{bmatrix} = \begin{bmatrix} {}^1b & {}^2b & \ldots & {}^mb\end{bmatrix}$$

with ${}^{i}b = i^{th}$ column of B, so that

$$M(i)\ {}^{i}B = \left[{}^{1}M(i)\ \dots\ {}^{n_i}M(i)\right]\left[0\ \dots\ 0\quad 1\right]^T = {}^{n_i}M(i) = {}^{i}b\ . \tag{3.33}$$

Equation (3.33) and the last of equations (3.32) then show that $A^{n_1}\ {}^{1}b$ is a linear combination of the independent vectors $A^k\ {}^{1}b$ with k varying from 1 to $n_1 - 1$, which fixes the dimension n_1 of the block A_{11} and the last row of the companion matrix. Equations 1) to n - 1) of (3.32), with ${}^{n_1}M(1) = {}^{1}b$, allow us also to determine the n_1 columns of the block M(1) of M.

In this development, we have assumed the dimension n_1 of the first subsystem to be less than n. (If $n_1 = n$, the system (A, ${}^{1}b$) is completely controllable and we have the case dealt with in chapter 2).

To continue generating the complete space, let us consider the second of equations (3.30), which corresponds to i = 2 and can be written:

$$M(2)\ A_{22} + M(1)\ A_{12} = AM(2) \tag{3.34}$$

and display as before the columns of M(1), M(2), A_{12} in the form

$$M(1) = \left[{}^{1}M(1)\ \dots\ {}^{n_1}M(1)\right],\quad M(2) = \left[{}^{1}M(2)\ \dots\ {}^{n_2}M(2)\right],\quad A_{12} = \left[{}^{1}A_{12}\ \dots\ {}^{n_2}A_{12}\right]$$

Equation (3.34) can be developed into n_2 vector equations

$$\begin{aligned}
{}^{n_2-1}M(2) &= (A + a^2_{n_2-1})\ {}^{n_2}M(2) - M(1)\ {}^{n_2}A_{12} \\
{}^{n_2-2}M(2) &= \left[A^2 + Aa^2_{n_2-1} + a^2_{n_2-2}\ I\right]\ {}^{n_2}M(2) - M(1)\ {}^{n_2-1}A_{12} - AM(1)\ {}^{n_2}A_{12} \\
&\ \vdots \\
{}^{1}M(2) &= \left[A^{n_2-1} + A^{n_2-2}\ a^2_{n_2-1} + \dots + Ia^2_1\right]\ {}^{n_2}M(2) - \\
&\quad - M(1)\ {}^{2}A_{12} - \dots - A^{n_2-2}\ M(1)\ {}^{n_2}A_{12}
\end{aligned} \tag{3.35}$$

and

$$\begin{aligned}
A^{n_2}\ {}^{n_2}M(2) &= -\left[A^{n_2-1}\ a^2_{n_2-1} + \dots + Ia^2_0\right]\ {}^{n_2}M(2) + \\
&\quad + M(1)\ {}^{1}A_{12} + \dots + A^{n_2-1}\ M(1)\ {}^{n_2}A_{12}
\end{aligned}$$

The quantity outside the bracket in the last equation is a linear combination of vectors of the form

$$A^{i-1}\,{}^{k}M(1) \quad k = 1 \text{ à } n_1 \quad i = 1 \text{ à } n_2$$

hence a linear combination of the ${}^{k}M(1)$, and we can always satisfy the preceding relations by setting

$${}^{2}A_{12} = {}^{3}A_{12} = \ldots = {}^{n_1}A_{12} = 0 \qquad {}^{1}A_{12} \neq 0$$

Taking account of the relation ${}^{n_2}{}^{2}M(2) = {}^{2}b$, the last of equations (3.35) can be written

$$A^{n_2}\,{}^{2}b = -\left[a^{2}_{n_2-1}\,A^{n_2-1} + \ldots + a^{2}_{0}\,I\right]{}^{2}b + \left[{}^{1}M(1) \;\ldots\; {}^{n_1}M(1)\right]{}^{1}A_{12}$$

which gives the rules of formation of the second subsystem.

We construct the sequence of vectors:

$${}^{2}b,\; A\,{}^{2}b,\; \ldots,\; A^{n_2-1}\,{}^{2}b,\; A^{n_2}\,{}^{2}b \;\ldots$$

until the last vector $A^{n_2}\,{}^{2}b$ is found to be linearly dependent on the preceding vectors ${}^{2}b$, $A\,{}^{2}b,\ldots$, and the vectors ${}^{1}M(1),\ldots{}^{n_1}M(1)$ determined in the first stage. (These latter n_1 vectors can evidently be replaced, as regards independence, by the vectors ${}^{1}b,\ldots A^{n_1-1}\,{}^{1}b$). Then

- n_2 fixes the order of the subsystem;
- the coefficients of the linear relation between the vectors ${}^{2}b,\ldots A^{n_2-1}\,{}^{2}b$, give the entries in the last row of the companion matrix A_{22}, while those associated with the vectors ${}^{1}M(1),\ldots{}^{n_1}M(1)$ are equal to the elements of ${}^{1}A_{12}$;
- equations (3.35) in turn allow us to determine the n_2 columns of the block M(2) of M.

In the above, we have assumed the existence of two independent subspaces, of dimensions n_1 and n_2, generated by ${}^{1}b$ and ${}^{2}b$. If $n_1 + n_2 < n$, the same procedure can be followed, by considering equation (3.30) for $i = 3$, so as to generate the subspace constructed on ${}^{3}b$, independent of the other two. The assumption of controllability ensures that the space so generated will be complete (of dimension n_i) for $i \leqslant r \leqslant m$. (1)

C - Remarks

1) The decomposition into subsystems is clearly not unique. It can depend, in particular, on the order in which the columns ${}^{i}b$ of B are taken when constructing the basis. We have used here the sequential order, but any other order is obviously possible.

2) In particular, it may happen that the structural decomposition leads to a number of subsystems equal to the number of inputs (r = m). In this case, u^* no longer exists, and each column of $\tilde{B}$ contains only one nonzero element, which is equal to 1. However, this possibility arises only occasionally, and we shall see in the following subsection a method of systematically decomposing the total system into m subsystems. This will impose a different structure.

3) It is also clear that, instead of testing the linear dependence by using the vectors ${}^{k}M(i)$, we can do it with the vectors $A^k\,{}^{i}b$. This makes the procedure easier, but we cannot then directly determine the elements of A_{ij} for $j \neq i$.

4) When the system is of high dimension, the calculations cannot be done by hand. A Fortran program for this purpose is given in the appendix.

D - Example

Let the system be

$$\dot{x} = \begin{bmatrix} -1 & & \\ & -2 & \\ & & 3 \end{bmatrix} x + \begin{bmatrix} 1 & 1 \\ -1 & 0 \\ -1 & 1 \end{bmatrix} u, \qquad y = \begin{bmatrix} 1 & 1 & 0 \\ 1 & 0 & 1 \end{bmatrix} x$$

(1) In the case that the system is not completely controllable, the above method can easily be generalised, cf. refs. 3.14 and 3.20.

If we take the columns of B in sequential order, ${}^1b$, ${}^2b$, then, since the system is completely controllable by u_1, only one space will appear. We have then:

$$\tilde{A} = \begin{bmatrix} 0 & 1 & 0 \\ 0 & 0 & 1 \\ 6 & 7 & 0 \end{bmatrix} \qquad \tilde{B} = \begin{bmatrix} 0 & -0{,}3 \\ 0 & +0{,}1 \\ 1 & -0{,}7 \end{bmatrix} \qquad M = \begin{bmatrix} -6 & -1 & 1 \\ 3 & 2 & -1 \\ -2 & -3 & -1 \end{bmatrix}$$

Suppose we now take the columns of B in the order, ${}^2b$, ${}^1b$. We must then form the sequence:

$${}^2b = \begin{bmatrix} 1 \\ 0 \\ 1 \end{bmatrix} \qquad A\,{}^2b = \begin{bmatrix} -1 \\ 0 \\ 3 \end{bmatrix} \qquad A^2\,{}^2b = \begin{bmatrix} 1 \\ 0 \\ 9 \end{bmatrix}$$

and we have

$$A^2\,{}^2b = 3\,{}^2b + 2\,A\,{}^2b \ .$$

Subsystem 1, constructed from ${}^2b$, is of dimension 2, and the characteristic polynomial of A_{11} is

$$s^2 - 2s - 3, \quad a_0 = -3, \quad a_1 = -2,$$

whence

The 2 columns of M corresponding to M(1) are:

$$\left[(A + a_1^1\, I)\,{}^2b \quad {}^2b\right] = \begin{bmatrix} -3 & 1 \\ 0 & 0 \\ 1 & 1 \end{bmatrix}$$

To generate the rest of the space, we consider:

$${}^1b = \begin{bmatrix} 1 \\ -1 \\ -1 \end{bmatrix} \qquad A\,{}^1b = \begin{bmatrix} -1 \\ 2 \\ -3 \end{bmatrix} .$$

We have

$$A\,{}^1b = -2\,{}^1b - 1{,}5\,A\,{}^2b - 0{,}5\,{}^2b$$

a relation which reveals the second subspace, of dimension 1, corresponding to the matrix $A_{22} = -2$.

Hence

$$\tilde{A} = \begin{bmatrix} -2 & 0 & 0 \\ -1,5 & 0 & 1 \\ -3,5 & 3 & 2 \end{bmatrix}, \quad \tilde{B} = \begin{bmatrix} 1 & 0 \\ 0 & 0 \\ 0 & 1 \end{bmatrix}, \quad M = \begin{bmatrix} 1 & -3 & 1 \\ -1 & 0 & 0 \\ -1 & 1 & 1 \end{bmatrix}$$

Remark: we note in $\tilde{B}$ the permutation of the columns ${}^1b$, ${}^2b$, in comparison with the form indicated in paragraph A, which is connected with the permutation of the components u_1, u_2 in the generation of the subspaces.

3.2.2 Decomposition of a system into a number of controllable subsystems equal to the number of inputs (1)

A - Structure of the decomposition

Given a system (S), assumed to be controllable, it is possible to decompose it into m single-input controllable subsystems S_i (m being the number of inputs), ordered so that their dimensions n_i satisfy the conditions $n_i \leq n_{i+1}$. More precisely, denoting by u_j ($j = 1, 2, \ldots m$) the components of the input:

S_1 is controlled by the single component u_1,
S_2 is controlled by $u_2 + \alpha_2 u_1$,
S_k is controlled by $u_k + \alpha_k u_{k-1} + \ldots + \rho_k u_1$.

In other words, at the expense of multiplying the control vector u by an appropriate scalar matrix, we can define a control u^* such that u_j^* is the only input to S_j, this subsystem being completely controllable by means of this component.

This decomposition will be possible, if, given the system

$$\dot{x} = Ax + Bu \qquad y = Cx$$

(1) cf. D.G. Luenberger, refs. 3.22-3.23,
and C.D. Johnson, ref. 3.20.

it is possible to find a transformation matrix M such that if

$$\tilde{A} = M^{-1} AM \qquad \tilde{B} = M^{-1} B$$

$\tilde{A}$ and $\tilde{B}$ have respectively the forms

$$\tilde{A} = \begin{bmatrix} A(1,1) & \dots & A(1,m) \\ \\ A(m,1) & \dots & A(m,m) \end{bmatrix} \qquad \tilde{B} = \begin{bmatrix} B(1) \\ \vdots \\ B(m) \end{bmatrix}$$

where A(i,i) is a block of dimensions (n_i, n_i) which we shall suppose to be in companion form, A(i,j), with i ≠ j, is a block of dimensions (n_i, n_j) having the structure

$$A(i,j) = \begin{bmatrix} 0 & 0 & \dots & 0 \\ 0 & & \dots & 0 \\ a_{ij}^{1} & & \dots & a_{ij}^{n_j} \end{bmatrix}$$

and

$$B(i) = \begin{bmatrix} 0 \\ 0 \\ \vdots \\ ({}^{i}b)^{T} \end{bmatrix}$$

We note that, in the expressions for A(i,j) and B(i), only the last rows are non-null.

B - Determination of the transformation matrix

The equation defining $\tilde{A}$ can be written

$$M^{-1} A = \tilde{A}M^{-1}$$

so that, after transposition,

$$A^T (M^{-1})^T = (M^{-1})^T \tilde{A}^T$$

which can be put in the form

$$A^T P = P\tilde{A}^T$$

by defining

$$P = (M^{-1})^T.$$

If P is decomposed into column blocks, compatible with the dimensions of the various blocks A(i,j), the matrix equation

$$A^T \left[P(1) \ \dots \ P(m)\right] = \left[P(1) \ \dots \ P(m)\right] \begin{bmatrix} A^T(1,1) & \dots & A^T(m,1) \\ \vdots & & \\ A^T(1,m) & \dots & A^T(m,m) \end{bmatrix}$$

decomposes into m equations of lower order:

$$A^T P(i) = \sum_{k=1}^{m} P(k)\, A^T(i,k) \quad \text{for } i = 1 \text{ to } m \tag{3.36}$$

If ${}^jP(i)$ denotes the jth column of the block P(i) (which contains n_i columns), each of the above equations can be written out as

$$A^T \left[{}^1P(i)\ {}^2P(i) \ \dots \ {}^{n_i}P(i)\right] = \sum_{\substack{k=1 \\ k \neq i}}^{m} \left[{}^1P(k) \ \dots \ {}^{n_k}P(k)\right] \begin{bmatrix} 0 \dots 0 & a_{ik}^1 \\ \vdots & \vdots \\ 0 \quad 0 & a_{ik}^{n_k} \end{bmatrix} +$$

$$+ \left[{}^1P(i) \ \dots \ {}^{n_i}P(i)\right] \begin{bmatrix} 0 & 0 & & a_{ii}^1 \\ 1 & 0 & & \cdot \\ 0 & 1 & & \cdot \\ \cdot & \cdot & & \cdot \\ \cdot & \cdot & & \cdot \\ 0 & \cdot & 1 & a_{ii}^{n_i} \end{bmatrix}$$

and identification of the n_i columns leads to

$$A^T\, {}^1P(i) = {}^2P(i) \qquad {}^2P(i) = A^T\, {}^1P(i)$$

$$A^T\, {}^2P(i) = {}^3P(i) \qquad {}^3P(i) = (A^T)^2\, {}^1P(i)$$

$$\vdots \qquad \vdots$$

$$A^T\, {}^{n_i-1}P(i) = {}^{n_i}P(i) \qquad {}^{n_i}P(i) = (A^T)^{n_i-1}\, {}^1P(i) \tag{3.37}$$

$$A^T\, P^{n_i}(i) = \sum_{k=1}^{m} \sum_{j=1}^{n_i} {}^jP(k)\, a_{ik}^j$$

We can thus calculate the last n_i-1 columns of $P(i)$, given the first one. Since, besides, we are imposing no constraints on the elements $\tilde{a}^j_{ik}$, we see that the conditions specified for $\tilde{A}$ allow an arbitrary choice of the m vectors ${}^1P(i)$.

Taking account of the condition

$$M^{-1} B = \tilde{B} \qquad \text{or} \qquad B^T P = \tilde{B}^T$$

we have

$$B^T \left[P(1) \ \ldots \ P(m)\right] = \left[0 \ \ldots \ 0 \ \ {}^1\tilde{b}, \ 0 \ \ldots \ {}^2\tilde{b} \ \ldots \ 0 \ \ldots \ {}^m\tilde{b}\right] \qquad (3.38)$$

That is, showing explicitly the various columns of each block $P(i)$,

$$B^T \left[{}^1P(i) \ \ldots \ {}^{n_i}P(i)\right] = \left[0 \ \ldots \ {}^i\tilde{b}\right]$$

$$\begin{aligned} B^T \, {}^1P(i) &= 0 \\ B^T \, {}^2P(i) &= 0 \\ \ldots&\ldots\ldots \\ B^T \, {}^{n_i-1}P(i) &= 0 \\ B^T \, {}^{n_i}P(i) &= {}^i\tilde{b} \end{aligned} \qquad (3.39)$$

which, in view of (3.37), can be written:

$$\begin{aligned} B^T \, {}^1P(i) &= 0 \\ (AB)^T \, {}^1P(i) &= 0 \\ \ldots&\ldots\ldots \\ \left[A^{n_i-2} B\right]^T \, {}^1P(i) = 0, &\quad \left[A^{n_i-1} B\right]^T \, {}^1P(i) = {}^1\tilde{b} \end{aligned} \qquad (3.40)$$

In this way, we are expressing the fact that the vector ${}^1P(i)$, which is sought, belongs to the null-space of the matrices B^T, $(AB)^T$, $\ldots(A^{n_i-2}B)^T$, that is to say, it is orthogonal to each vector of the sequence:

$${}^1b, \ {}^2b, \ \ldots, \ {}^mb, \ A^T \, {}^1b, \ \ldots, \ (A^T)^{n_i-2} \, {}^mb.$$

Further, the jth component of ${}^i\tilde{b}$ is the scalar product of ${}^1P(i)$ with $A^{n_i-1} \, {}^jb$.

The above sequence of vectors is taken from the controllability matrix of the system. Let $A^{n_{i1}} \, {}^{i_1}b$ be the first vector of this sequence which is dependent on those which precede it (it follows that all the vectors $A^k \, {}^{i_1}b$ for $k > n_{i1}$ will be likewise dependent): we shall say that the column ${}^{i_1}b$ of B generates the first subsystem, of dimension n_{i1}. If $A^{n_{i2}} \, {}^{i_2}b$ is the second vector found to be dependent on the preceding ones, we shall say, similarly, that ${}^{i_2}b$ generates the second subsystem, of dimension n_{i2}, and so on.

The controllability assumption ensures, besides, that it is possible to find in this way m subsystems such that

$$\sum_{k=1}^{m} n_{ik} = m \quad .$$

In order to maintain the ordering of the subsystems, we shall be led to make a substitution among the components of the input vector u, that is, to renumber the columns of B in the order

$$\left[{}^{i_1}b \quad {}^{i_2}b \ldots {}^{i_m}b\right] \rightarrow \left[{}^{1}b \quad {}^{2}b \ldots {}^{m}b\right]$$

which we shall suppose done, so as to avoid complicating the notation. Let

$$Q = \left[{}^{1}b \ldots {}^{m}b \quad A\,{}^{1}b \ldots A\,{}^{m}b \ldots A^{n_1-1}\,{}^{m}b \quad A^{n_1}\,{}^{2}b \ldots A^{n_m-1}\,{}^{r}b\right].$$

be the nonsingular matrix composed of the n independent vectors isolated by the above algorithm. We have seen that ${}^{1}P(1)$ should be orthogonal to all the columns of Q from ${}^{1}b$ up to $A^{n_1-2}\,{}^{m}b$, hence, to the first $m(n_1-1)$ vectors of Q. We can thus take for ${}^{1}P(1)^T$ the $\left[m(n_1-1)+1\right]^{th}$ row of Q^{-1}. For this choice, we have moreover

$${}^{1}\tilde{b} = \left[A^{n_1-1}\,B\right]^T\;{}^{1}P(1) = \left[1 \quad 0 \ldots 0\right]^T$$

and the first subsystem will be controlled by the single component u_1 of u.

Similarly, ${}^{1}P(2)$ should be orthogonal to all the columns of B, AB,... A^{n_2-2} B, that is to say, to the first $mn_1 + (m-1)(n_2-n_1-1)$ vectors of Q. We can hence take for ${}^{1}P(2)^T$ the $\left[(m-1)(n_2-1)+n_1+1\right]^{th}$ row of Q^{-1}, and we shall then have

$${}^{2}\tilde{b} = \left[A^{n-2}\,B\right]^T\;{}^{2}P(1) = \left[\alpha \quad 1 \quad 0 \ldots 0\right]^T$$

α being the component of $A^{n_2-1}\,{}^{1}b$ in $A^{n_2-1}\,{}^{2}b$. The second subspace is controlled by $\alpha u_1 + u_2$.

More generally, we can take for ${}^{1}P(k)^T$ the $\left[mn_1 + \ldots + (m-k+1)(n_k-n_{k-1}-1) + 1\right]^{th}$ row of Q^{-1}, and, in particular, ${}^{1}P(m)^T$ will be equal to the last row of Q^{-1}.

In conclusion, this decomposition allows us to convert the original system

$$\dot{x} = Ax + Bu \qquad y = Cx$$

into the form

$$\dot{\tilde{x}} = \tilde{A}\tilde{x} + \tilde{B}\tilde{u} \qquad y = \tilde{C}\tilde{x}$$

where $\tilde{A}$ and $\tilde{B}$ have the following structures:

$$\tilde{A} = \left[\begin{array}{c|c|c|c} [A(11)] & \alpha_{12} & & \alpha_{1m} \\ \hline \alpha_{21} & [A(22)] & & \alpha_{2m} \\ \hline & & & \\ \hline \alpha_{m1} & \alpha_{m2} & & [A(mm)] \end{array}\right] \qquad \tilde{B} = \left[\begin{array}{c|c|c|c} 0 & & & \\ 0 & & & \\ \vdots & & & \\ 1 & & & \\ \hline \vdots & 0 & & \\ \vdots & \vdots & & \\ 0 & \vdots & & \\ \times & 1 & & \\ \hline & & & \\ \hline 0 & 0 & & \vdots \\ \vdots & \vdots & & \vdots \\ \vdots & \vdots & & 0 \\ \times & \times & \times & 1 \end{array}\right]$$

$\times$ representing a nonzero scalar, $A(ii)$ a companion matrix and α_{ij} a non-null row of dimension n_j. This form of decomposition will turn out to be particularly advantageous in problems of compensation (cf. chapter 4).

C - Example

Let the system be defined by the equations

$$\dot{x} = \begin{bmatrix} 1 & 0 & 0 & 0 \\ 0 & 1 & 0 & 1 \\ 1 & 1 & 0 & 0 \\ 0 & 0 & 1 & 1 \end{bmatrix} x + \begin{bmatrix} 1 & 0 \\ 1 & 0 \\ 0 & 1 \\ 1 & 1 \end{bmatrix} u \qquad y = \begin{bmatrix} 0 & 1 & 1 & 0 \\ 0 & 0 & 1 & 1 \end{bmatrix} x \quad ;$$

Applying the above algorithm, we obtain

$$Q = \begin{bmatrix} 1 & 0 & 1 & 0 \\ 1 & 0 & 2 & 1 \\ 0 & 1 & 2 & 0 \\ 1 & 1 & 1 & 2 \end{bmatrix} \qquad Q^{-1} = \frac{1}{4}\begin{bmatrix} 5 & -2 & -1 & 1 \\ 2 & -4 & 2 & 2 \\ -1 & 2 & +1 & -1 \\ -3 & 2 & -1 & 1 \end{bmatrix}$$

$${}^{1}P(1) = \left[-1/4,\ 1/2,\ 1/4,\ -1/4\right]^T \text{ (third row of } Q^{-1}\text{)}, \quad {}^{2}P(1) = A^T\, {}^{1}P(1),$$

$${}^{1}P(2) = \left[-3/4,\ 1/2,\ -1/4,\ 1/4\right]^T \text{ (last row of } Q^{-1}\text{)}, \quad {}^{2}P(2) = A^T\, {}^{1}P(2),$$

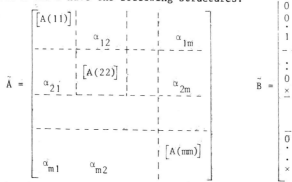

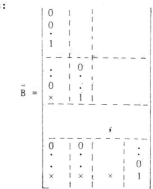

$$P = \begin{bmatrix} -1/4 & 0 & -3/4 & -1 \\ 1/2 & 3/4 & 1/2 & +1/4 \\ 1/4 & -1/4 & -1/4 & 1/4 \\ -1/4 & 1/4 & 1/4 & 3/4 \end{bmatrix} = (M^{-1})^T \text{and } M = \frac{1}{4}\begin{bmatrix} -1 & 4 & -5 & 0 \\ 3 & 4 & -1 & 0 \\ 5 & 0 & -7 & 4 \\ -4 & 4 & -4 & 4 \end{bmatrix}$$

$$\tilde{A} = M^{-1}AM = \left[\begin{array}{cc|cc} 0 & 1 & 0 & 0 \\ -1/4 & 5/4 & -5/4 & 5/4 \\ \hline 0 & 0 & 0 & 1 \\ 1/2 & 3/4 & -3/2 & 7/4 \end{array}\right] \quad \text{and} \quad \tilde{B} = M^{-1}B = \left[\begin{array}{cc} 0 & 0 \\ 1 & 0 \\ \hline 0 & 0 \\ 0 & 1 \end{array}\right]$$

3.2.3 Decomposition of a system into single-output observable subsystems(1)

A - Structure of the decomposition

In the previous decompositions, the fundamental hypothesis concerned the controllability of the system. We are now going to utilise the properties of observability. The consideration of the observation matrix, in place of the control matrix, thus suggests, a priori, a different approach. Whereas the decompositions into controllable subsystems were envisaged from the point of view of control, this new decomposition is intended to facilitate the problem of reconstruction of the state vector, and will form a basis for the realisation of a minimal degree observer, as we shall see in chapter 4.

We thus propose to decompose the system S into a set of p single-output observable subsystems, arranged in such a way that the coupling between them occurs solely through the outputs, in the way shown in figure 3.17.

(1) Cf. ref. 3.22, D.G. Luenberger

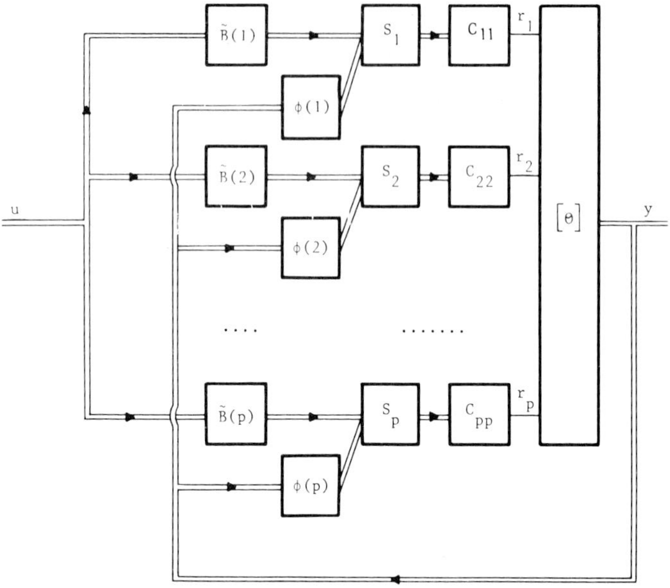

Fig. 3.17 The principle of the decomposition of a multivariable system into p subsystems observable from single outputs.

The equation of each subsystem will then be of the form

$$\dot{\tilde{x}}_i = A(i,i)\,\tilde{x}_i + \phi(i)\,y + \tilde{B}(i)\,u$$
$$r_i = C_{ii}\,\tilde{x}_i \qquad (3.41)$$

where the matrices $A(i,i)$, $\tilde{B}(i)$, $\phi(i)$, C_{ii} have the following dimensions:

$A(i,i)$	dimension	n_i, n_i
$\tilde{B}(i)$	"	n_i, m
$\phi(i)$	"	n_i, p
C_{ii}	"	$1, n_i$.

B - Construction of the transformation

To achieve this, we investigate the possibility of finding a transformation matrix M ($\tilde{x} = Mx$) which converts the original system

$$\dot{x} = Ax + Bu \qquad y = Cx$$

into the form

$$\dot{\tilde{x}} = \tilde{A}\tilde{x} + \tilde{B}u \qquad y = \tilde{C}\tilde{x}$$

satisfying the required conditions. For this, taking account of the form imposed on each subsystem, we must have

$$y = \begin{bmatrix} y_1 \\ \vdots \\ y_p \end{bmatrix} = \theta \begin{bmatrix} r_1 \\ \vdots \\ r_p \end{bmatrix} = \theta \begin{bmatrix} C_{11} & \tilde{x}_1 \\ \vdots & \\ C_{pp} & \tilde{x}_p \end{bmatrix} = \theta \begin{bmatrix} C_{11} & & \\ & \ddots & \\ & & C_{pp} \end{bmatrix} \tilde{x} = \theta C_d \, \tilde{x} = \tilde{C}\tilde{x} \tag{3.42}$$

designating by C_d the row-diagonal matrix

$$C_d = \begin{bmatrix} C_{11} & & & \\ & C_{22} & & \\ & & \ddots & \\ & & & C_{pp} \end{bmatrix} \tag{3.43}$$

Also, the total system constituted from the p subsystems (3.41) can be written in the form

$$\dot{\tilde{x}} = \begin{bmatrix} A(1,1) & & \cdot \\ & \ddots & \\ & & A(p,p) \end{bmatrix} \tilde{x} + \begin{bmatrix} \phi(1) \\ \phi(2) \\ \vdots \\ \phi(p) \end{bmatrix} y + \begin{bmatrix} \tilde{B}(1) \\ \tilde{B}(2) \\ \vdots \\ \tilde{B}(p) \end{bmatrix} u \tag{3.44}$$

$$\dot{\tilde{x}} = A_d \, \tilde{x} + \phi\tilde{C}\tilde{x} + \tilde{B}u = (A_d + \phi\tilde{C}) \, \tilde{x} + \tilde{B}u = \tilde{A}\tilde{x} + \tilde{B}u$$

with

$$\tilde{A} = A_d + \phi\tilde{C}$$

$$\phi = \begin{bmatrix} \phi(1) \\ \vdots \\ \phi(p) \end{bmatrix} \qquad A_d = \begin{bmatrix} A(1,1) & & \\ & \ddots & \\ & & A(p,p) \end{bmatrix} \qquad \tilde{B} = \begin{bmatrix} \tilde{B}(1) \\ \vdots \\ \tilde{B}(p) \end{bmatrix} \tag{3.45}$$

Now, the transformation matrix M is defined by:

$$M^{-1} AM = \tilde{A}$$

so that, since $C = \tilde{C}M^{-1}$:

$$M^{-1} A = \tilde{A}M^{-1} = \left[A_d + \phi\tilde{C}\right] M^{-1} = A_d \, M^{-1} + \phi C$$

Transposing both sides of the last equation, we have

$$A^T (M^{-1})^T = C^T \phi^T + (M^{-1})^T A_d^T$$

and, setting

$$(M^{-1})^T = P \quad C^T = H \quad A^T = F \quad \phi^T = \psi \quad A_d^T = F_d \tag{3.46}$$

we get

$$FP = PF_d + H\psi \tag{3.47}$$

where F, F_d and P are matrices of dimensions (n, n),

$$H = C^T \quad \text{of dimension} \quad n, p$$

$$\psi = \phi^T \quad \text{of dimension} \quad p, n.$$

If we decompose the matrices P and ψ into column blocks P(i), ψ(i), of dimensions (n, n_i), (p, n_i), respectively, associated with the subsystems S_i, the last equation can be written:

$$F\left[P(1) \ \dots \ P(p)\right] = \left[P(1) \ \dots \ P(p)\right] \begin{bmatrix} A^T(11) & & \\ & \ddots & \\ & & A^T(pp) \end{bmatrix} + H\left[\psi(1) \ \dots \ \psi(p)\right]$$

and decomposes into p matrix equations of orders n_i

$$FP(i) = P(i)\, A^T(i, i) + H\psi(i) \tag{3.48}$$

Displaying explicitly the n_i columns of these blocks, and supposing each block A^T(i,i) to be in companion form

$$A^T(ii) = \begin{bmatrix} -a^i_{n_i-1} & \dots & \dots & \dots -a^i_1 & -a^i_0 \\ 1 & 0 & . & 0 & 0 \\ 0 & 1 & 0 & . & 0 \\ \vdots & \vdots & \vdots & \vdots & \vdots \\ 0 & \vdots & \vdots & 1 & 0 \end{bmatrix}$$

we have

$$F\left[{}^1P(i) \ \dots \ {}^{n_i}P(i)\right] = \left[{}^1P(i) \ \dots \ {}^{n_i}P(i)\right] A^T(i, i) + H\left[{}^1\psi(i) \ \dots \ {}^{n_i}\psi(i)\right]$$

that is to say, p(n_i-1) vector equations of the form

$$F\,{}^kP(i) = -\,{}^1P(i)\, a^i_{n_i-k} + {}^{k+1}P(i) + H\,{}^k\psi(i) \tag{3.49}$$

and p equations of the form

$$F^{n_i}P(i) = -a_0^i \; {}^1P(i) + H\,{}^{n_i}\psi(i) \tag{3.50}$$

which can easily be solved iteratively, starting from ${}^1P(i)$, in the form:

$$\begin{aligned}
{}^2P(i) &= \left[F + Ia^i_{n_i-1}\right] {}^1P(i) - H\,{}^1\psi(i)\\
{}^3P(i) &= \left[F^2 + Fa^i_{n_i-1} + Ia^i_{n_i-2}\right] {}^1P(i) - FH\,{}^1\psi(i) - H\,{}^2\psi(i)\\
&\cdots\cdots\cdots\cdots\cdots\cdots\cdots\cdots\cdots\cdots\\
{}^kP(i) &= \left[F^{k-1} + F^{k-2}a^i_{n_i-2} + \ldots + Ia^i_{n_i-k+1}\right] {}^1P(i) - \sum_{j=1}^{k-1} F^{j-1} H\,{}^{k-j}\psi(i)
\end{aligned} \tag{3.51}$$

the last p equations being expressible in the form:

$$\left[F^{n_i} + \ldots + a_1^i F\right] {}^1P(i) - \sum_{j=1}^{n_i-1} F^j H\,{}^{n_i-j}\psi(i) = -a_0^i\,{}^1P(i) + H\,{}^{n_i}\psi(i)$$

or

$$\left[F^{n_i} + F^{n_i-1}a^i_{n_i-1} + \ldots + Fa_1^i + Ia_0^i\right] {}^1P(i) = \sum_{j=0}^{n-1} F^j H\,{}^{n-j}\psi(i) \tag{3.52}$$

If we set

$$\begin{aligned}
{}^1P(1) &= {}^IP\\
{}^1P(2) &= {}^{n_1+1}P = {}^{N_2}P\\
&\vdots\\
{}^1P(i) &= {}^{n_1+n_2+\ldots+(n_i-1)+1}P = {}^{N_i}P
\end{aligned} \tag{3.53}$$

the preceding equation can be written:

$$F^{n_i}\,{}^{N_i}P = \sum_{j=0}^{n_i-1} F^j \left[H\,{}^{n_i-j}\psi(i) - a_j^i\,{}^{N_i}P\right] \tag{3.54}$$

The form of this equation suggests a possible solution of the problem. Let

$${}^1H \quad {}^2H \ldots {}^pH$$

be the p column vectors of the matrix H, and let us look in the sequence

$${}^1H\;{}^2H \ldots {}^pH \quad F\,{}^1H \ldots F\,{}^pH \ldots F^{n_i-1}\,{}^1H \ldots F^{n_i}\,{}^iH \ldots$$

for the first n vectors independent of their predecessors, the existence of n such vectors being assured if the proposed system is observable, that is to say, if

$$\text{rank}\begin{bmatrix} H & FH & \dots & F^{n-1}H \end{bmatrix} = n$$

Let us suppose that, in the formation of this sequence, we find that $F^{n_i}\,{}^iH$ is dependent on the preceding vectors (whence it follows that $F^{\nu}\,{}^iH$ for $\nu > n_i$ is likewise dependent). Displaying the coefficients of this linear dependence, we can write:

$$F^{n_i}\,{}^iH = \sum_{k=0}^{n_i} \sum_{j=1}^{p} F^k \xi_k^j(i)\,{}^jH$$

with $\xi_k^j(i) = 0$ for $k = n_i$ if $j \geq i$.

If, further, we agree to set $\xi_k^j(i) = 0$ for $k > n_i$, we can replace the index n_i in the above summation by the observability index ν, and write:

$$F^{n_i}\,{}^iH = \sum_{k=0}^{\nu} \sum_{j=1}^{p} F^k \xi_k^j(i)\,{}^jH \tag{3.55}$$

Note: in the triply-indexed quantity $\xi_k^j(i)$, the index i refers to the subsystem number, j to the number of the column vector ${}^jH$ associated with the kth power of F.

If we compare equations (3.54) and (3.55), it is clear that a solution can be found in the form:

$$N_i\,p = {}^iH - \sum_{j=1}^{i-1} \xi_{n_i}^j(i)\,{}^jH \tag{3.56}$$

C - Determination of the companion matrices A(i,i) and the matrices $\psi(i)$

Equations (3.55) and (3.52), with the columns of H displayed, allow us to write:

$$F^{n_i}\,{}^iH - F^{n_i}\sum_{j=1}^{i-1} \xi_{n_i}^j(i)\,{}^jH = \sum_{k=0}^{n_i-1} F^k a_k^i \left[\sum_{j=1}^{i-1} \xi_{n_i}^j(i)\,{}^jH - {}^iH\right] +$$

$$+ \sum_{k=0}^{n_i-1} F^k \left[{}^1H \dots {}^pH\right]\,{}^{n_i-k}\psi(i)$$

$$= -\sum_{k=0}^{n_i-1} F^k a_k^i \, {}^iH + \sum_{k=0}^{n_i-1} F^k a_k^i \sum_{j=1}^{i-1} \xi^j_{n_i}(i) \, {}^jH +$$

$$+ \sum_{k=0}^{n_i-1} F^k \sum_{\rho=1}^{p} {}^{\rho}H \, {}^{n_i-k}_{\rho}\psi(i)$$

denoting by ${}^{n_i-k}_{\rho}\psi(i)$ the ρth element of the $(n_i-k)^{th}$ column of the block $\psi(i)$.

We thus have, finally, the equation

$$F^{n_i}\left[{}^iH - \sum_{j=1}^{i-1} \xi^j_{n_i}(i) \, {}^jH\right] = \sum_{k=0}^{n_i-1} F^k \left[\sum_{\rho=1}^{p} {}^{\rho}H \, {}^{n_i-k}_{\rho}\psi(i) - a_k^i \, {}^iH + \right.$$

$$\left. + a_k^i \sum_{j=1}^{i-1} \xi^j_{n_i}(i) \, {}^jH \right] \qquad (3.57)$$

If, in this last equation, we equate the terms in $F^j \, {}^kH$, we obtain pn_i equations which can be put in the form, for each j from 0 to n_i-1:

$$k < i \qquad \xi^k_j(i) = a^j_k \, \xi^k_{n_i}(i) + {}^{n_i-j}_{i}\psi(i) \qquad (3.58)$$

$$k = i \qquad \xi^i_j(i) = -a^i_j + {}^{n_i-j}_{i}\psi(i) \qquad (3.59)$$

$$k > i \qquad \xi^k_j(i) = {}^{n_i-j}_{i}\psi(i) \qquad (3.60)$$

To solve this set of equations, we can set

$${}^{\rho}_{i}\psi(i) = 0 \qquad \rho = 1, \ldots, n_i$$

and we then have

$$a^i_j = -\xi^i_j(i)$$

$${}^{j}_{k}\psi(i) = \xi^k_{n_i-j}(i) \qquad \text{if } k > i$$

$${}^{j}_{k}\psi(i) = \xi^k_{n_i-j}(i) - a^i_{n_i-j} \, \xi^k_{n_i}(i) \qquad \text{if } k < i$$

$${}^{j}_{i}\psi(i) = 0$$

The matrices A(i,i) and ψ, hence ϕ, are thus determined.

D - Determination of the transformation matrix

This matrix is determined via the matrix $P = (M^{-1})^T$ which, as we recall, is decomposed into the form

$$P = \left[P(1) \quad P(2) \ldots P(p)\right]$$

each block P(i) being itself in the form

$$P(i) = \left[{}^{1}P(i) \ldots {}^{n_i}P(i)\right]$$

with

$${}^{1}P(i) = {}^{N_i}P$$

We have from equation (3.56)

$${}^{N_i}P = {}^{1}P(i) = {}^{i}H - \sum_{j=1}^{i-1} \xi^{j}_{n_i}(i)\ {}^{j}H$$

and, in particular,

$${}^{1}P = {}^{1}H$$

$${}^{N_1}P = {}^{1}P(2) = {}^{2}H - \xi^{1}_{n_2}(2)\ {}^{1}H$$

$$\ldots\ldots$$

The other columns in each block referring to subsystems i are defined by equations (3.51), which can be rewritten in the form

$${}^{j}P(i) = F^{j-1}\ {}^{1}P(i) - \sum_{k=1}^{j-1} F^{k-1}\left[H\ {}^{j-k}\nu(i) + \xi^{i}_{n_i-j+k}(i)\ {}^{1}P(i)\right] \tag{3.61}$$

E - Determination of the matrices θ and C_d

The relation $\tilde{C} = CM$ can be written, after transposition, in the form

$$P\tilde{C}^T = H$$

where $H = C^T$ and $\tilde{C}^T$ are matrices of dimensions (n, p) whose columns can be displayed in the form

$$H = \left[{}^{1}H \ldots {}^{p}H\right] \qquad \tilde{C}^T = \left[{}^{1}\tilde{C}^T \ldots {}^{p}\tilde{C}^T\right]$$

thus giving p vector equations of the form

$${}^{i}H = P\ {}^{i}\tilde{C}^T$$

and np scalar equations

$$ {}^{i}_{k}H = \sum_{j=1}^{n} {}^{j}_{k}P \, {}^{i\sim}_{j}C = {}^{1}_{k}P \, {}^{i\sim}_{1}C + \sum_{j=2}^{n} {}^{j}_{k}P \, {}^{i\sim}_{j}C $$

${}^{j}_{k}P$ being the $(k,j)^{th}$ element of P and ${}^{i\sim}_{j}C$ the jth element of the column vector ${}^{i\sim}C^{T}$.

It is easy to see, taking account of the expressions for P, that the above equations can be satisfied by taking $\tilde{C}$ in the form

$$ \tilde{C} = \begin{bmatrix} 1 & 0 & \cdots\cdots\cdots\cdots & & & \\ \xi^{1}_{n_2}(2) & 0 \,..\, 1 & \cdots\cdots\cdots\cdots & & & \\ \xi^{1}_{n_3}(3) & 0 & \cdots & \xi^{2}_{n_3}(3) & \cdots & 1 \quad 0 \\ \cdots\cdots\cdots\cdots & & & & & \end{bmatrix} $$

whence

$$ C_{ii} = [1, 0, \ldots 0] \quad (n_i \text{ columns}). $$

The matrix θ is found directly from the relation $\theta C_d = \tilde{C}$,

$$ \theta = \begin{bmatrix} 1 & 0 & 0 & \cdots & \\ \xi^{1}_{n_2}(2) & 1 & 0 & \cdots & \\ \xi^{1}_{n_3}(3) & \xi^{2}_{n_3}(3) & 1 & \cdots & \\ \xi^{1}_{n_p}(p) & \xi^{2}_{n_p}(p) & \cdots & \cdots & 1 \end{bmatrix} \tag{3.62} $$

The decomposition into subsystems is thus completely established.

F - Example

Let the system be defined by

$$ \dot{x} = Ax + Bu \qquad\qquad y = Cx $$

wi

$$ A = \begin{bmatrix} 1 & 1 & 2 \\ 2 & 0 & -2 \\ 4 & 2 & -5 \end{bmatrix}, \quad B = \begin{bmatrix} 1 & 0 \\ 2 & 1 \\ 1 & 3 \end{bmatrix}, \quad C = \begin{bmatrix} 1 & 0 & 0 \\ 0 & 1 & 0 \end{bmatrix} $$

and let us follow, step by step, the procedure indicated above:

$$F = A^T = \begin{bmatrix} 1 & 2 & 4 \\ 1 & 0 & 2 \\ 2 & -2 & -5 \end{bmatrix}, \quad H = C^T = \begin{bmatrix} 1 & 0 \\ 0 & 1 \\ 0 & 0 \end{bmatrix}$$

We thus form the sequence of vectors

$$\begin{matrix} {}^1H & {}^2H & F\,{}^1H & F\,{}^2H & F^2\,{}^1H \\ \begin{bmatrix} 1 \\ 0 \\ 0 \end{bmatrix} & \begin{bmatrix} 0 \\ 1 \\ 0 \end{bmatrix} & \begin{bmatrix} 1 \\ 1 \\ 2 \end{bmatrix} & \begin{bmatrix} 2 \\ 0 \\ -2 \end{bmatrix} & \begin{bmatrix} 11 \\ 5 \\ -10 \end{bmatrix} \end{matrix}$$

and test them for linear dependence:

$$F\,{}^2H = 3\,{}^1H + {}^2H - F\,{}^1H$$

$$F^2\,{}^1H = 16\,{}^1H + 10\,{}^2H - 5\,F\,{}^1H$$

We deduce:

1) that subsystem 2 (formed from ${}^2H$) has degree 1, that subsystem 1 (formed from ${}^1H$) has degree 2, and further that

$$\xi_0^1(2) = 3 \qquad \xi_0^2(2) = 1 \qquad \xi_1^1(2) = -1$$

$$\xi_0^1(1) = 16 \qquad \xi_0^2(1) = 10 \qquad \xi_1^1(1) = -5$$

2) the companion matrices

$$A^T(1,1) = \begin{bmatrix} -a_1^1 & -a_0^1 \\ 1 & 0 \end{bmatrix} \qquad \begin{matrix} -a_1^1 = \xi_1^1(1) = -5 \\ -a_0^1 = \xi_0^1(1) = 16 \end{matrix} \qquad A_{11} = \begin{bmatrix} -5 & 1 \\ 16 & 0 \end{bmatrix}$$

$$A^T(2,2) = -a_0^2 \qquad -a_0^2 = \xi_0^2(2) = 1 \qquad A_{22} = 1$$

3) the matrices θ and C_{ii}

$$\theta = \begin{bmatrix} 1 & 0 \\ \xi_1^1(2) & 0 \end{bmatrix} = \begin{bmatrix} 1 & 0 \\ -1 & 1 \end{bmatrix} \qquad \begin{matrix} C_{11} = \begin{bmatrix} 1 & 0 \end{bmatrix} \\ C_{22} = 1 \end{matrix}$$

4) the matrices ϕ_i

$$\phi^T = \psi = \begin{bmatrix} \psi(1) & \psi(2) \end{bmatrix} = \begin{bmatrix} {}^1\psi(1) & {}^2\psi(1) & \vdots & {}^1\psi(2) \end{bmatrix}$$

with

$${}^1\psi(1) = \begin{bmatrix} 0 \\ \xi_1^1(1) \end{bmatrix} = \begin{bmatrix} 0 \\ 0 \end{bmatrix} \qquad \psi(1) = \begin{bmatrix} 0 & 0 \\ 0 & 10 \end{bmatrix} \qquad \phi(1) = \begin{bmatrix} 0 & 0 \\ 0 & 10 \end{bmatrix}$$

$${}^2\psi(1) = \begin{bmatrix} 0 \\ \xi_0^2(1) \end{bmatrix} \quad \begin{bmatrix} 0 \\ 10 \end{bmatrix}$$

$${}^1\psi(2) = \begin{bmatrix} \xi_0^1(2) - a_0^2\ \xi_1^1\ (2) \\ 0 \end{bmatrix} = \begin{bmatrix} 2 \\ 0 \end{bmatrix} \qquad \phi(2) = \begin{bmatrix} 2 & 0 \end{bmatrix}$$

5) the transformation matrix

$$P = \left[M^{-1}\right]^T = \left[P(1) \quad P(2)\right] = \left[{}^1P(1) \quad {}^2P(1) \;\vdots\; {}^1P(2)\right] = \left[{}^IP \quad {}^2P(1) \quad {}^{III}P\right]$$

with

$${}^Ip = {}^1H = \begin{bmatrix} 1 \\ 0 \\ 0 \end{bmatrix}, \quad {}^2P(1) = F\ {}^1P(1) - H\ {}^1\psi(1) - \xi_1^1(1)\ {}^1P(1) = \begin{bmatrix} 6 \\ 1 \\ 2 \end{bmatrix}$$

$${}^1P(2) = {}^{III}P = {}^2H - \xi_1^1(2)\ {}^1H = \begin{bmatrix} 1 \\ 1 \\ 0 \end{bmatrix}$$

$$(M^{-1})^T = \begin{bmatrix} 1 & 6 & 1 \\ 0 & 1 & 1 \\ 0 & 2 & 0 \end{bmatrix} \qquad M = \begin{bmatrix} 1 & 0 & 0 \\ -1 & 0 & 1 \\ -2,5 & 0,5 & -0,5 \end{bmatrix}$$

and

$$\tilde{B} = M^{-1}\ B = \begin{bmatrix} 1 & 0 \\ 10 & 7 \\ 3 & 1 \end{bmatrix}$$

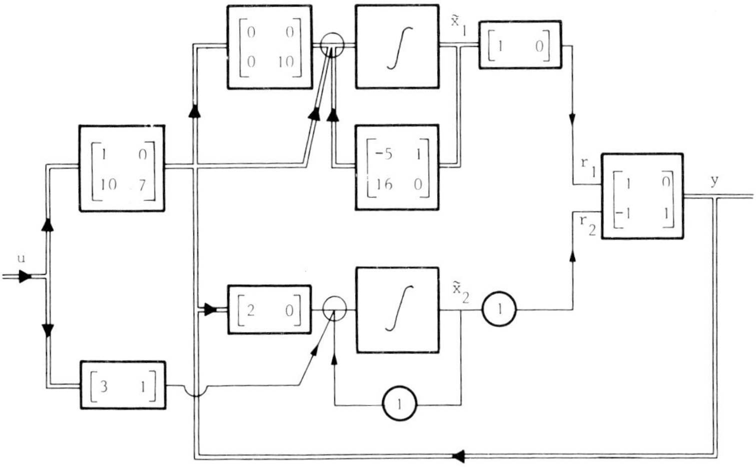

Fig. 3.18 An example of the decomposition

In the form of a signal-flow diagram, the system is represented by figure 3.18. It may be verified that

$$\tilde{A} = M^{-1}AM = A_d + \phi\theta C_d = \begin{bmatrix} -5 & 1 & 0 \\ 16 & 0 & 0 \\ 0 & 0 & 1 \end{bmatrix} + \begin{bmatrix} 0 & 0 \\ 0 & 10 \\ 2 & 0 \end{bmatrix} \begin{bmatrix} 1 & 0 \\ -1 & 1 \end{bmatrix} \begin{bmatrix} 1 & 0 & 0 \\ 0 & 0 & 1 \end{bmatrix}$$

$$= \begin{bmatrix} -5 & 1 & 0 \\ 6 & 0 & 10 \\ 2 & 0 & 1 \end{bmatrix} = \begin{bmatrix} 1 & 0 & 0 \\ 6 & 1 & 2 \\ 1 & 1 & 0 \end{bmatrix} \begin{bmatrix} 1 & 1 & 2 \\ 2 & 0 & -2 \\ 4 & 2 & -5 \end{bmatrix} \begin{bmatrix} 1 & 0 & 0 \\ -1 & 0 & 1 \\ -2{,}5 & 0{,}5 & -0{,}5 \end{bmatrix}$$

Chapter 4

Interactive control

The design of a control system for an industrial process or, more simply, for a complicated mechanical device, whether it concerns a satellite, a reactor or a boiler, evidently presents many problems. Among these, we have seen that the problem of setting up the equations, in the wide sense (i.e. not only the representation of the system to be controlled by a suitable set of equations, but possibly the linearisation of these equations or their simplification by elimination of fast stable modes) is a problem which justifies the importance given to chapters 2 and 3. This first stage having been covered, the role of the control engineer will be, finally, to design, around this model, a control system to achieve the performance required by the user.

In the case of single-variable systems, there are available a certain number of tools whose efficacy no longer requires demonstration, such as Nyquist and Nichols charts , the root-locus method,etc. The situation appears quite differently in the case of multivariable systems, for which, until recently, only rather empirical tools were available and one was often reduced to choosing the compensator parameters by tentative simulation trials. Is it worthwhile to indicate the drawbacks of such a method, which not only rests on a compensator structure chosen a priori, but which remains, even at best, haphazard and wasteful of time, in consequence of the cross-coupling inherent in all multivariable systems?

The use, still very widespread in the industrial context, of the transfer function matrix representation, obliges us to look again at certain problems connected with this type of representation and with the concepts of controllability and observability. We shall, indeed, see the fundamental practical importance of these ideas, whose underestimation has often led to the design of control schemes which were stable only on paper.

We shall propose, subsequently, a systematic solution of the compensation problem using a collection of methods based on the state-space representation. This representation has the advantage of being mathematically exact, i.e. of taking account of the uncontrollabilities and (or) unobservabilities which can appear as a result of the connection of the process and the compensators. It has both the advantage and the drawback of not revealing the internal physical structure of the system: an undesirable drawback at the practical level of realisation and technology, but an advantage at the level of synthesis, where the multiplicity of state-space forms allows us to choose that best adapted to the solution of a given type of problem (cf. chapter 3), without anticipating the final structure of the control scheme. This approach by state-space methods is clearly an intermediate stage in the calculation: we shall specify various means of passing from a compensation by means of state variables, more theoretical than practical, to a compensation which uses only accessible and measurable signals.

Let us note, finally, that the control problems considered here are being looked at from an interactive viewpoint. The particular problems raised by non-interactive control will be examined in chapter 5.

4.1 GENERAL PROBLEMS OF COMPENSATION AND THE CONNECTION WITH FUNDAMENTAL CONCEPTS

4.1.1 The problem of compensation

Let there be a process S, assumed to be linear and time-invariant, which, under the action of m independent inputs u, generates p outputs y. To control the system involves acting on the input variables u in such a way that it achieves the required aim, which, in general, can be expressed in the form of a certain performance index g(u,y), to be maximised or minimised. Although, in a wider perspective, this index may correspond to maximum productivity, maximum economy,..., we shall take here a more limited view, considering particularly the case where the purpose of control is to ensure satisfactory performance and suitable dynamics of the outputs y in comparison with some reference inputs e. In reference to figures 4.1., typical of the structure of a classical servo-system, the problem will be, e.g. to find controllers C(or R) such that:

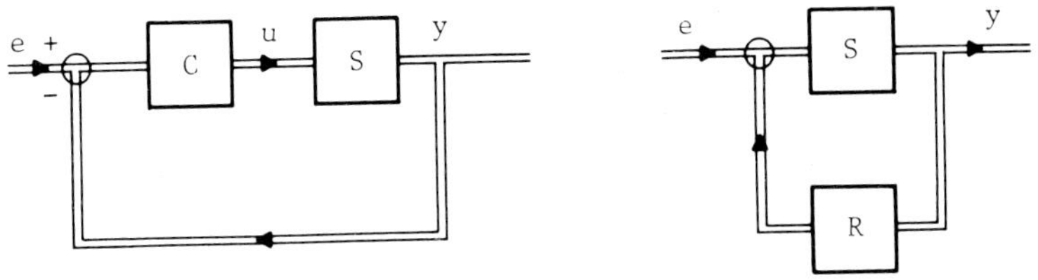

Fig. 4.1 Structure of control systems
(a) controller in the forward path
(b) controller in the feedback path

(a) to any desired output vector function y(t), there corresponds a possible control input e(t);

(b) to a change of input e, there corresponds a change in the output y which can be expressed in terms of damping or response time.

Of the two requirements (a) and (b), the first corresponds to a rather theoretical conception of control; the second, less restrictive, to a more practical idea (although there is often a tendency to confuse them). By way of illustration, we can consider an example of application to the control of a jet engine. For this system, very schematically represented in figure 4.2., the practical problem can be formulated as follows: the measurable quantities being ΔP_4 (compressor outlet pressure), ΔN (running speed) and ΔT (a quantity associated with the temperature of the turbine outflow), to find a control system (e.g. the compensator R) such that the responses to a step change of set-point δ are characterised, for both ΔN and ΔT, by a response time of less than 1.5 seconds and an overshoot of less than 10%.

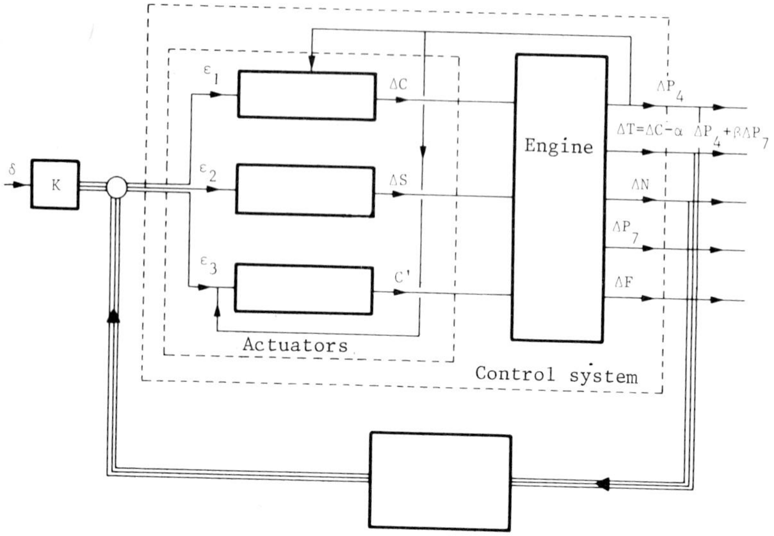

Fig. 4.2 Schematic diagram of the control of a jet engine

There are, of course, other considerations to be taken into account in the realisation of a control system. To mention only three: the problems of noise, sensitivity and complexity.

4.1.2 The structural viewpoint, connections with fundamental concepts

A - Controllability, observability, reproducibility

Given a system represented by its state-space equations

$$\dot{x} = Ax + Bu$$
$$y = Cx + Du \tag{4.1}$$

where u is the input vector (dimension m) and y the output vector (dimension p), the first problem which presents itself is to know if y can be controlled, in the practical sense of this term, in a suitable manner. This is what Rosenbrock calls the problem of the structural design of control systems (cf. ref. 4.13), a problem intimately connected with the fundamental ideas defined in chapter 1.

These ideas are, we recall, three in number:

(a) controllability means the existence of an input u allowing us to bring the state-vector x (or the output y) from an initial condition x_0 (or y_0) to a final condition x_1 (or y_1) in a given time t_1. The first condition relates to state controllability, the second to controllability with regard to the outputs.

(b) observability means the possibility, starting from the observation of the output vector during the interval (t_0, t_1), of reconstructing the initial state x_0.

(c) reproducibility expresses the possibility, given a vector function y(t) in the neighbourhood of an admissible output $y_a(t)$, of finding an input u(t) which, when applied to the system, gives rise to the output y(t) (cf. chapter 1 and ref. 4.22).

Although these notions are involved, to various extents, in all compensation problems, it is nevertheless important to realise that, in practice, they will be neither necessary nor sufficient:

(a) State controllability is involved in the sense that any uncontrollable mode, which is observable, will cause terms to arise, in the output, whose dynamics will be determined by the system, whose amplitude will depend on the initial conditions, and on which no input whatever will have any effect. If such a mode has slow dynamics with respect to the dominant modes of the control system, its contribution to the transient regime may be important. This contribution will even be disastrous, if it is unstable; but, if it stable with fast dynamics, its influence may be practically negligible.

Complete state controllability is thus, from the practical point of view, necessary only where it concerns unstable modes. It nevertheless remains a necessary and sufficient condition for the poles of the closed-loop system to be specifiable arbitrarily (cf. paragraph 4.2.2.A), and hence leaves open the possibility of thus establishing satisfactory control.

Controllability of the outputs (cf. chapter 1 and ref. 4.7) is, on the other hand, not a sufficient condition. If an unstable uncontrollable mode is unobservable, the system may still be output controllable. There nevertheless remains, as we saw in chapter 3, the fact that such an internal instability, even though invisible in the outputs, can have dramatic effects,whether in the failure of the model to reflect the physical reality of the process or in the behaviour of the materials involved.

(b) Observability of the system is also not a necessary and sufficient condition in every case. It is necessary, for the same reasons as mentioned above, in regard to unstable modes. For stable modes, it depends on whether or not state feedback compensation leads to a compensator which makes use of these modes (cf. paragraph 4.2.3.B).

(c) As to the concept of reproducibility(1), it is important to realise that it is, in a certain sense, only a local property. The fact that a system is reproducible by no means implies that every vector function y(t) can be chosen as the output of the system for some suitable input u(t): this is only true if y(t) is in the neighbourhood of an admissible function. If the latter has "bad" initial behaviour (non-minimum phase), this regrettable characteristic will affect all the outputs which we can hope to obtain.

B - Controllability (L) in the sense of Rosenbrock (2)

Despite their importance and their influence in all control problems, none of the above concepts allows us thus to solve the fundamental problem of structural design, namely: given a system S, is it justifable to seek a control system such that the output y of S can be made an arbitrarily given function y(t)? The answer to this question is to found in a concept introduced by Rosenbrock (cf. ref. 4.13) who called it "controllability (L)". We shall give here only the fundamental points of this theory, referring the reader to the original reference for the details.

(1) When we speak here of reproducibility, it refers to functional reproducibility, in the sense of Mesarovic and Brockett (ref. 4.2), and not to pointwise reproducibility, which is nothing but output controllability (refs.4.2 and 4.7) or what Rosenbrock calls "controllability(F)"(ref. 4.13).

(2) A notion defined by H.H. Rosenbrock (ref. 4.13).

Let us recall, to begin with, that the criteria of reproducibility, given in chapter 1, in terms of the state-representation matrices (A, B, C, D), can be more simply expressed in terms of the transfer function matrix $Z = \mathcal{Z}/z$ of the system (1). Indeed, the fact that y(t) can be regarded as the output of the system S, can be expressed by the condition that Z has a right inverse, i.e.

$$\text{rank } \mathcal{Z} = p \qquad \forall j\omega \text{ (2)}$$

This condition also implies that, if we decompose into Smith-Macmillan form (cf. chapter 2)

$$\mathcal{Z}(s) = V(s)\Gamma(s)W(s) \tag{4.2}$$

all the elements of the diagonal matrix Γ are nonzero. In the case that the rank of $\mathcal{Z}$ is $\rho < p$, the matrix Γ will contain ρ nonzero diagonal entries, the other $p-\rho$ being zero.

If the system is not functionally reproducible in its entirety, one can show that it is possible to isolate a subsystem with ρ inputs and ρ outputs which is: we shall say in this case that we have <u>partial reproducibility</u> of order ρ. (This property will be denoted by the symbol R_ρ).

The system S, with transfer function matrix Z, will be called controllable (L_q) (cf. ref. 4.13) if it is reproducible (R_ρ) ($q \leq \rho \leq p$) and if, among the ρ nonzero elements of Γ, q elements have all their zeroes in the left half-plane and $\rho-q$ elements do not. Rosenbrock has shown that, in this case, only q of the p outputs can have the characteristic response behaviour of a minimum-phase system, the $p-q$ others having that of a non-minimum-phase system. This result is essential in that it shows the importance which the determinant of the process assumes in any control system (we shall return to this point with a more thorough treatment in chapter 5).

(1) We suppose here that all the elements of the transfer function matrix have been reduced to a common denominator z.

(2) We shall assume, in what follows, that p = m.

<u>Example</u>: Let a three-input, three-output system be defined by

$$Z(s) = \begin{bmatrix} \dfrac{1}{3s+2} & \dfrac{2}{(s+1)(3s+2)} & \dfrac{1}{(s+1)(3s+2)} \\ \dfrac{1}{s+2} & \dfrac{5}{(s+2)(3s+2)} & \dfrac{1}{(s+1)(s+2)} \\ \dfrac{3s+1}{(s+2)(3s+2)} & \dfrac{5s+3}{(s+1)(s+2)(3s+2)} & \dfrac{3s+1}{(s+1)(s+2)(3s+2)} \end{bmatrix}$$

We have $Z = \dfrac{\mathcal{Z}}{z}$, with $z = (s+1)(s+2)(3s+2)$,

$$\mathcal{Z} = \begin{bmatrix} (s+1)(s+2) & 2(s+2) & s+2 \\ (s+1)(3s+2) & 5(s+1) & 3s+2 \\ (s+1)(3s+1) & 5s+3 & 3s+1 \end{bmatrix}$$

which can be written in Smith-Macmillan form

$$\mathcal{Z} = V(s)\Gamma(s)W(s)$$

with

$$V(s) = \begin{bmatrix} s+2 & 0 & 1 \\ 3s+2 & 1 & s \\ 3s+1 & 1 & s \end{bmatrix}, \quad W(s) = \begin{bmatrix} s+1 & 2 & 1 \\ 0 & 1 & 0 \\ s+1 & s & 1 \end{bmatrix} \quad \text{and} \quad \Gamma(s) = \begin{bmatrix} 1 & 0 & 0 \\ 0 & 1-s & 0 \\ 0 & 0 & 0 \end{bmatrix}.$$

The system in question is reproducible (R_2) and controllable (L_1). Only one of the outputs can be satisfactorily controlled at a time.

<u>C - Controllability, observability and compensation</u>

The purpose of this chapter being to explain a number of methods of compensation using state-space techniques, the compensation problem will not be considered specifically from the transfer function matrix viewpoint. In this case, moreover, the problem presents no difficulties other than calculational ones, with due regard, of course, (and this is fundamental) to the remarks made in the preceding paragraph and those which follow.

We saw in chapter 3, dealing directly with state-space representations, that the cascade connection of two subsystems, each controllable and observable, can give rise to a system containing uncontrollable and/or unobservable parts. Such an eventuality will clearly be less likely if the systems thus connected are selected arbitrarily. But this is precisely not the case in a compensation problem: the connection takes place between two systems, the process and the compensator, one of which is specifically designed by comparison with the other.

Let Z be the transfer function matrix of the process, which we shall write in the form

$$Z(s) = \frac{\mathcal{Z}(s)}{z(s)} \qquad (4.3)$$

being a polynomial matrix and z a polynomial, the common denominator of the elements of Z.

Similarly, let $C(s) = \mathcal{C}(s)/c(s)$ be the transfer function matrix of the compensator, placed upstream of the process as in figure 4.1a. We shall suppose, as is legitimate, that the process and the compensator are correctly represented by their respective transfer function matrices (i.e. we are dealing with minimal systems), and that the compensator is defined in such a way that

$$Z(s)C(s) = N(s) \qquad (4.4)$$

(N will represent, e.g., the desired open-loop transfer function matrix). From the relation (4.4) it follows that:

$$C = Z^{-1}\ N = \frac{[\mathrm{adj}\ Z]\ N}{|Z|}$$

i.e.

$$C = \frac{z\,[\mathrm{adj}\,\mathcal{Z}]\,\mathcal{N}}{|\mathcal{Z}|\,n} \qquad (4.5)$$

Since there may be cancellations between the numerator and denominator on the right-hand side of this equation, while, because of the minimality assumption, no cancellation is possible between $\mathcal{C}$ and c, we are led to write equation (4.5) in the form: (1)

$$C = \frac{z\,[\mathrm{adj}\,\mathcal{Z}]\,\mathcal{N}}{|\mathcal{Z}|\ n} = \frac{\mathcal{C}\ g}{c\ g} \qquad (4.6)$$

(1) These considerations were, to our knowledge, first expressed by E.G. Gilbert, cf. ref. 4.5.

where g comprises all the factors capable of being simplified by cancellation. Based on this last relation for C, the effective expression for the resultant overall transfer function matrix is

$$N^* = ZC = \frac{\mathcal{N}^*}{n^*} = \frac{\mathcal{Z}}{z} \frac{z\,[\mathrm{adj}\,\mathcal{Z}]\,\mathcal{N}\,g^{-1}}{|\mathcal{Z}|\,ng^{-1}}$$

$$N^* = \frac{z\,\mathcal{Z}\,[\mathrm{adj}\,\mathcal{Z}]\,\mathcal{N}\,g^{-1}}{z\,|\mathcal{Z}|\,ng^{-1}} = \frac{z\,|\mathcal{Z}|\,g^{-1}\mathcal{N}}{z\,|\mathcal{Z}|\,g^{-1}\,n}, \qquad (4.7)$$

Let us emphasise that, in this expression, we have taken care not to allow those factors which we could have considered cancelling, to disappear from either the numerator or the denominator.

The cascade thus constructed will not contain uncontrollable or unobservable modes if there is no cancellation possible between $\mathcal{N}^*$ and n^*, i.e. in view of (4.7) with $\mathcal{N}$ and n having no cancellation, if g contains all the factors of $z|\mathcal{Z}|$.

From (4.6),

$$\mathcal{C} = z\,[\mathrm{adj}\,\mathcal{Z}]\,\mathcal{N}\,g^{-1}$$

$$c = |\mathcal{Z}|\,ng^{-1}$$

The preceding condition will thus only be satisfied if

- n includes the factors of z
- $(\mathrm{adj}\,\mathcal{Z})\,\mathcal{N}$ includes the factors of $|\mathcal{Z}|$ (4.8)

Remark 1

In practice, conditions (4.8), which turn out to be quite stringent, can be somewhat softened, and it is often sufficient to ensure the controllability and observability of only the unstable modes, so that (4.8) will frequently be used in the form:

- n includes the unstable factors of z
- $(\mathrm{adj}\,\mathcal{Z})\,\mathcal{N}$ includes the unstable factors of $|\mathcal{Z}|$ (4.9)

It is important always to bear in mind constraints of this kind (to which we shall return in chapter 5, devoted to non-interaction), since experience shows that the great majority of errors made in multivariable compensation find their origin here. Put another way, this indicates that, given a process S characterised by its transfer function matrix Z, one cannot necessarily impose arbitrarily an overall open-loop (or closed-loop) transfer function matrix for the control system, without causing uncontrollable or unobservable modes to appear.

Remark 2

In certain problems, conditions (4.9), although necessary since the stability of the system depends on them, are not sufficient. A stable uncontrollable mode, which however has slow dynamics compared with the controllable and observable modes of the system, may well cause disappointing behaviour in the transient dynamic response, so that it is always useful to detect the existence of such a mode when choosing the structure of the controller.

Let us consider, e.g., the following very simple case of an ideal process S on which no perturbation acts and which is subject only to the influence of the inputs u; let

$$Z(s) = \frac{1}{s+1}\begin{bmatrix} -2 & 3 \\ 4 & 8s+2 \end{bmatrix}$$

be the transfer function matrix of the process, and

$$H(s) = \frac{1}{s+1}\begin{bmatrix} -2 & 0 \\ 0 & 8s+2 \end{bmatrix}$$

be the desired transfer function matrix of the control system.

Three control structures are possible a priori, corresponding to the schemes of figure 4.3. The first is an open-loop structure, while the other two have closed-loop structures differing only in the position of the coupling. Arguing directly from the transfer function matrices and using the defining relations

$$H = ZC_0$$

$$H = \left[I + ZC_1 R_1\right]^{-1} ZC_1$$

$$H = \left[I + ZR_2\right]^{-1} ZC_2$$

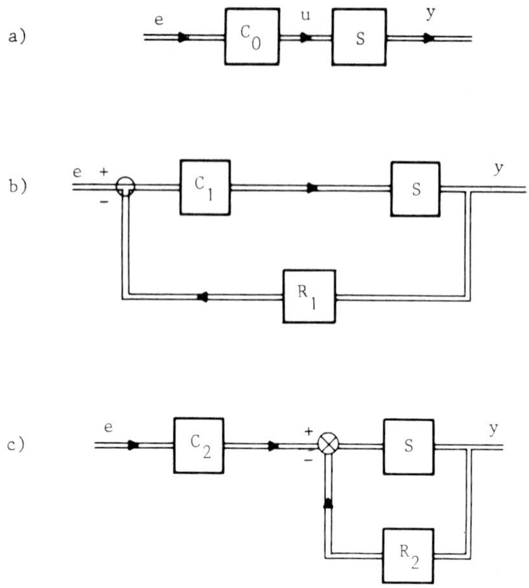

Fig. 4.3 Various structures corresponding to the same overall transfer function matrix.

we find for the various compensators the following expressions:

$$\text{a)}\quad C_0 = \frac{s+0.25}{s+1}\begin{bmatrix} 1 & 1.5 \\ -1/2(s+0.25) & 1 \end{bmatrix}$$

$$\text{b)}\quad C_1 = I,\; R_1 = \frac{1}{8}\begin{bmatrix} -3 & -1.5 \\ -2 & 3/(4s+1) \end{bmatrix}$$

$$\text{c)}\quad C_2 = \frac{1}{s+1}\begin{bmatrix} (4s+9)/4 & 3(4s+1)/8 \\ -0.5 & 9(4s+1)/4 \end{bmatrix},\; R_2 = \begin{bmatrix} -1 & 0 \\ 0 & 1 \end{bmatrix}$$

Although structures a) and b) are minimal, it is easy to see that this property does not hold for structure c), whose state representation is of order 3, e.g. of the form

$$\dot{x} = -\frac{1}{3}\begin{bmatrix} 7 & 8 & 0 \\ 0 & 3 & 0 \\ 0 & 0 & 3 \end{bmatrix} x + \begin{bmatrix} -2 & 0 \\ 1 & 0 \\ 0 & 1 \end{bmatrix} e$$

$$y = \frac{1}{9}\begin{bmatrix} 9 & 0 & 0 \\ -2 & -4 & -54 \end{bmatrix} x - \frac{1}{9}\begin{bmatrix} 0 & 0 \\ 4 & 54 \end{bmatrix} e .$$

Here, there appears an uncontrollable mode corresponding to the factor $3s + 7$.

4.2 COMPENSATION BY STATE FEEDBACK

4.2.1 Introduction

The following paragraphs are devoted to a particular type of compensation, called compensation by state feedback. Given the system S

$$S : \quad \dot{x} = Ax + Bu \qquad y = Cx \tag{4.10}$$

where x, u, y are respectively the state, input and output vectors, of respective dimensions n, m, p, the principle of the method is to apply a control law of the form (cf. figure 4.4)

$$u = e + Kx \tag{4.11}$$

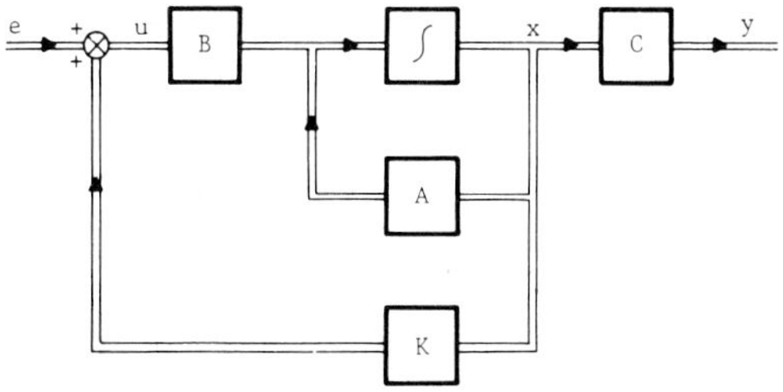

Fig. 4.4 Principle of state feedback

thus replacing the open-loop transfer function matrix

$$Z_0 = C(sI-A)^{-1}B \tag{4.12}$$

by the closed-loop transfer function matrix

$$Z_c = C(sI-A-BK)^{-1}B \tag{4.13}$$

In particular, the modes of the closed-loop system being the roots of the characteristic equation

$$|sI-A-BK| = 0 \tag{4.14}$$

we appreciate that a suitable choice of K may allow us to improve the stability and the dynamical properties of the system.

We shall show first that the assumption of controllability for S is equivalent to the possibility of arbitrarily specifying the modes of the closed-loop system, i.e. the roots of equation (4.14).

Then, we shall study the construction of a feedback matrix by systematic procedures, utilising an adequate structural decomposition of S. We shall also show that it is possible, not only to fix the modes completely, but to specify, to a certain extent, the dynamics of each output y.

Clearly, not all the states of the system will, in practice, be accessible and measurable. It is thus important, finally, to discover methods of transforming a theoretical state feedback into a feedback which makes use only of the inputs and outputs of the system S. This will form the subject of the second part of this chapter.

4.2.2 Controllability and compensation

A - Controllability and the closed-loop system characteristic equation

Given the controllable system

$$\dot{x} = Ax + Bu$$

we know that it is always possible (1) to find a transformation $x = M\tilde{x}$ which puts the matrix A in a block-diagonal companion form

(1) cf. subsection 3.2.1.

$$\tilde{A} = \begin{bmatrix} A(0) & 0 & \cdots & 0 \\ 0 & A(1) & & \vdots \\ \vdots & & \ddots & 0 \\ 0 & & 0 & A(r) \end{bmatrix}, \text{ with } A(i) = \begin{bmatrix} 0 & 1 & 0 & \cdots & 0 \\ 0 & 0 & 1 & & 0 \\ \vdots & & & & \vdots \\ -a_0(i) & -a_1(i) & & & -a_{n_i-1}(i) \end{bmatrix} \quad (4.15)$$

This transformation corresponds to a decomposition of the space R_X into cyclic subspaces R_{X_i} (and is thus not unique). In the new basis, the compensated system takes the form

$$\dot{\tilde{x}} = M^{-1}AM\,\tilde{x} + M^{-1}\,B(KM\tilde{x} + e)$$

$$\dot{\tilde{x}} = (\tilde{A} + \tilde{B}\tilde{K})\,\tilde{x} + \tilde{B}e, \text{ with } \tilde{K} = KM \quad (4.16)$$

In view of the form chosen for A, we see that the problem can be solved if it is possible to find a matrix $\tilde{K}$ such that $\tilde{B}\tilde{K}$ satisfies the relations

$$\tilde{B}\tilde{K} = \begin{bmatrix} \Delta(1) & & & \\ & \cdot & & \\ & & \cdot & \\ & & & \cdot \\ & & & \Delta(r) \end{bmatrix} = \Delta, \text{ with } \Delta(i) = \begin{bmatrix} 0 & 0 & 0 \\ 0 & 0 & 0 \\ \vdots & \vdots & \vdots \\ -\Delta_0(i) & -\Delta_1(i) \cdots & -\Delta_{n-1}(i) \end{bmatrix}$$

In fact, under these conditions, the characteristic polynomial can be written

$$\phi(s) = \prod_{i=1}^{r} \phi_i(s)$$

$$\phi_i(s) = s^{n_i} + \left[a_{n_i-1}(i) + \Delta_{n_i-1}(i)\right] s^{n_i-1} + \ldots + \left[a_0(i) + \Delta_0(i)\right]$$

and we can specify its n roots.

For this, we must have

$$\tilde{B}\tilde{K} = M^{-1}\,B\tilde{K} = \Delta\,,$$

thus

$$B\tilde{K} = M\Delta\,.$$

If we decompose $\tilde{K}$ and M into blocks, in the form:

$$B\tilde{K} = B\left[\tilde{K}(1)\ \ldots\ \tilde{K}(r)\right] \qquad M = \left[M(1)\ M(2)\ \ldots\ M(r)\right]$$

the preceding equation decomposes into:

$$B\tilde{K}(i) = M(i)\ \Delta(i)$$

which is, in terms of columns,

$$B\left[{}^{1}\tilde{K}(i)\ \ {}^{2}\tilde{K}(i)\ \dots\ {}^{n_i}\tilde{K}(i)\right] = \left[{}^{1}M(i)\ \dots\ {}^{n_i}M(i)\right]\begin{bmatrix} 0 & 0 & 0 \\ 0 & 0 & 0 \\ -\Delta_0(i) & \dots\dots & \end{bmatrix}$$

$$B\ {}^{j}\tilde{K}(i) = -\ \Delta_{j-1}(i)\ {}^{n_i}M(i) \qquad j = 1, \dots, n_i$$

If we consider two columns, with indices j and k, belonging to the same block i, we then have

$$B\ {}^{k}\tilde{K}(i) = \frac{\Delta_{k-1}(i)}{\Delta_{j-1}(i)}\ B\ {}^{j}\tilde{K}(i)$$

and it follows that the columns ${}^{j}\tilde{K}(i)$ belonging to the same block are proportional,

$${}^{j}\tilde{K}(i) = \frac{\Delta_{j-1}(i)}{\Delta_{k-1}(i)}\ {}^{k}\tilde{K}(i)$$

and that

$${}^{n_i}M(i) = -\ Bd(i), \qquad \text{avec} \qquad d(i) = -\ {}^{k}\tilde{K}(i)/\Delta_{k-1}(i)\ .$$

On the other hand, we must also satisfy the relation:

$$M^{-1}\ AM = \tilde{A}$$
$$AM = M\tilde{A}$$

The same procedure as before shows that we have:

$${}^{n_i-1}M(i) = \left[A + Ia_{n_i-1}(i)\right]\ {}^{n_i}M(i)$$

$${}^{n_i-2}M(i) = \left[A^2 + Aa_{n_i-1}(i) + Ia_{n_i-2}(i)\right]\ {}^{n_i}M(i)$$

....

$${}^{1}M(i) = \left[A^{n_i-1} + A^{n_i-2}\ a_{n_i-1}(i) + \dots + Ia_1\right]\ {}^{n_i}M(i)$$

$$\left[A^{n_i} + a_{n_i-1}(i)\ A^{n_i-1} + \dots + a_0\ I\right]\ {}^{n_i}M(i) = 0\ .$$

Since the matrix M must be regular, the existence of the d(i) depends on the condition:

$$\text{rank}\left[Bd(1)\ \dots\ A^{n_1-1}\ Bd(1)\quad Bd(2)\ \dots\ A^{n_r-1}\ Bd(r)\right] = n$$

i.e.

$$\text{rank}\begin{bmatrix}B & AB & \dots & A^{n_r-1} & B\end{bmatrix}\begin{bmatrix}\mathcal{D}(1) & & & \\ & \mathcal{D}(2) & & \\ & & \ddots & \\ & & & \mathcal{D}(r)\end{bmatrix} = n = \text{rank}\begin{bmatrix}\mathcal{A} & \mathcal{D}\end{bmatrix}$$

$$\mathcal{D}(i)\begin{bmatrix}d(i) & & & \\ & d(i) & & \\ & & \ddots & \\ & & & d(i)\end{bmatrix}$$

Given that the rank of $\mathcal{D}$ is n, Sylvester's inequality shows that a necessary and sufficient condition for this is

$$\text{rank}\begin{bmatrix}\mathcal{D}\end{bmatrix} = n \qquad (4.17)$$

Hence, if a system is controllable, it is possible to find a feedback matrix K,

$$u = Kx + u_0$$

such that the roots of the characteristic equation of the closed-loop system can be specified a priori.

B - Controllability of the closed-loop system

We have seen that the state equation of the closed-loop system takes the form:

$$S_b : \begin{cases} \dot{x} = (A + BK)\,x + Be \\ y = Cx \end{cases} \qquad (4.18)$$

It is clearly important to make sure that the system S_b is also controllable, K being a matrix of dimensions (m, n). This property has been established by Wonham, to whom we refer the reader (cf. ref. 4.15). We shall incidentally demonstrate it in the following paragraphs during the practical determination of K, showing that the characteristic equation of sI-A-BK can be arbitrarily fixed.

C - Case where the system S is not completely controllable

If the system S is not completely controllable, it is not possible to specify all the roots of the characteristic equation of the closed-loop system. In this case, we should at least like the closed-loop system to be stable. If we go back to the classical decomposition into two subsystems, one completely controllable and the other uncontrollable, we see from the above development that state feedback will only allow us to act on the controllable modes of the system and will have no effect on the others. The closed-loop system will thus only be stabilisable if these latter modes are stable to begin with (cf. ref. 4.15).

4.2.3 Theoretical determination of state feedback matrices

We have seen in the preceding paragraphs that, if a system is controllable, it is possible to find a state feedback matrix K so as to fix arbitrarily the modes of the closed-loop system. The problem now consists of specifying a method, in the form of precise algorithms, for actually constructing such feedback matrices.

The techniques which will be exhibited in the following paragraphs are based on the utilisation of particular structural decompositions of the system. We know indeed that, since the physical nature and structure of the system are not preserved in a state-space representation (which is defined only to within a basis transformation), we can always, by a suitable transformation, put the system in a mathematical form specially adapted to the problem to be solved. Taking this viewpoint, the problem will be separated into three stages:

a - casting the system into an appropriate form;
b - finding a compensator associated with this form;
c - transforming the compensator to the original coordinates.

Stage a) constitutes the essential step in the problem, at least from the point of view of complexity and calculational effort. The advantage of this method is that this stage can be completely automated thanks to certain computational subroutines; stage b) can then be carried out systematically on the basis of simple tests.

A - Case of single-input systems

We shall first consider the case of a single-input system defined by

$$\dot{x} = Ax + Bu$$
$$y = Cx \qquad (4.19)$$

completely controllable from its only input u, and suppose that all the states are accessible and measurable. The problem consists of finding a row vector K such that

$$\dot{x} = (A + BK)\,x + be \qquad (4.20)$$

has an arbitrarily given characteristic equation of degree n. In fact, we have seen that the solution depends directly on the controllability of the system, and that, if it is completely controllable, it is possible to fix arbitrarily the modes of the compensated system.

1 - Transformation matrix:

Since the system is controllable from its sole input, we know (1) that there exists a transformation matrix M ($x = M\tilde{x}$) which converts the matrix A into a companion matrix $\tilde{A}$, and the matrix B into $\tilde{B}$:

$$\tilde{B} = \begin{bmatrix} 0 & 0 & \dots & 1 \end{bmatrix}^T ,$$

with

$$\tilde{A} = M^{-1}\,AM \qquad \tilde{B} = M^{-1}\,B$$

$$\tilde{A} = \begin{bmatrix} 0 & 1 & \dots\dots & 0 \\ 0 & 0 & 1 \;\dots & 0 \\ \vdots & & & \vdots \\ \vdots & & & 1 \\ -a_0 & \dots\dots & & -a_{n-1} \end{bmatrix} \qquad (4.21)$$

In this form, the characteristic polynomial manifests itself directly as

$$|sI-A| = s^n + a_{n-1}s^{n-1} + \dots + a_0 = |sI-\tilde{A}| . \qquad (4.22)$$

This transformation also converts K into $\tilde{K}$

$$\tilde{K} = KM . \qquad (4.23)$$

Let us try to specify the transformation: the equation defining $\tilde{A}$, rewritten in the form

$$M\tilde{A} = AM$$

(1) cf. chapter 3. We emphasise that the transformation used here affects simultaneously both the evolution and control matrices.

can be replaced by n vector equations if we display the n columns of M in the form $M = \left[{}^1M \; {}^2M \ldots {}^nM\right]$. In fact, we have

$$\left[{}^1M \quad {}^2M \ldots {}^nM\right] \begin{bmatrix} 0 & 1 & \ldots & 0 \\ & & 1 & \\ & & & \ddots \\ & & & 1 \\ -a_0 & & & -a_{n-1} \end{bmatrix} = A \left[{}^1M \quad {}^2M \ldots {}^nM\right]$$

which decomposes into

$$\begin{aligned} {}^{n-1}M &= \left[A + Ia_{n-1}\right] {}^nM \\ {}^{n-2}M &= \left[A^2 + a_{n-1} A + Ia_{n-2}\right] {}^nM \\ &\cdots\cdots\cdots\cdots\cdots\cdots \\ {}^1M &= \left[A^{n-1} + a_{n-1} A^{n-2} + \ldots + Ia_1\right] {}^nM \\ 0 &= \left[A^n + a_{n-1} A^{n-1} + \ldots + Ia_0\right] {}^nM \,. \end{aligned} \tag{4.23b}$$

The first n-1 equations allow us to express the first n-1 columns of M as functions of the last one and of the coefficients a_i in the characteristic polynomial; the last equation shows that $A^n\,{}^nM$ is linearly dependent on the vectors $A^k\,{}^nM$ for k = 1,... n-1, the coefficients of the linear dependence being precisely those of the last row of the companion matrix.

On the other hand, we have the equation

$$M^{-1} B = \tilde{B} \rightarrow B = M\tilde{B}$$

which can be written in the form

$$B = \left[{}^1M \quad {}^2M \ldots {}^nM\right] \left[0 \quad 0 \ldots 1\right]^T = {}^nM$$

and we can simply replace ${}^nM$ by B in equations (4.23b). This decomposition is possible if the matrix M thus formed is indeed a transformation matrix, that is to say, if its rank is n. For this, it is necessary and sufficient that:

$$\text{rank} \left[B \quad AB \ldots A^{n-1} B\right] = n$$

hence, that the system is controllable.

2 - Form of the compensator:

The simplest way of imposing the characteristic equation of the compensated system, e.g.

$$s^n + \alpha_{n-1}s^{n-1} + \ldots + \alpha_0 = 0$$

is to put $\tilde{A} + \tilde{B}\tilde{K}$ in companion form

$$\tilde{A} + \tilde{B}\tilde{K} = \begin{bmatrix} 0 & 1 & 0 & \ldots & 0 \\ 0 & 0 & 1 & \ldots & 0 \\ \vdots & & & & \\ 0 & & & & 1 \\ -\alpha_0 & \ldots & \ldots & \ldots & -\alpha_{n-1} \end{bmatrix} \tag{4.24}$$

i.e. taking account of the form of $\tilde{A}$, to put $\tilde{B}\tilde{K}$ in the form

$$\tilde{B}\tilde{K} = (\tilde{A} + \tilde{B}\tilde{K}) - \tilde{A}$$

$$= \begin{bmatrix} 0 & 0 & \ldots\ldots & 0 \\ \vdots & & & \\ 0 & 0 & & 0 \\ a_0-\alpha_0 & a_1-\alpha_1 & \ldots & a_{n-1}-\alpha_{n-1} \end{bmatrix} = \begin{bmatrix} 0 \\ 0 \\ \vdots \\ 1 \end{bmatrix} \begin{bmatrix} \tilde{k}_0 & \tilde{k}_1 & \ldots & \tilde{k}_{n-1} \end{bmatrix}$$

whence we deduce immediately

$$\tilde{K} = \begin{bmatrix} \tilde{k}_0 & \tilde{k}_1 & \tilde{k}_{n-1} \end{bmatrix}, \quad \tilde{k}_i = a_i - \alpha_i \tag{4.25}$$

and

$$K = \tilde{K}M^{-1} \tag{4.26}$$

3 - Example:

Let a single-input system be defined by

$$\dot{x} = \begin{bmatrix} -1 & & \\ & -2 & \\ & & +3 \end{bmatrix} x + \begin{bmatrix} 1 \\ -1 \\ -1 \end{bmatrix} u, \quad y = \begin{bmatrix} 1 & 1 & 0 \\ 1 & 0 & 1 \end{bmatrix} x .$$

The system is unstable in consequence of the existence of the mode $s = 3$ in the right half-plane. We should like the compensated system to have the modes $s = -2$, $s = -3$ and $s = -5$, corresponding to the characteristic equation $s^3 + 10s^2 + 31s + 30 = 0$.

To put the system in canonical form, we compute the column vectors

B	AB	A^2B	A^3B
1	− 1	1	− 1
− 1	2	− 4	8
− 1	− 3	− 9	− 27

We can check that, consistently with the controllability of the system, the first three vectors are independent, the fourth being dependent on them through the relation

$$A^3 B = 6 b + 7 AB$$

corresponding to the characteristic equation

$$s^3 - 7s - 6 = 0 \quad (a_0 = -6, a_1 = -7).$$

The transformation matrix is defined by:

$$M = \left[(A^2 + a_1 I) B , AB , B\right] = \begin{bmatrix} -6 & -1 & 1 \\ 3 & 2 & -1 \\ -2 & -3 & -1 \end{bmatrix}$$

The required feedback matrix is given by

$$\tilde{K} = \left[a_0 - \alpha_0 \quad a_1 - \alpha_1 \quad a_2 - \alpha_2\right] ,$$

with $\alpha_0 = 30$, $\alpha_1 = 31$, $\alpha_2 = 10$,

so that

$$\tilde{K} = \left[-36 \quad -38 \quad -10\right] \quad \text{et} \quad K = \tilde{K}M^{-1} = \left[2 \quad 0 \quad 12\right] .$$

B - General case of a multi-input system

Consider the general case of a system having several inputs. The control matrix B is then of dimensions (n, m), m being the number of inputs, and the system is assumed to be controllable from the complete set of its inputs, i.e.

$$\text{rank} \left[B, AB, \ldots A^{n-1} B\right] = n.$$

Two cases can arise:

1 - The system is completely controllable from a single input u_i, i.e. if ${}^iB$ is the ith column of B (corresponding to the input u_i), the matrix $\left[{}^iB, A\, {}^iB, \ldots A^{n-1}\, {}^iB\right]$ is of rank n.

The method of the previous paragraph is then directly applicable. If we take the vector ${}^{i}B$ as generator, only one subspace will arise, and $\tilde{A}$ and $\tilde{B}$ will appear in the forms

$$\tilde{A} = \begin{bmatrix} 0 & 1 & 0 & \cdots & 0 \\ 0 & 0 & 1 & \cdots & \\ & & & 1 & \\ & & & & 1 \\ -a_0 & \cdots & & \cdots & -a_{n-1} \end{bmatrix} \qquad \tilde{B} = \begin{bmatrix} {}^{1}\tilde{B}, {}^{2}\tilde{B}, {}^{i-1}\tilde{B} & \begin{matrix} 0 \\ 0 \\ \vdots \\ \vdots \\ 1 \end{matrix} & {}^{i+1}\tilde{B}, {}^{m}\tilde{B} \end{bmatrix} .$$

From the preceding paragraph, it will thus be sufficient to take $\tilde{K}$ in the form

$$\tilde{K} = \begin{bmatrix} 0 & \cdots & 0 \\ \vdots & & \vdots \\ a_0 - \alpha_0 & & a_{n-1} - \alpha_{n-1} \\ \vdots & & \vdots \\ 0 & \cdots & 0 \end{bmatrix} \leftarrow i^{th} \text{ row} \qquad (4.27)$$

in order for A + BK to have the characteristic equation

$$s^n + \alpha_{n-1} s^{n-1} + \ldots + \alpha_0 = 0 .$$

<u>2 - The system is controllable from the whole set of inputs</u>. In general, the system will not be completely controllable except through the action of several inputs. We shall thus not be able, as in the above method, to generate the whole space on the basis of a single input component: we are thus led to seek a new structural form of decomposition.

We shall make use of a property set forth in the chapter relating to structural decompositions of a system. (1)

a - <u>Decomposition into single-input controllable subsystems</u>

We have seen, indeed, that if a system is controllable, it is always possible to decompose it into a number r of subsystems, such that:
- each of them is controllable from a single input
- they are hierarchically organised in such a way that

(1) cf. chapter 3 and refs. (3.14), (4.8), (4.9).

S_r reacts on subsystems $S_{r-1}, \dots S_2, S_1,$

$\vdots$

S_3 reacts on S_2 and S_1,

S_2 reacts only on S_1.

Schematically, the decomposition of the system is as indicated in figure 4.5, which shows the one-sided nature of the interactions, S_i (of degree n_i) being controllable from the input component u_i; if $r < m$, the $m - r$ remaining components of the input, denoted by $u^* = [u_{r+1}, \dots u_m]^T$, may possibly act on all the subsystems.

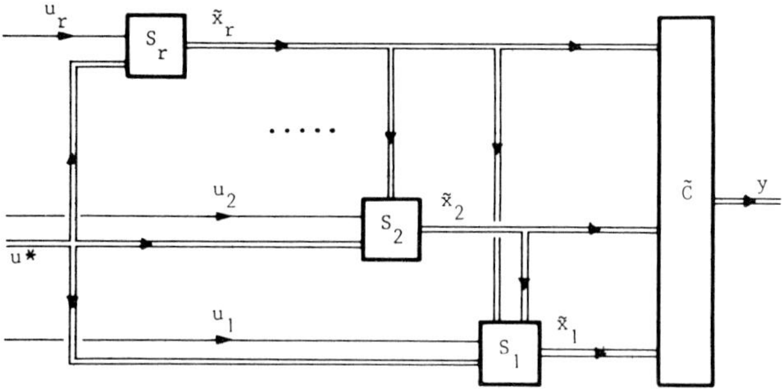

Fig. 4.5 Decomposition of a system into single-input controllable subsystems.

The equation of subsystem S_i being of the form:

$$\dot{\tilde{x}}_i = A_{ii}\,\tilde{x}_i + \left[A_{i,i+1}\,\tilde{x}_{i+1} + \dots + A_{ir}\,\tilde{x}_r\right] + {}^{i}B\,u_i + \beta_i\,u^* \quad (4.28)$$

with the term in brackets representing the interaction of subsystems $S_{i+1}, \dots S_r$ on S_i, we have seen that an appropriate transformation matrix M

$$\tilde{A} = M^{-1} AM \qquad \tilde{B} = M^{-1} B$$

allows us to put $\tilde{A}$ and $\tilde{B}$ in the forms:

$$\tilde{A} = \left[\begin{array}{c|c|c} A_{rr} & 0 & 0 \\ \hline \square & A_{22} & 0 \\ \hline \square & \square & A_{11} \end{array}\right] \quad (4.29) \qquad \tilde{B} = \left[\begin{array}{c|c|c|c} 0 & 0 & {}^{r}B & \times \\ \hline 0 & {}^{2}B & 0 & \times \\ \hline {}^{1}B & 0 & 0 & \times \end{array}\right] \quad (4.30)$$

where the matrices A_{ii} and ${}^{i}B$ have the forms

$$A_{ii} = \begin{bmatrix} 0 & 1 & \cdots\cdots & 0 \\ 0 & 0 & 1 \cdots & 0 \\ \vdots & & & \vdots \\ \vdots & & & 1 \\ -a_0^i & & & -a_{n_i-1}^i \end{bmatrix} \quad (4.31) \qquad {}^{i}B = \begin{bmatrix} 0 \\ 0 \\ \vdots \\ 1 \end{bmatrix} \quad (4.32)$$

and where the vertical rectangles of formula (4.29) represent non-null column vectors.

The block-triangular form of the matrix $\tilde{A}$, together with the structure of the A_{ii}, enables us to write the complete characteristic polynomial directly in the form

$$\psi(s) = \prod_{i=1}^{r} \psi_i(s), \text{ with } \psi_i(s) = s^{n_i} + a^i_{n_i-1} s^{n_i-1} + \ldots + a^i_0 . \quad (4.33)$$

We have also seen that the decomposition into subsystems is not unique. It will depend on the order in which the columns of B are taken during the construction of the basis. If, in fact, the controllability indices corresponding to the ${}^{i}B$ are different, a change in the ordering of the columns will lead to a modification of the number and order of the blocks A_{ii} appearing in $\tilde{A}$. In particular, it can happen that the above structural decomposition leads to the number of systems being equal to m. In this case, u^* no longer exists and each column of $\tilde{B}$ contains only one nonzero element, which is unity.

b - Construction of the compensator

The problem is, we recall, that, given the system

$$\dot{x} = Ax + Bu$$

with characteristic equation

$$\psi(s) = s^n + a_{n-1}s^{n-1} + \ldots + a_0 = 0 \quad (4.34)$$

we have to find a feedback compensator K, such that the compensated system, with evolution matrix A + BK, should have the characteristic equation

$$\phi(s) = s^n + \alpha_{n-1}s^{n-1} + \ldots + \alpha_0 = 0 \quad (4.35)$$

where the α_i are specified arbitrarily.

Making use of the decomposition set up in paragraph a), we see that if the desired characteristic polynomial ϕ is decomposed into factors

$$\phi(s) = \prod_{i=1}^{r} \phi_i(s)$$

with the degrees of ϕ_i and ψ_i being equal, a simple procedure can be envisaged. It suffices indeed, taking account of the companion form of the blocks A_{ii} and the structure of $\tilde{B}$, to modify the modes of each subsystem by taking for the ith row of $\tilde{K}$:

$$\left[0 \ldots 0 \quad \left[a_0^i - \alpha_0^i \ldots a_{n_i-1}^i - \alpha_{n_i-1}^1\right] \quad 0 \ldots 0\right] \qquad (4.36)$$

the last m − r rows being null. We have in fact

$\tilde{B}\tilde{K} =$

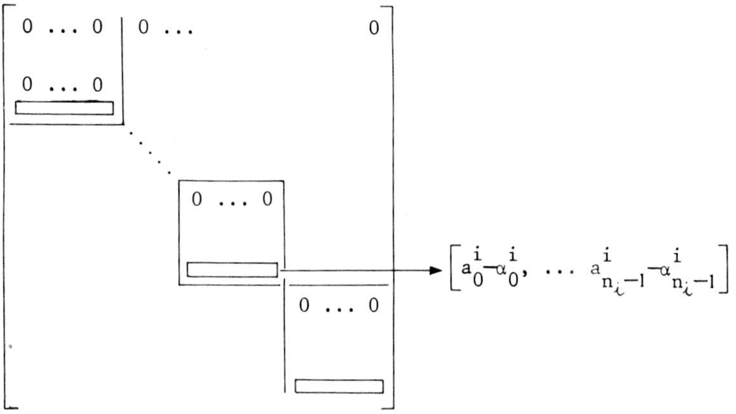

and

$\tilde{A} + \tilde{B}\tilde{K} =$

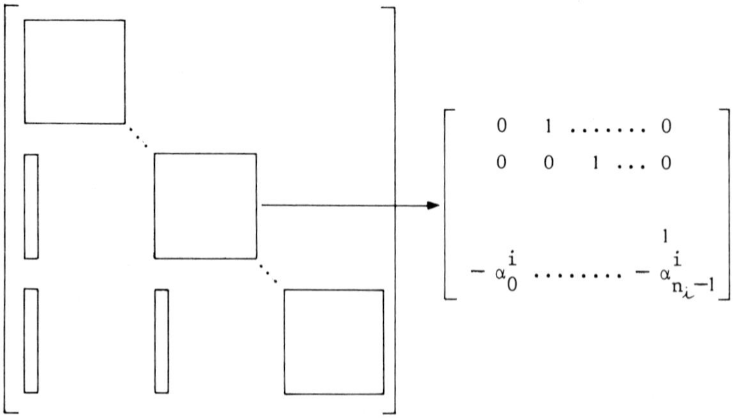

The compensator, in terms of the original state variables, will be obtained by the inverse transformation $\tilde{K} = KM$, $K = \tilde{K}M^{-1}$.

Remark

We see that, in the general case, a number of compensators K will thus be possible. We saw in the preceding paragraph that a first choice lies in the non-uniqueness of the decomposition into controllable subsystems. A second choice resides in the distribution of the desired modes among the various polynomials $\phi_i(s)$ associated with the subsystems, as the following example shows.

c - Example

Let the system be

$$\dot{x} = \begin{bmatrix} -1 & & \\ & -2 & \\ & & 3 \end{bmatrix} x + \begin{bmatrix} 1 & 1 \\ -1 & 0 \\ -1 & 1 \end{bmatrix} u, \qquad y = \begin{bmatrix} 1 & 1 & 0 \\ 1 & 0 & 1 \end{bmatrix} x .$$

Suppose we wish to compensate it in such a way that the closed-loop system possesses the modes $s = -2$, $s = -3$ and $s = -5$.

Two decompositions are possible, the system being completely controllable by u_1 and partially controllable by u_2.

If then we take the columns of B in the order ${}^1B$, ${}^2B$, only one subspace will be generated:

$$\tilde{A} = \begin{bmatrix} 0 & 1 & 0 \\ 0 & 0 & 1 \\ 6 & 7 & 0 \end{bmatrix}, \quad \tilde{B} = \begin{bmatrix} 0 & -0{,}3 \\ 0 & 0{,}1 \\ 1 & -0{,}7 \end{bmatrix}, \quad M = \begin{bmatrix} -6 & -1 & 1 \\ 3 & 2 & -1 \\ -2 & -3 & -1 \end{bmatrix}$$

we have then

$$\phi(s) = s^3 + 10s^2 + 31s + 30$$

In accordance with the theory, we take $\tilde{K}$ to have a null row:

$$\tilde{K} = \begin{bmatrix} -36 & -38 & -10 \\ 0 & 0 & 0 \end{bmatrix} \qquad K = \tilde{K}M^{-1} = \begin{bmatrix} 2 & 0 & 12 \\ 0 & 0 & 0 \end{bmatrix} \qquad \text{(a)}$$

If, on the contrary, we take the columns of B in the order ^2B, ^1B, we shall cause two subsystems to appear:

$$\tilde{A} = \left[\begin{array}{c|cc} -2 & 0 & 0 \\ \hline -1{,}5 & 0 & 1 \\ -3{,}5 & 3 & 2 \end{array}\right] \qquad \tilde{B} = \left[\begin{array}{c|c} 1 & 0 \\ \hline 0 & 0 \\ 0 & 1 \end{array}\right] \qquad M = \begin{bmatrix} 1 & -3 & 1 \\ -1 & 0 & 0 \\ -1 & 1 & 1 \end{bmatrix}$$

We then have

$$\phi(s) = \phi_1(s)\,\phi_2(s).$$

Depending on how we allocate the modes to ϕ_1 and ϕ_2, we have the three possible solutions:

$$\phi_1 = s^2 + 8s + 15,\ \phi_2 = s+2,\quad \tilde{K} = \begin{bmatrix} 0 & 0 & 0 \\ 0 & -18 & -10 \end{bmatrix}, K = \begin{bmatrix} 0 & 0 & 0 \\ 2 & 14 & -12 \end{bmatrix}, \quad \text{(b)}$$

$$\phi_1 = s^2 + 7s + 10,\ \phi_2 = s+3,\quad \tilde{K} = \begin{bmatrix} -1 & 0 & 0 \\ 0 & -13 & -9 \end{bmatrix}, K = \begin{bmatrix} 0 & 1 & 0 \\ 1 & 11 & -10 \end{bmatrix}, \quad \text{(c)}$$

$$\phi_1 = s^2 + 5s + 6,\ \phi_2 = s+5,\quad \tilde{K} = \begin{bmatrix} -3 & 0 & 0 \\ 0 & -9 & -7 \end{bmatrix}, K = \begin{bmatrix} 0 & 3 & 0 \\ 0.5 & 8 & -7.5 \end{bmatrix}. \quad \text{(d)}$$

Remarks

1 - It will be important to make the best use of these various possibilities in order to determine the compensator which is easiest to realise in practice. We can, e.g., choose the one which allows us to assign the smallest coefficients to those states which are hardest to reconstruct or most affected by noise (cf. subsection 4.3.6). Thus, in the above example, if x_2 is unmeasurable or very "noisy", there will be an advantage in choosing solution a) which allows us to dispense with x_2. On the other hand, solution d) will be the one to keep if the difficult variable is x_3.

2 - We may also remark that, as a general rule, in the case where there is no valid reason to favour one state rather than another, it is advantageous to decompose the system into the largest possible number of subsystems. Such a precaution will usually enable us to reduce the magnitude of the gains in the feedback matrix.

Figure 4.6 also shows that decomposition into a number of subsystems much smaller than the number of inputs leads to the exercise of adequate control only over this small number of inputs, thereby even losing many of the possibilities of the system.

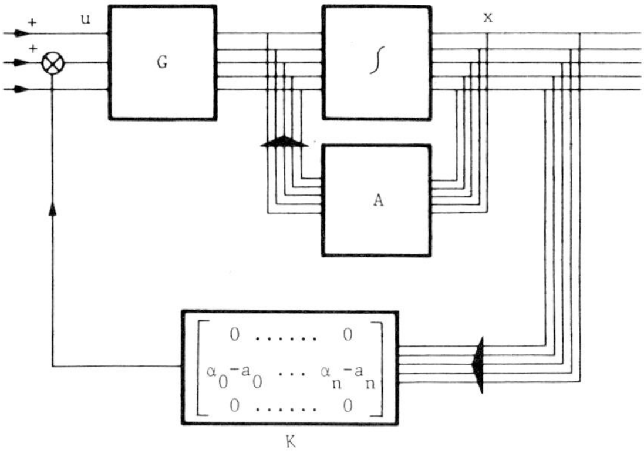

Fig. 4.6 The case of decomposition into a number of subsystems less than the number of inputs.

<u>3 - Remarks on the factorisation of the desired characteristic equation.</u> We have implicitly assumed above, that we can assign the modes of the desired characteristic equation to the various subsystems of dimensions n_i. In fact, if we are limited to real compensators, some difficulties can arise. Let us consider, by way of example, the case of a system of dimension 4, which can only be decomposed into two subsystems, of respective dimensions 1 and 3, and suppose that the desired characteristic polynomial is the product of two damped quadratic factors. For example:

$$\tilde{A} = \left[\begin{array}{c|ccc} 0 & 0 & 0 & 0 \\ \hline 0 & 0 & 1 & 0 \\ 1 & 0 & 0 & 1 \\ 0 & 1 & 1 & 1 \end{array}\right] \qquad \tilde{B} = \left[\begin{array}{cc} 0 & 1 \\ \hline 0 & 0 \\ 0 & 0 \\ 1 & 0 \end{array}\right]$$

$$\phi(s) = (s^2 + 1.4s + 1)(s^2 + 3s + 4)$$

The above method would lead to compensators with complex coefficients (!). This difficulty can be overcome in the following manner, pointed out by Wonham (cf. ref. 4.15).

We look for a feedback matrix $\tilde{K}$ in the form

$$\tilde{K} = \tilde{K}_1 + \tilde{K}_2 \tag{4.37}$$

where $\tilde{K}_1$ (which has the structure indicated previously) is subjected to the sole condition that $\tilde{A} + \tilde{B}\tilde{K}_1$ has distinct roots, which is always possible in view of the assumption of controllability. Since the n-dimensional space is cyclic with respect to this matrix, it is possible to find a vector $\tilde{b}$ such that $(\tilde{A} + \tilde{B}\tilde{K}_1, \tilde{b})$ is controllable and a vector g such that

$$\tilde{b} = \tilde{B}g \tag{4.38}$$

We can then find a control law of the form

$$u = \tilde{K}_1 \tilde{x} + g\alpha + W \tag{4.39}$$

and the closed-loop system is defined by

$$\dot{\tilde{x}} = (\tilde{A} + \tilde{B}\tilde{K}_1)\tilde{x} + \tilde{b}\alpha + \tilde{B}W$$

Taking $\alpha = H^T\tilde{x}$, we have

$$\dot{\tilde{x}} = (\tilde{A} + \tilde{B}\tilde{K}_1 + \tilde{b}H^T)\tilde{x} + \tilde{B}W$$

whence

$$\tilde{B}\tilde{K}_2 = \tilde{b}H^T = \tilde{B}gH^T$$

$$\tilde{K} = \tilde{K}_1 + gH^T \tag{4.40}$$

Let us apply this technique to the preceding example; we determine $\tilde{K}_1$ such that $\tilde{A} + \tilde{B}\tilde{K}_1$ has distinct eigenvalues, for example, s = 2, s = 1, $s^2 + s + 1 = 0$, so that

$$\tilde{K}_1 = \begin{bmatrix} 0 & 0 & -1 & -1 \\ 2 & 0 & 0 & 0 \end{bmatrix},$$

$\tilde{A} + \tilde{B}\tilde{K}_1$ is controllable from $\tilde{b} = [1, 0, 0, 0]^T$, whence

$$g = [0 \quad 1]^T .$$

If we take $H^T = [h_1, h_2, h_3, h_4]$, the characteristic equation of $\tilde{A} + \tilde{B}(\tilde{K}_1 + gH^T)$ becomes

$$s^4 - (2+h_1)s^3 - h_3s^2 - (1 + h_2)s + (2 + h_1 - h_4) = 0$$

which has the desired roots if

$$H^T = \begin{bmatrix} -6,4 & -9,6 & -9,2 & -8,4 \end{bmatrix},$$

whence

$$K = \tilde{K}_1 + gH^T = \begin{bmatrix} 0 & 0 & -1 & -1 \\ -4,4 & -9,6 & -9,2 & -8,4 \end{bmatrix}.$$

C - Decomposition into a number of subsystems equal to the number of inputs

1 - Structure:

It is important to notice that in the preceding paragraphs, we have sought only to achieve an overall compensation, involving the characteristic equation as a whole. Although this constitutes an important first step, we appreciate nevertheless that this approach is insufficient in practice, where we shall in general wish to specify:

- the modes affecting each output;
- the modes involved in the transmission from each input.

With regard to these objectives, the method of decomposition envisaged up to now lends itself badly to computation, for two reasons:

a - the existence in the matrix $\tilde{B}$ of m - r columns having no particular structural properties: this leads us to take the last m - r rows of $\tilde{K}$ to be null, and puts all the effort of compensation into only r channels;

b - the fact that the nonzero elements of the blocks A_{ij} of $\tilde{A}$ appear in columns, which considerably complicates the calculation of the input-output transfer functions:

$${}_iC(sI - \tilde{A} - \tilde{B}\tilde{K})^{-1}\tilde{B}.$$

Indeed, even in the case that r = m, if we write the matrix $\tilde{K}$ in the form

$$\tilde{K} = \begin{bmatrix} {}_m\tilde{k}_1 & \cdots\cdots & {}_1\tilde{k}_1 \\ \vdots & & \vdots \\ {}_m\tilde{k}_m & \cdots\cdots & {}_1\tilde{k}_m \end{bmatrix}$$

where ${}_i\tilde{k}_1$ is a row of n_i elements, the matrix $\tilde{A} + \tilde{B}\tilde{K}$ will have the following structure:

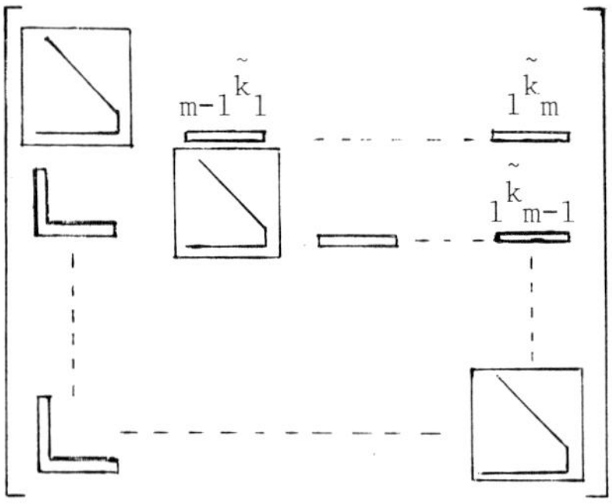

(4.41)

where the diagonal blocks are in companion form and the blocks below the diagonal have first columns corresponding to the non-null first columns of the A_{ij} and last rows corresponding to the ${}_i\tilde{k}_j$.

The inversion of such a matrix, even if

$${}_i\tilde{k}_j = 0 \text{ for } i < j$$

and

$${}_ia_i + {}_i\tilde{k}_i = {}_i\alpha_i$$

where ${}_ia_i$ denotes the last row of the uncompensated companion block A_{ii} and ${}_i\alpha_i$ is the corresponding row of the compensated block, is not too easy.

This difficulty will be resolved if we decompose the system into a number of subsystems systematically equal to the number m of inputs. We saw in chapter 3 that such a decomposition is possible, under the assumption of controllability, and that each subsystem is controlled by a linear combination of the input components.

More precisely, the matrices $\tilde{A}$ and $\tilde{B}$ will appear in the forms: (1)

$$\tilde{A} = \left[\begin{array}{c:c:c:c} A_{11} & \vdash\!\dashv & \vdash\!\dashv & \vdash\!\dashv \\ \hdashline \vdash\!\dashv & A_{22} & \vdash\!\dashv & \vdash\!\dashv \\ \hdashline {}_ia_1 & {}_ia_2 & A_{ii} & {}_ia_m \\ \hdashline \vdash\!\dashv & \vdash\!\dashv & \vdash\!\dashv & A_{mm} \end{array}\right] \qquad (4.42)$$

$$\tilde{B} = \left[\begin{array}{c:c:c:c} 0 & \vdots & & 0 \\ 0 & \vdots & & \vdots \\ \vdots & \vdots & & \vdots \\ 1 & 0 & & 0 \\ \hdashline 0 & 0 & & 0 \\ \vdots & \vdots & & \vdots \\ & 0 & & \\ \rho_{21} & 1 & & 0 \\ \hdashline 0 & 0 & & 0 \\ \vdots & \vdots & & \vdots \\ \rho_{i,i-1} & \cdots & 1 & 0 \\ \hdashline & 0 & & 0 \\ & \vdots & & \vdots \\ \rho_{m,m-1} & \vdots & \rho_{m,1} & 1 \end{array}\right] \qquad (4.43)$$

In comparison with the previous decomposition, the following differences will be noticed:

- 1) the matrix A is no longer block-triangular, which means that the characteristic equation of the system will not now be obtained so simply;

- 2) the matrices A_{ij}, for $i \neq j$, are null except for their last rows ${}_ia_j$ (whereas previously the non-nullity occurred in the first columns for $i > j$);

- 3) the last rows of each block of $\tilde{B}$ contain more than one non-zero element; on the other hand, all the columns of $\tilde{B}$ have the same structure.

We are trying to find in what form the compensator $\tilde{K}$ should be sought so that the characteristic polynomial may be arbitrarily chosen (of degree n) and we suppose, to begin with, that the ρ_{ik} are all zero.

The simplest possibility will be to end up with a matrix $\tilde{A} + \tilde{B}\tilde{K}$ in

(1) In expression (4.42) the symbol $\vdash\!\dashv$ represents the last row, which alone is non-null, of each matrix A_{ij} $(i \neq j)$. Note also that the ordering of the blocks in this decomposition is not the same as that used in the previous one.

block-triangular form, each diagonal block being in companion form: indeed, we then have the characteristic equation directly.

If we start with a matrix $\tilde{K}$ written in the form

$$\tilde{K} = \begin{bmatrix} {}_1\tilde{k}_1 & \cdots & {}_1\tilde{k}_m \\ {}_m\tilde{k}_1 & \cdots & {}_m\tilde{k}_m \end{bmatrix}$$

where ${}_i\tilde{k}_j$ is a row vector having n_j elements, we shall have the following structure:

$$\tilde{A} + \tilde{B}\tilde{K} = \left[\begin{array}{c:c:c:c} A^*_{11} & \underline{{}_1a^*_2} & \cdots & \underline{{}_1a^*_m} \\ \hdashline \underline{{}_2a^*_1} & A^*_{22} & \cdots & \underline{{}_2a^*_m} \\ \hdashline & & & \\ & & & \\ \hdashline \underline{{}_ma^*_1} & \underline{{}_ma^*_2} & \cdots & A^*_{mm} \end{array}\right] \qquad (4.44)$$

with

$${}_ka^*_j = {}_ka_j + \sum_{\nu=0}^{k-1} \rho_{k\nu}\, {}_{k-\nu}\tilde{k}_j \quad \text{and} \quad \rho_{k,0} = 1 . \qquad (4.45)$$

The system will have the desired characteristic equation if $\tilde{K}$ is chosen so that $\tilde{A} + \tilde{B}\tilde{K}$ appears in block-triangular form, i.e. if we specify that the blocks above the diagonal are null:

$${}_ka^*_j = 0 \quad \text{for } j > k \qquad (4.46)$$

and that

$${}_ja^*_j = {}_j\alpha_j = {}_ja_j + \sum_{\nu=0}^{j-1} \rho_{j\nu}\, {}_{j-\nu}\tilde{k}_j \qquad (4.47)$$

In doing this, we prescribe, for determining the nm coefficients of the matrix $\tilde{K}$, a number of relations

$$n + \sum_{k=1}^{m-1} (m - k)\, n_{m+1-k}$$

which is less than nm, for $m > 1$. The remaining degrees of freedom can be simply used to specify, to some extent, the form of the outputs.

2 - Shaping of the outputs:

A first possibility is to impose

$$ {}_k a_j^* = 0 \qquad \text{for } j \neq k $$

in place of equation (4.45). The matrix $\tilde{A} + \tilde{B}\tilde{K}$ will then appear in block-diagonal form, each block being the companion matrix of a polynomial D_i of degree n_i. In consequence of the form of $\tilde{B}$, we then have, by inversion,

$$ Y(s) = \tilde{C}(sI-\tilde{A}-\tilde{B}\tilde{K})^{-1}\tilde{B}\,E(s) = \tilde{C}\begin{bmatrix} {}^1\ell E_1/D_1 \\ \vdots \\ {}^m\ell E_m/D_m \end{bmatrix}, \quad {}^i\ell = \begin{bmatrix} 1 \\ s \\ \vdots \\ s^{n_i-1} \end{bmatrix} $$

whence

$$ y_i = \sum_{k=1}^{n} \frac{\tilde{C}_{ik}\,{}^k 1 u_k}{D_k} \tag{4.48} $$

with

$$ {}_i\tilde{C} = \begin{bmatrix} \tilde{C}_{i1} & \ldots & \tilde{C}_{im} \end{bmatrix}. $$

Thus, in the expression for each of the system outputs, every input component is associated with a set of given, arbitrarily fixed, modes, the same for each output.

In the general case, we have seen that the ρ_{ik} are not all zero. It is, however, easy to reduce this to the case considered above, by putting, upstream of the system, a precompensator R, of dimensions (m,m), such that

$$ \tilde{B}R = \tilde{B}^* $$

where $\tilde{B}^*$ has the previous form.

The presence of such a compensator, which is moreover triangular, does not affect the controllability of the system, seeing that the controllability matrix of $(\tilde{B}^*, \tilde{A})$ is equal to

$$ Q_c^* = \begin{bmatrix} \tilde{B} & \tilde{A}\tilde{B} & \ldots & \tilde{A}^{n-1} & \tilde{B} \end{bmatrix} \begin{bmatrix} R & & & \\ & R & & \\ & & \ddots & \\ & & & R \end{bmatrix} $$

and Gauss' condition

$$ n + nm - nm \leqslant \text{rank } Q_c^* \leqslant \text{minimum } (n, nm) $$

entails

$$\text{rank } Q_c^* = n.$$

The structure of the system is then as indicated in figures 4.7 and 4.8.

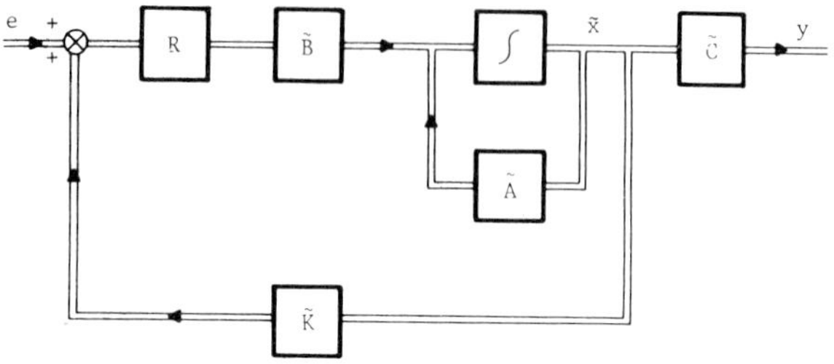

Fig. 4.7 Use of a precompensator

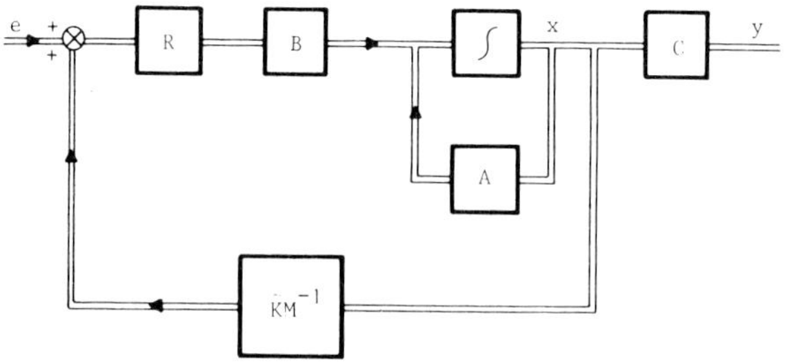

Fig. 4.8 General structure of a system with state feedback

3 - Example

Let the system be

$$\dot{\tilde{x}} = \begin{bmatrix} 0 & 1 & 0 \\ 0 & 0 & 1 \\ 2 & 1 & -3 \end{bmatrix} \tilde{x} + \begin{bmatrix} 0 & 0 \\ 1 & 0 \\ 2 & 1 \end{bmatrix} \tilde{u}, \qquad \tilde{y} = \begin{bmatrix} 1 & 2 & 3 \\ 1 & 1 & 2 \end{bmatrix} \tilde{x} .$$

We want the closed-loop system to have the characteristic equation

$$(s + 1)(s^2 + 3s + 4) = 0$$

the input e_2 affecting only the mode $s = -1$, while the input e_1 affects only the complex modes corresponding to $s^2 + 3s + 4$. We put $\tilde{B}$ into the form

$$\tilde{B}^* = \begin{bmatrix} 0 & 0 \\ 1 & 0 \\ 0 & 1 \end{bmatrix}$$

by taking

$$R = \begin{bmatrix} 1 & 0 \\ -2 & 1 \end{bmatrix}$$

The matrix $\tilde{K}$ is then defined by the equations (cf. equations 4.46, 4.47)

$$_1\tilde{k}_2 = -1$$

$$_2\tilde{k}_1 = [-2, -1]$$

$$[0,0] + {}_1\tilde{k}_1 = [-4, -3]$$

$$-3 + {}_2\tilde{k}_2 = -1$$

whence

$$\tilde{K} = \begin{bmatrix} -4 & -3 & -1 \\ -2 & -1 & 2 \end{bmatrix}$$

and

$$Y_1 = \frac{(1+2s)E_1}{s^2+3s+4} + \frac{3E_2}{s+1} \quad , \quad Y_2 = \frac{(s+1)E_1}{s^2+3s+4} + \frac{2E_2}{s+1}$$

4 - Use of a precompensator

The use of the precompensator R also enables us to impose supplementary constraints on the form of the outputs, as a way of trying to exploit the m(n + m) adjustable parameters.

Let us first determine $\tilde{K}$ so that $\tilde{A} + \tilde{B}\tilde{K}$ should simply be triangular (the nonzero subdiagonal elements then appear in rows and the calculation of input-output transfer functions is not appreciably altered). For greater simplicity of exposition, let us confine our attention to the case of a two-input system. We then have:

$$\tilde{A} = \left[\begin{array}{c|c} A_{11} & {}_1\underline{a}_2 \\ \hline {}_2\underline{a}_1 & A_{22} \end{array}\right] \qquad \tilde{B} = \left[\begin{array}{c|c} 0 & 0 \\ r_{11} & r_{12} \\ \hline 0 & 0 \\ \vdots & \vdots \\ r_{21} & r_{22} \end{array}\right] \qquad \tilde{K} = \left[\begin{array}{c|c} {}_1\tilde{k}_1 & {}_1\tilde{k}_2 \\ \hline {}_2\tilde{k}_1 & {}_2\tilde{k}_2 \end{array}\right]$$

to give $\tilde{A} + \tilde{B}\tilde{K}$ in the form

$$\tilde{A} + \tilde{B}\tilde{K} = \left[\begin{array}{c|c} A^*_{11} & 0 \\ \hline -{}_2\underline{m}_1 & A^*_{22} \end{array}\right], \text{ where } A^*_{ii} = \begin{bmatrix} 0 & 1 & 0 & \dots \\ 0 & 0 & 1 & \dots \\ & & & \ddots \quad 1 \\ & {}_i\underline{a}^*_i & & \end{bmatrix}$$

We have to satisfy the equations

$$\begin{aligned} {}_1a^*_1 &= {}_1a_1 + r_{11}\,{}_1\tilde{k}_1 + r_{12}\,{}_2\tilde{k}_1 \\ {}_2a^*_2 &= {}_2a_2 + r_{21}\,{}_1\tilde{k}_2 + r_{22}\,{}_2\tilde{k}_2 \\ 0 &= {}_1a_2 + r_{11}\,{}_1\tilde{k}_2 + r_{12}\,{}_2\tilde{k}_2 \\ -{}_2m_1 &= {}_2a_1 + r_{21}\,{}_1\tilde{k}_1 + r_{22}\,{}_2\tilde{k}_1 \end{aligned} \qquad (4.49)$$

With $\tilde{C} = [C(1),\ C(2)]$, we have the expressions for Y:

$$Y(s) = \left[C(1)\Big|^{n_1} \frac{r_{11}}{D_1} + \frac{r_{21}}{D_2}C(2)\Big|^{n_2} - \frac{r_{11}}{D_1 D_2}C(2)\Big|^{n_2} {}_2\underline{m}_1 \Big|^{n_1}\right] E_1 +$$

$$+ \left[\frac{r_{12}}{D_1}C(1)\Big|^{n_1} + \frac{r_{22}}{D_2}C(2)\Big|^{n_2} - \frac{r_{12}}{D_1 D_2}C(2)\Big|^{n_2} {}_2\underline{m}_1 \Big|^{n_1}\right] E_2$$

using the abbreviated notation

$$\Big|^{n_1} = \begin{bmatrix} 1 \\ s \\ \vdots \\ s^{n_i - 1} \end{bmatrix}$$

This form of Y can be analysed, to reveal a number of practically interesting cases.

a) One of the outputs y_i depends only on D_2,

$$Y_i = \frac{f(E_1, E_2)}{D_2}$$

This condition can occur:

i) - if $r_{11} = r_{12} = 0$, which in general will be incompatible with equations (4.49);

ii) - if $D_2\,{}_iC_1\big|^{n_1} - {}_iC_2\big|^{n_2}\,{}_2m_1\big|^{n_1}$ is divisible by D_2; since ${}_iC_2\big|^{n_2}\,{}_2m_1\big|^{n_1}$ is a polynomial of degree $n_1 + n_2 - 2 = n - 2$, this condition can arise if the degree of D_1 is $\leqslant n - 2$.

Example

$$\dot{\tilde{x}} = \begin{bmatrix} 0 & 1 & 0 \\ 0 & 0 & 1 \\ 2 & 1 & -3 \end{bmatrix}\tilde{x} + \begin{bmatrix} 0 & 0 \\ 1 & 0 \\ 2 & 1 \end{bmatrix} u \qquad y = \begin{bmatrix} 1 & 2 & 3 \\ 1 & 1 & 2 \end{bmatrix}\tilde{x} .$$

We desire the characteristic equation to be

$$D_1 D_2 = (s^2 + 6s + 4)(s + 1) = 0$$

and further that y_1 should respond like a first-order system with time-constant equal to 1. We have the equations

$$[-4 \quad -6] = [0 \quad 0] + r_{11}\ {}_1\tilde{k}_1 + r_{12}\ {}_2\tilde{k}_1$$

$$-1 = -3 + r_{21}\ {}_1\tilde{k}_2 + r_{22}\ {}_2\tilde{k}_2$$

$$1 + r_{11}\ {}_1\tilde{k}_2 + r_{12}\ {}_2\tilde{k}_2 = 0$$

$$-{}_2m_1 = [2 \quad 1] + r_{21}\ {}_1\tilde{k}_1 + r_{22}\ {}_2\tilde{k}_1$$

with

$$\tilde{K} = \left[\begin{array}{c|c} {}_1\tilde{k}_1 & {}_1\tilde{k}_2 \\ \hline {}_2\tilde{k}_1 & {}_2\tilde{k}_2 \end{array}\right] = \left[\begin{array}{cc|c} k_{11} & k_{12} & k_{13} \\ \hline k_{21} & k_{22} & k_{23} \end{array}\right] .$$

We also have to satisfy

$$r_{21}\ k_{12} + r_{22}\ k_{22} = 2$$

$$r_{21}\ k_{11} + r_{22}\ k_{21} = 1/3 .$$

We can take, for example,

$$r_{11} = r_{22} = 1 \qquad r_{12} = r_{21} = 0 \qquad K = \begin{bmatrix} -4 & -6 & -1 \\ 1/3 & 2 & 2 \end{bmatrix}$$

and

$$R = \begin{bmatrix} 1 & 0 \\ -2 & 1 \end{bmatrix}, \quad Y_1 = \frac{2E_1 + 3E_2}{s+1}, \quad Y_2 = \frac{(s^2+8s+17/3)E_1}{(s+1)(s^2+6s+4)} + \frac{2E_2}{s+1}$$

b) One of the outputs y_i depends only on D_1. This condition is satisfied if $\left[r_{21}\, D_1 - r_{11}\; {}_2m_1^{\eta_1}\right]_i C_2^{\eta_2}$ and $\left[r_{22}\, D_1 - r_{12}\; {}_2m_1^{\eta_1}\right]_i C_2^{\eta_2}$ are divisible by D_2. This implies

$$r_{21}/r_{11} = r_{22}/r_{12} = \rho \qquad r_{21}\; {}_1\tilde{k}_2 + r_{22}\; {}_2\tilde{k}_2 = {}_2a_2^{*} - {}_2a_2 = -\rho_1\, a_2$$

a condition which is in general impossible to satisfy. In fact, this impossibility arises only because of the structural form chosen (lower triangular).

Let us again take the previous example, and suppose that we want to find a compensator such that the overall characteristic polynomial should be $(s+1)(s^2+6s+4)$ and that

$$Y_1 = \frac{f(E_1, E_2)}{s^2+6s+4}$$

Since conditions b) cannot be satisfied, we return to case a) by rewriting the system in the form:

$$\dot{\xi} = \begin{bmatrix} -3 & 2 & 1 \\ 0 & 0 & 1 \\ 1 & 0 & 0 \end{bmatrix} \xi + \begin{bmatrix} 2 & 1 \\ 0 & 0 \\ 1 & 0 \end{bmatrix} u , \quad y = \begin{bmatrix} 3 & 2 & 1 \\ 2 & 1 & 1 \end{bmatrix} \xi$$

with

$$\tilde{x} = \begin{bmatrix} 0 & 1 & 0 \\ 0 & 0 & 1 \\ 1 & 0 & 0 \end{bmatrix} \xi$$

We then find immediately

$$Y_1 = \frac{3(s+3)E_1 + (2s+1)E_2}{s^2+6s+4} , \quad Y_2 = \frac{(s+1)E_1}{s^2+6s+4} + \frac{(2s^2+9s+5)E_2}{(s+1)(s^2+6s+4)}$$

$$K_\xi = \begin{bmatrix} 2 & -2 & -1 \\ -4 & -4 & -6 \end{bmatrix} , \quad R = \begin{bmatrix} 0 & 1 \\ 1 & -2 \end{bmatrix}$$

This amounts in fact to putting $\tilde{A} + \tilde{B}\tilde{K}$ in the form

$$\begin{bmatrix} A_{11} & \longmapsto \\ 0 & A_{22} \end{bmatrix} \quad \text{in place of} \quad \begin{bmatrix} A_{11} & 0 \\ \longmapsto & A_{22} \end{bmatrix} .$$

4.3 OBSERVER THEORY

4.3.1 Consequences of the non-measurability of all the states

We have shown, in the preceding paragraphs, that, given the assumption of controllability, it is possible, by state feedback techniques, to compensate a system and to specify the characteristic equation of the closed-loop system. In fact, the methods proposed so far remain somewhat theoeretical since they assume all the states of the system to be accessible and measurable, which will evidently not be the case in practice, where we shall have access only to the outputs (i.e. to some combinations of states) and possibly to some other state.

A priori, then, every inaccessible state entails a supplementary constraint on the structure of the feedback matrix, namely: the corresponding column of K must be null. If this is so, the possibilities of compensating the system are considerably limited.

Let us consider, e.g., the system

$$\dot{x} = \begin{bmatrix} -1 & & \\ & -2 & \\ & & 3 \end{bmatrix} x + \begin{bmatrix} 1 & 1 \\ -1 & 0 \\ -1 & 1 \end{bmatrix} u \qquad y = \begin{bmatrix} 1 & 1 & 1 \\ 0 & 1 & 0 \end{bmatrix} x$$

(which differs from the case considered in paragraph 4.2.3.B only in respect of the observation matrix), and suppose that we wish to achieve the characteristic equation

$$(s^2 - \beta_2 s - \beta_1)(s - \gamma) = 0$$

If we use a compensator of type b), we have

$$\tilde{K}_b = \begin{bmatrix} \gamma + 2 & 0 & 0 \\ 0 & \beta_1 - 3 & -2 + \beta_2 \end{bmatrix}$$

$$K_b = \frac{1}{20} \begin{bmatrix} 5(\gamma + 2) & 4(\gamma + 2) & \gamma + 2 \\ 5(\beta_1 - \beta_2 - 1), & 8(1 + \beta_1 - 2\beta_2), & -3(\beta_1 + 3\beta_2 - 9) \end{bmatrix} .$$

Let us suppose that, for practical reasons, x_1 and x_2 are accessible, and x_3 is not. We must then impose:

$$\gamma + 2 = 0 \qquad \beta_1 + 3\beta_2 - 9 = 0$$

and the characteristic equation obtained is of the form

$$(s + 2)(s^2 - \beta_2 s + 3\beta_2 - 9) = 0$$

which always corresponds to an unstable system.

It is thus important, if we want the above methods to remain useful, to free ourselves from these drawbacks. This can be achieved by two routes, which, as we shall see, are merely two different ways of interpreting the problem:

- either, by reconstructing the non-measurable states directly, through an appropriate dynamical system;
- or, by finding a way of transforming the state feedback into an output feedback.

In both cases, it is clear that we shall have to bring in the ideas of observability (through the observation matrix C, not considered up to now). We shall see that this modification expresses itself in both cases through an increase in the order of the system. The only difference is that while, in the first case, the extra modes created are unobservable (and will not appear in the expression for the overall transfer function matrix), they will remain observable in the second case.

The first solution will be considered in the following paragraphs, the second in section 4.4.

4.3.2 Theory of observers

The reconstruction of those states of a dynamical system which are not directly measurable could, a priori, be envisaged in two different ways:

A first method would consist of constructing a model of the system and utilising the states of the model in the control scheme, in place of the true system states. The validity of such a method depends on the establishment of a complete identification, sufficiently rapidly to follow the possible variations of the system states.

A second method, which would avoid the delays in the process of identification, consists of differentiating the outputs a certain number of times and thence reconstructing the state vector. It is clear that in this case, problems of physical realisation would make themselves sharply felt, and that, in general, all differentiation processes should be ruled out.

Between these two methods, both extreme from the dynamical point of view, we can imagine a certain compromise, i.e. a direct and sufficiently rapid means of reconstructing the states of the system, which does not suffer from the disadvantages inherent in the second solution.

It is from this viewpoint that the theory of observers, which will be presented in what follows, was conceived. (1)

A - Definitions and principle

By an observer is meant a dynamical system capable of reproducing the non-measurable states of a system S, starting from a knowledge of the inputs, the outputs and possibly the measurable states (cf. the scheme of figure 4.9).

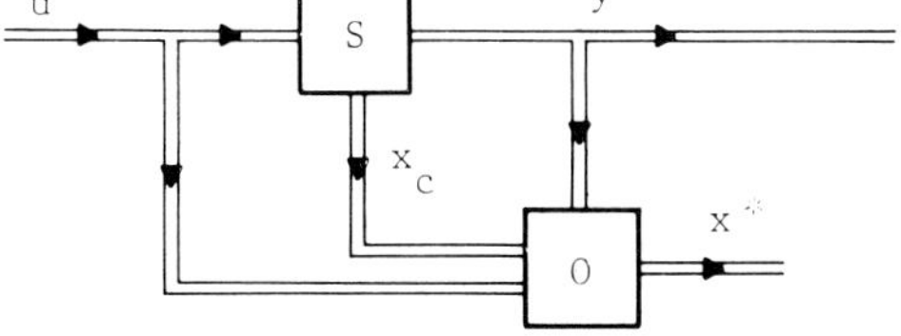

Fig. 4.9 Illustration of the principle of an observer

(1) The theory of observers was invented by D.G. Luenberger, and we may have recourse in particular to refs. (3.22, 3.23, 4.8, 4.9), to which this presentation owes a great deal.

If S is the linear process:

$$\dot{x} = Ax + Bu \qquad y = Cx \tag{4.50}$$

it is usual to try to represent the observer by another linear model, defined by:

$$\dot{z} = \alpha z + \beta_1' y + \beta_1'' x_c + \beta_2 u \qquad x^* = \gamma z + \delta_1' y + \delta_1'' x_c \tag{4.51}$$

where:

u and y represent the inputs and outputs of S,

x and z represent the state vectors of S and O,

x_c represents the measurable states of S (figure 4.10).

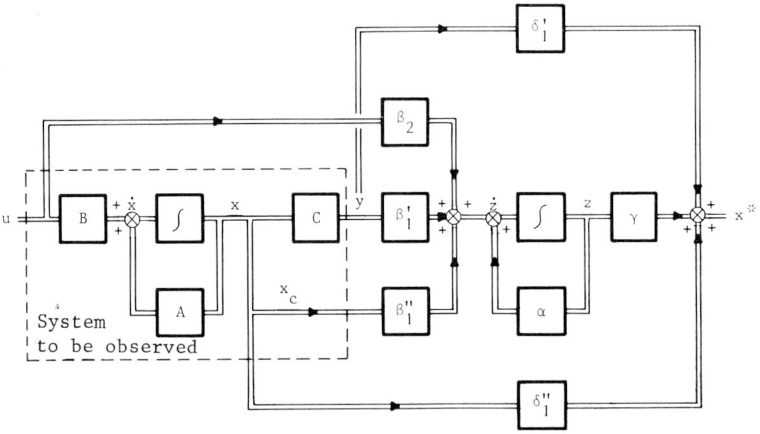

Fig. 4.10 Structure of an observer

In order to lighten the notation, we will suppose that the accessible states of $S(x_c)$ are included in the output y, taking, if necessary, $[y^T, x_c^T]^T$ as a new output vector. In the following, we shall assume that this has been done, and define the observer by:

$$\dot{z} = \alpha z + \beta_1 y + \beta_2 u \qquad x^* = \gamma z + \delta y . \tag{4.52}$$

B - Fundamental relations

Since both system and observer are assumed to be linear, so is the relation connecting x and z, and we can write it in the form

$$z = Tx + \varepsilon \ . \qquad (4.53)$$

For T to exist, it is nevertheless necessary, in view of (4.53), that the systems (4.50) and (4.52) should be compatible. Substituting equation (4.53) into (4.50) and (4.52), we have then:

$$(TA - \alpha T - \beta_1 C)\, x + (TB - \beta_2)\, u + \dot{\varepsilon} - \alpha\varepsilon \equiv 0 \qquad \forall u,\ \forall x$$

whence follow the conditions

$$\dot{\varepsilon} = \alpha\varepsilon \qquad (4.56)$$

$$TA - \alpha T = \beta_1 C \qquad (4.54)$$

$$TB = \beta_2 \qquad (4.55)$$

The last condition implies that

$$\varepsilon = z - Tx = e^{\alpha(t-t_0)}(z_0 - Tx_0) \qquad (4.57)$$

Put another way, there will be a linear relation between the states of the system and those of the observer if:

1) - the equation $TA - \alpha T = \beta_1 C$ has a solution, for arbitrary choice of the matrix β_1;

2) - the initial conditions on the system states, x_0, and on the observer states, z_0, satisfy:

$$z_0 = Tx_0 .$$

The possibility of satisfying these two conditions is examined below.

The first is satisfied if the matrices A and α have no common eigenvalue.

The second seems more difficult to satisfy. Accepting that T does indeed exist, it assumes that the initial conditions x_0 of the system can be measured in order to initialise the states of the observer, which is impossible. In fact, returning to equations (4.53) and (4.57), we see that if the observer dynamics (characterised by the eigenvalues of α) are more rapid than those of the system, the discrepancy between z and Tx diminishes quickly in the course of time and, after a certain transient period, can be considered negligible.

This remark does not completely solve the problem, since it raises that of the choice of observer dynamics, i.e. we have to answer the following questions:

Can we choose the observer dynamics arbitrarily, and under what conditions? It is clear, indeed, that, the more we try to give the observer fast dynamics, the more delicate its realisation is likely to be (the enlargement of the pass-band raising the problem of noise, both at the inputs and at the outputs of S).

How will the dynamics of the observer affect those of the complete system?

C - Existence conditions

Returning to equation (4.52), we have

$$x^* = \gamma z + \delta y = \gamma Tx + \delta Cx + \gamma\varepsilon$$
$$= \begin{bmatrix}\gamma & \delta\end{bmatrix}\begin{bmatrix}T\\C\end{bmatrix} x + \gamma\varepsilon = \begin{bmatrix}\gamma & \delta\end{bmatrix}\left[\begin{bmatrix}T\\C\end{bmatrix} x + \varepsilon^*\right]$$

with

$$\varepsilon^* = \begin{bmatrix}\varepsilon\\0\end{bmatrix} = \begin{bmatrix}e^{\alpha(t-t_0)}\,\varepsilon(t_0)\\0\end{bmatrix}$$

The eigenvalues of α should obviously be chosen to have negative real parts. In the steady-state régime, the relation z = Tx then gives:

$$x^* = (\gamma T + \delta C)x$$

so we need

$$[\gamma,\ \delta]\begin{bmatrix}T\\C\end{bmatrix} = I_n\ .$$

This will be possible only if:

$$\text{rank}\ [\gamma, \delta] \geqslant n, \qquad \text{rank} \begin{bmatrix} T \\ C \end{bmatrix} \geqslant n,$$

i.e. since the number of columns of T limits the rank of $\begin{bmatrix} T \\ C \end{bmatrix}$ to n,

$$\text{rank} \begin{bmatrix} T \\ C \end{bmatrix} = n.$$

It follows, in particular, that, in the most favourable case, with rank C = p, the minimum number of rows of T, i.e. the number of observer states, is n - p. (We recall that, in view of the assumptions made at the beginning, p is the number of outputs augmented by the number of states which may be directly measured).

D - Structure of the matrix T and its consequences

We have seen that the transformation matrix T is defined by relation (4.54). Successively multiplying both sides by A, and using the equation itself, we get:

$$TA^2 - \alpha^2 T = \beta_1 CA + \alpha\beta_1 C$$

$$\vdots$$

$$TA^n - \alpha^n T = \beta_1 CA^{n-1} + \alpha\beta_1 A^{n-2} + \ldots + \alpha^{n-1} \beta_1 C .$$

If these equations are multiplied by the scalars $a_1, \ldots, a_n$, where the a_i are the coefficients in the characteristic polynomial of A,

$$\Phi_0(s) = |sI-A| = \sum_{i=1}^{n} a_i s^i \quad (a_n=1)$$

and are added to each other and to the equation

$$TA^0 - \alpha^0 T = 0$$

multiplied by a_0, we obtain

$$T\Phi_0(A) - \Phi_0(\alpha) T = \sum_{i=1}^{n} \beta_1 C\Phi_i(A)$$

with

$$\Phi_i(s) = \frac{1}{s} \left[\Phi_{i-1}(s) - a_{i-1}\right] \quad (i = 1,\ldots n).$$

Bongiorno and Youla (cf. ref. 4.1) have shown that this equation can be solved in the form

$$T = \Phi_0^{-1}(\alpha)\ \Gamma_0\ \Omega\nabla$$

where $\nabla = [C^T, A^T C^T, \ldots (A^{n-1})^T C^T]^T$ is the observability matrix of the process, $\Gamma_0 = [\beta_1, \alpha\beta_1, \ldots \alpha^{n-1}\ \beta_1]$ contains the controllability matrix of the observer, and Ω is a "block-triangular" matrix whose blocks each have dimensions (p, p):

$$\Omega = \begin{bmatrix} a_1 Is & a_2 Is \cdots\cdots & a_n Is \\ a_2 Is & \cdot^{\cdot^{\cdot}} & 0 \\ \vdots & & \vdots \\ a_n Is & 0 \cdots\cdots & 0 \end{bmatrix}$$

It follows in particular that, for an observer to be compatible, i.e. capable of reconstructing all the states of the system, condition (4.58) implies the necessity of the observability of the process and the controllability of the observer. In particular: for single-output systems (p = 1), the necessary and sufficient condition for reconstruction of all the states, is that the process should be observable, and the observer controllable, from the sole output of the process.

4.3.3 Observers for single-output systems

Let us consider first the construction of an observer of order n for a single-output system.

A - Observer of order n

For the case of single-output systems (p = 1), let us try to determine T. The simplest solution would be to take $T = I_n$ in order to ensure identity (after the transient period) of the observer states with those of the system. Let us further suppose that, at the expense of a suitable basis change, defined by the transformation matrix M, the system equations are written in the form:

$$\dot{\tilde{x}} = \tilde{A}\tilde{x} + \tilde{B}u$$

$$y = \tilde{C}\tilde{x}$$

with

$$\tilde{C} = \begin{bmatrix} 1 & 0 & \cdots & 0 \end{bmatrix}$$

$$\tilde{A} = \begin{bmatrix} -a_{n-1} & 1 & 0 & \cdots & 0 \\ \vdots & & 1 & \cdots & 0 \\ \vdots & & & & \vdots \\ -a_1 & & & & 1 \\ -a_0 & 0 & \cdots & \cdots & 0 \end{bmatrix}$$

the characteristic equation of A being

$$s^n + a_{n-1}s^{n-1} + \ldots + a_1 s + a_0 = 0$$

Equation (4.54) then becomes:

$$\tilde{A} - \tilde{\alpha} = {}^1\tilde{\beta} \begin{bmatrix} 1 & \ldots & 0 \end{bmatrix} = \begin{bmatrix} \beta_{11} & 0 & \ldots & 0 \\ \vdots & & & \vdots \\ \beta_{n1} & 0 & \ldots & 0 \end{bmatrix}$$

the β_{j1} being the elements of the column vector ${}^1\tilde{\beta}$, and we deduce that

$$\alpha_{j1} = -(a_{n-j} + \beta_{j1})$$

$$\alpha_{jk} = \begin{cases} 1 & \text{for } j = k - 1 \\ 0 & \text{for } j \neq k - 1 \end{cases}$$

If these conditions are satisfied, $\tilde{x} = z$ and $x = Mz$.

A more elegant procedure consists of finding the matrix $\tilde{\alpha}$ directly in companion form

$$\tilde{\alpha} = \begin{bmatrix} -\alpha_{n-1} & 1 & \cdots & \cdots & 0 \\ \vdots & 0 & 1 & \cdots & \\ \vdots & & & & \\ -\alpha_1 & & & & 1 \\ -\alpha_0 & 0 & \cdots & \cdots & 0 \end{bmatrix}$$

and deducing the matrix ${}^1\tilde{\beta}$. It suffices indeed to write

$$\beta_{j1} = -a_{n-j} + \alpha_{n-j} \tag{4.59}$$

B - Construction of a minimal observer

Going back to condition (4.58), we appreciate that it is possible to take T of dimensions (n-1, n). Let us then see if we can satisfy equations (4.54) and (4.58), with a matrix T of n-1 rows, while specifying arbitrary dynamics for α.

Since the system is throughout assumed to be observable, let us write it in a basis where $\tilde{A}$ is in companion form and $\tilde{C}$ in the form $[1, 0, \ldots 0]$, and take $\tilde{\alpha}$ in the form

$$\tilde{\alpha} = \begin{bmatrix} -\alpha_{n-2} & 1 & 0 & 0 \ldots 0 \\ \vdots & & 1 & 0 \ldots \vdots \\ \vdots & & & \vdots \\ -\alpha_1 & & & 1 \\ -\alpha_0 & & & 0 \end{bmatrix}$$

$\tilde{\alpha}$ being in effect now only of dimensions (n-1, n-1). The equation

$$T\tilde{A} - \alpha T = {}^1\tilde{\beta}\tilde{C}$$

becomes

$$\begin{cases} -\sum_{k=1}^{n} t_{jk}\, a_{n-k} + \alpha_{n-1-j}\, t_{11} - t_{j+1,1} = +\beta_{j1} & j = 1 \text{ à } n-2 \\ -\sum_{k=1}^{n} t_{n-1,k}\, a_{n-k} + \alpha_0\, t_{11} = \beta_{n-1,1} & \end{cases} \qquad (4.60)$$

$$\begin{cases} t_{jk} = -\alpha_{n-1-j}\, t_{1,k+1} + t_{j+1,k+1} & j = 1, \ldots, n-2 \\ t_{n-1,k} = -\alpha_0\, t_{1,k+1} & \end{cases}$$

Since the β_{ij} are not fixed, we can choose the t_{ij} arbitrarily provided that the set (4.60) of n-1 equations, with n-1 unknowns, has a solution, and that the values chosen are consistent with (4.58). Among the infinity of possible choices, the most advantageous is, in our opinion, that recommended by Luenberger (1), which consists of taking T in the form

(1) cf. refs. (4.8, 4.9)

$$T = \begin{bmatrix} \omega_{n-2} & 1 & 0 & \dots & 0 \\ \vdots & & 1 & & \vdots \\ \omega_1 & & & & \vdots \\ \omega_0 & 1 & & & 1 \end{bmatrix} \tag{4.61}$$

The preceding set of equations can then be solved immediately. We have:

$$\begin{aligned} \omega_i &= -\alpha_i \\ \beta_{k1} &= \alpha_{n-k-1}(a_{n-1} - \alpha_{n-2}) - a_{n-k-1} + \alpha_{n-k-2} \\ \beta_{n-1,1} &= -\alpha_0(\alpha_{n-2} - a_{n-1}) - a_0 . \end{aligned} \tag{4.62}$$

It follows that, given a linear, single-output, observable system with n states, we can find an observer, of order n-1, whose dynamics can be arbitrarily chosen.

The expression for $\tilde{x}^*$ in terms of z and y can be obtained immediately from the equations

$$\tilde{x}^* = \gamma z + \delta y \qquad [\gamma \quad \delta]\begin{bmatrix} T \\ C \end{bmatrix} = I_n$$

whence

$$[\delta \quad \gamma] = \begin{bmatrix} C \\ T \end{bmatrix}^{-1} = \begin{bmatrix} 1 & 0 & & & 0 \\ -\alpha_{n-2} & 1 & & & \\ \vdots & 0 & 1 & 0 & \dots \\ \vdots & \vdots & & & \\ -\alpha_0 & & & & 1 \end{bmatrix}^{-1} = \begin{bmatrix} 1 & 0 & \dots\dots & 0 \\ \alpha_{n-2} & 1 & & \\ \vdots & & \ddots & \\ \alpha_0 & & & 1 \end{bmatrix}$$

so that

$$\gamma = \begin{bmatrix} 0 & 0 & \dots & 0 \\ 1 & 0 & & \\ & 1 & 0 & \\ 0 & & \ddots & 1 \end{bmatrix} \quad (4.63) \qquad \delta = \begin{bmatrix} 1 \\ \alpha_{n-2} \\ \vdots \\ \alpha_0 \end{bmatrix} \quad (4.64)$$

and hence

$$\begin{aligned} \tilde{x}_1^* &= y \\ \tilde{x}_2^* &= z_1 + \alpha_{n-2}\, y \\ &\vdots \\ \tilde{x}_n^* &= z_{n-1} + \alpha_0\, y \end{aligned}$$

We then find x^* from the transformation $x^* = M\tilde{x}^*$. The general scheme for the realisation of a minimal observer for a single-output system is illustrated in figure 4.11.

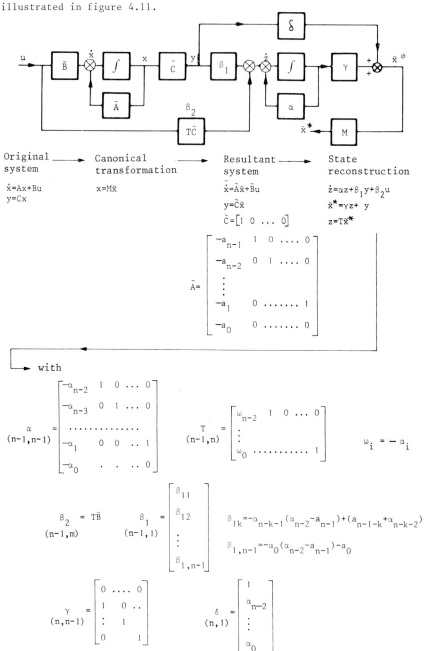

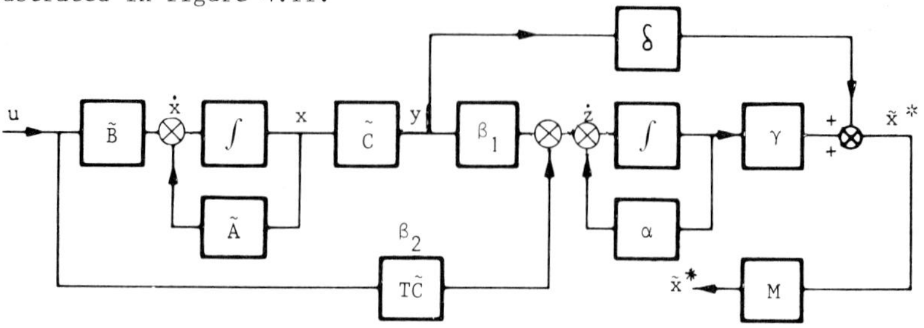

Original system → Canonical transformation → Resultant system → State reconstruction

Original system: $\dot{x}=Ax+Bu$, $y=Cx$

Canonical transformation: $x=M\tilde{x}$

Resultant system: $\dot{\tilde{x}}=\tilde{A}\tilde{x}+\tilde{B}u$, $y=\tilde{C}\tilde{x}$, $\tilde{C}=[1\ 0\ \dots\ 0]$

State reconstruction: $\dot{z}=\alpha z+\beta_1 y+\beta_2 u$, $\tilde{x}^*=\gamma z+\delta y$, $z=T\tilde{x}^*$

$$\tilde{A}=\begin{bmatrix} -a_{n-1} & 1 & 0 & \dots & 0 \\ -a_{n-2} & 0 & 1 & \dots & 0 \\ \vdots & & & & \\ -a_1 & 0 & \dots & \dots & 1 \\ -a_0 & 0 & \dots & \dots & 0 \end{bmatrix}$$

with

$$\underset{(n-1,n-1)}{\alpha}=\begin{bmatrix} -\alpha_{n-2} & 1 & 0 & \dots & 0 \\ -\alpha_{n-3} & 0 & 1 & \dots & 0 \\ \dots & & & & \\ -\alpha_1 & 0 & 0 & \dots & 1 \\ -\alpha_0 & \cdot & \cdot & \dots & 0 \end{bmatrix} \qquad \underset{(n-1,n)}{T}=\begin{bmatrix} \omega_{n-2} & 1 & 0 & \dots & 0 \\ \vdots & & & & \\ \omega_0 & \dots & & & 1 \end{bmatrix} \qquad \omega_i=-\alpha_i$$

$$\underset{(n-1,m)}{\beta_2}=T\tilde{B} \qquad \underset{(n-1,1)}{\beta_1}=\begin{bmatrix} \beta_{11} \\ \beta_{12} \\ \vdots \\ \beta_{1,n-1} \end{bmatrix}$$

$$\beta_{1k}=-\alpha_{n-k-1}(\alpha_{n-2}-a_{n-1})+(a_{n-1-k}+\alpha_{n-k-2})$$

$$\beta_{1,n-1}=-\alpha_0(\alpha_{n-2}-a_{n-1})-a_0$$

$$\underset{(n,n-1)}{\gamma}=\begin{bmatrix} 0 & \dots & 0 \\ 1 & 0 & \dots \\ \vdots & 1 & \\ 0 & & 1 \end{bmatrix} \qquad \underset{(n,1)}{\delta}=\begin{bmatrix} 1 \\ \alpha_{n-2} \\ \vdots \\ \alpha_0 \end{bmatrix}$$

Fig. 4.11 Procedure for the realisation of an observer

Example:

Let the system be defined by the equations

$$\dot{x} = \begin{bmatrix} 0 & 1 & 0 \\ 0 & 0 & 1 \\ -2 & -5 & -4 \end{bmatrix} x + \begin{bmatrix} 0 \\ 0 \\ 1 \end{bmatrix} u \qquad y = \begin{bmatrix} 1 & 0 & 2 \end{bmatrix} x \; .$$

Since the system is observable, we know that it is possible to find an observer of order 2, allowing us to reconstruct all the system states.

i) Transformation to canonical form: the transformation with $x = M\tilde{x}$

with
$$M = \frac{1}{9}\begin{bmatrix} -1 & 0 & 2 \\ 0 & -1 & 0 \\ 5 & 0 & -1 \end{bmatrix}$$

puts the system in the required form

$$\dot{\tilde{x}} = \begin{bmatrix} -4 & 1 & 0 \\ -5 & 0 & 1 \\ -2 & 0 & 0 \end{bmatrix} \tilde{x} + \begin{bmatrix} 2 \\ 0 \\ 1 \end{bmatrix} u \, , \qquad y = \begin{bmatrix} 1 & 0 & 0 \end{bmatrix} \tilde{x} \; .$$

ii) The eigenvalues of the system are -1, -1, -2. Suppose we require the observer dynamics to have a double pole at s = -3,

$$(s + 3)^2 = s^2 + 6s + 9 .$$

We have then

$$\alpha = \begin{bmatrix} -6 & 1 \\ -9 & 0 \end{bmatrix} \qquad T = \begin{bmatrix} -6 & 1 & 0 \\ -9 & 0 & 1 \end{bmatrix}$$

and we deduce

$$\gamma = \begin{bmatrix} 0 & 0 \\ 1 & 0 \\ 0 & 1 \end{bmatrix} \qquad \delta = \begin{bmatrix} 1 \\ 6 \\ 9 \end{bmatrix}$$

$$\beta_1 = \begin{bmatrix} -8 \\ -20 \end{bmatrix} \qquad \beta_2 = \begin{bmatrix} -12 \\ -17 \end{bmatrix}$$

whence we get the scheme of figure 4.12, where we explicitly display

$$\gamma' = M\gamma = \frac{1}{9}\begin{bmatrix} 0 & 2 \\ -1 & 0 \\ 0 & -1 \end{bmatrix} \qquad \delta' = M\delta = \frac{1}{9}\begin{bmatrix} 17 \\ -6 \\ -4 \end{bmatrix}.$$

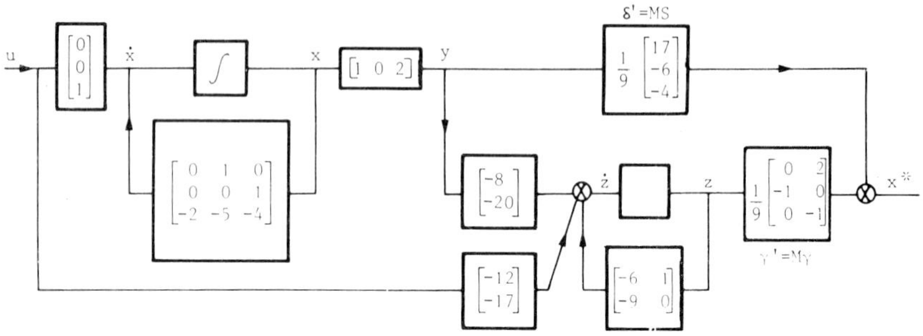

Fig. 4.12 Example of the construction of an observer for a single-output system.

4.3.4 Minimal observers for multi-output systems

In the general case where the output vector is of dimension p, the problem is more complicated. We nevertheless expect, in view of the results obtained in the single-output case, to be able to find an observer of order $n - p$. The purpose of this subsection is to show that this is actually possible if the system is observable, and to give a simple method for constructing an observer of minimal degree and arbitrary dynamics.

The realisation of a minimal observer (of order $n - p$) depends on the facts that:

- for a single-output system, of order n_i, the minimal observer is of order $n_i - 1$;

- if a multivariable system, with p outputs, is observable, it is possible to decompose it into p single-output observable subsystems S_i, according to the scheme of figure 4.13, corresponding to the equations

$$S_i : \begin{cases} \dot{\tilde{x}}_i = A_{ii}\,\tilde{x}_i + \phi(i)\,y + \tilde{B}(i)\,u \\ r_i = C_{ii}\,\tilde{x}_i \\ y = \theta r \qquad\qquad r = \begin{bmatrix} r_1 & r_2 & \dots & r_p \end{bmatrix}^T \end{cases} \tag{4.65}$$

(we refer to chapter 3 for all details concerning this transformation).

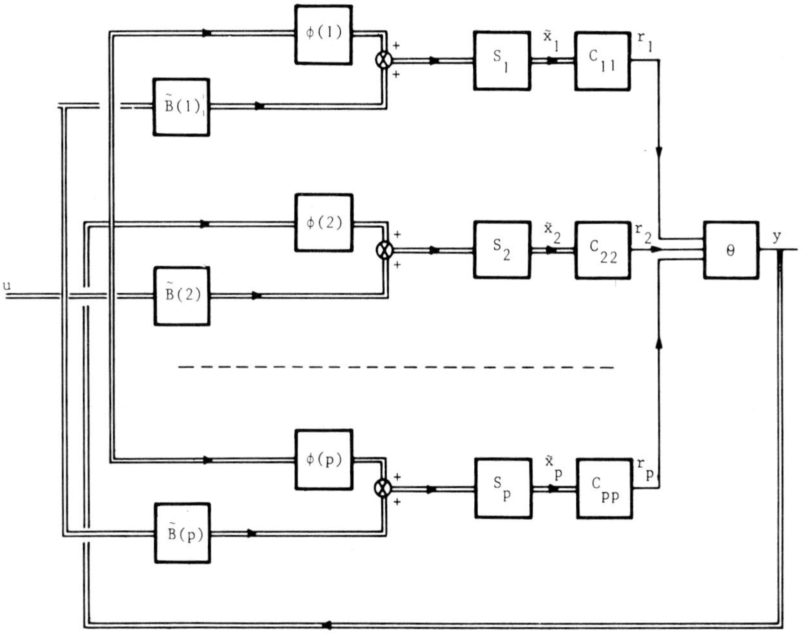

Fig. 4.13 Decomposition of a system into p observable subsystems

At the level of each single-output subsystem S_i, of order n_i, the reconstruction of the vector $\tilde{x}_i$ can be achieved, in view of the preceding subsection, by an observer of order n_i-1, defined by

$$\dot{z}(i) = \alpha(i)\,z(i) + \beta_1(i)\,r_i + \beta_2(i)\left[\phi(i)\,y + \tilde{B}(i)\,u\right]$$

$$\hat{x}^*(i) = \gamma(i)\,z(i) + \delta(i)\,r_i$$

where $\phi(i)y + \tilde{B}(i)u$ constitutes the input to S_i, and the matrices $\alpha(i)$, $\beta_1(i)$, $\beta_2(i)$, $\gamma(i)$, $\delta(i)$ are defined as previously indicated; the reconstruction of the complete state-vector $x = [\tilde{x}_1^T, \tilde{x}_2^T, \ldots \tilde{x}_p^T]^T$ can then be achieved by p observers of orders $n_i - 1$, i.e. by an observer of order

$$\sum_{i=1}^{p} (n_i - 1) = n - p .$$

The principle of this realisation is diagramatically illustrated in figure 4.14, where only one subsystem is considered.

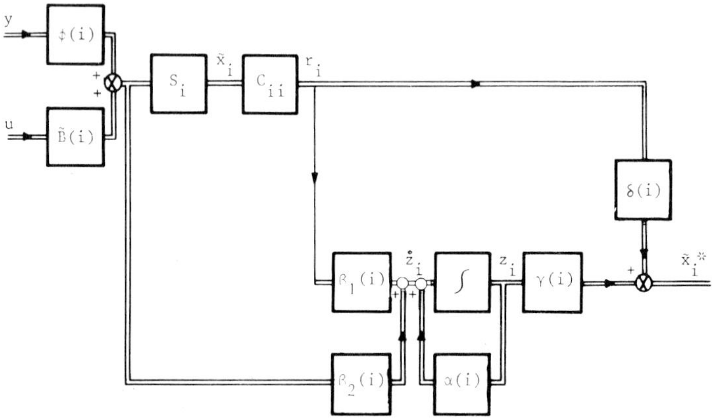

Fig. 4.14 Realisation of an observer for a decomposed multi-output system.

In the decomposition into subsystems which has been made, the matrix θ, which converts r into y, is always invertible. It will thus be possible, by applying an observation matrix θ^{-1} to the outputs, to have direct access to the components r_i. This enables us to write the elementary observer in the form indicated in figure 4.15, where z(i) is evaluated directly from the inputs u and outputs y of the system being observed.

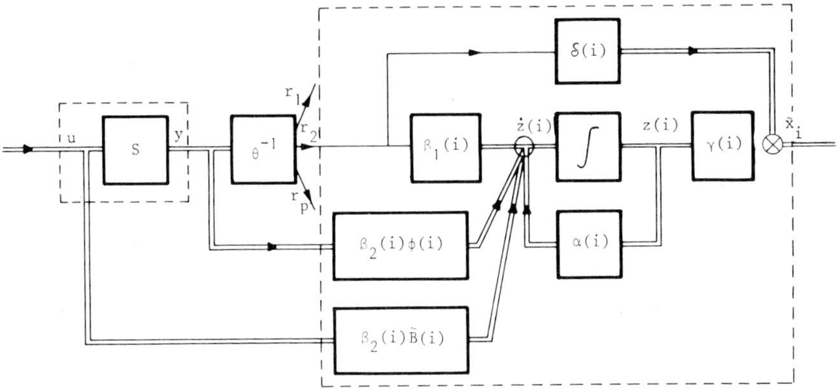

Fig. 4.15 Observer acting directly on the inputs and outputs of the observed system.

<u>Example</u>:

Let the system be

$$\dot{x} = Ax + Bu$$

$$y = Cx$$

with

$$A = \begin{bmatrix} -1 & & \\ & -2 & \\ & & 3 \end{bmatrix} \quad B = \begin{bmatrix} 1 & 1 \\ -1 & 0 \\ -1 & 1 \end{bmatrix} \quad C = \begin{bmatrix} 1 & 1 & 0 \\ 1 & 0 & 1 \end{bmatrix}.$$

We have seen that the transformation matrix

$$M = \begin{bmatrix} -1 & 1 & 0 \\ 2 & -1 & 0 \\ -3 & -1 & 1 \end{bmatrix}$$

enables us to decompose the system into two subsystems: a first subsystem of degree 2,

$$A_{11} = \begin{bmatrix} -3 & 1 \\ -2 & 0 \end{bmatrix} \quad \phi(1) = \begin{bmatrix} 0 & 0 \\ 0 & 0 \end{bmatrix} \quad \tilde{B}(1) = \begin{bmatrix} 0 & 1 \\ 1 & 2 \end{bmatrix}$$

and a second subsystem of degree 1,

$$A_{22} = 3 \qquad \phi(2) = \begin{bmatrix} -20 & 0 \end{bmatrix}, \quad \tilde{B}(2) = \begin{bmatrix} 0 & 6 \end{bmatrix}$$

while the matrix connecting y and r is

$$\theta = \begin{bmatrix} 1 & 0 \\ -4 & 1 \end{bmatrix}.$$

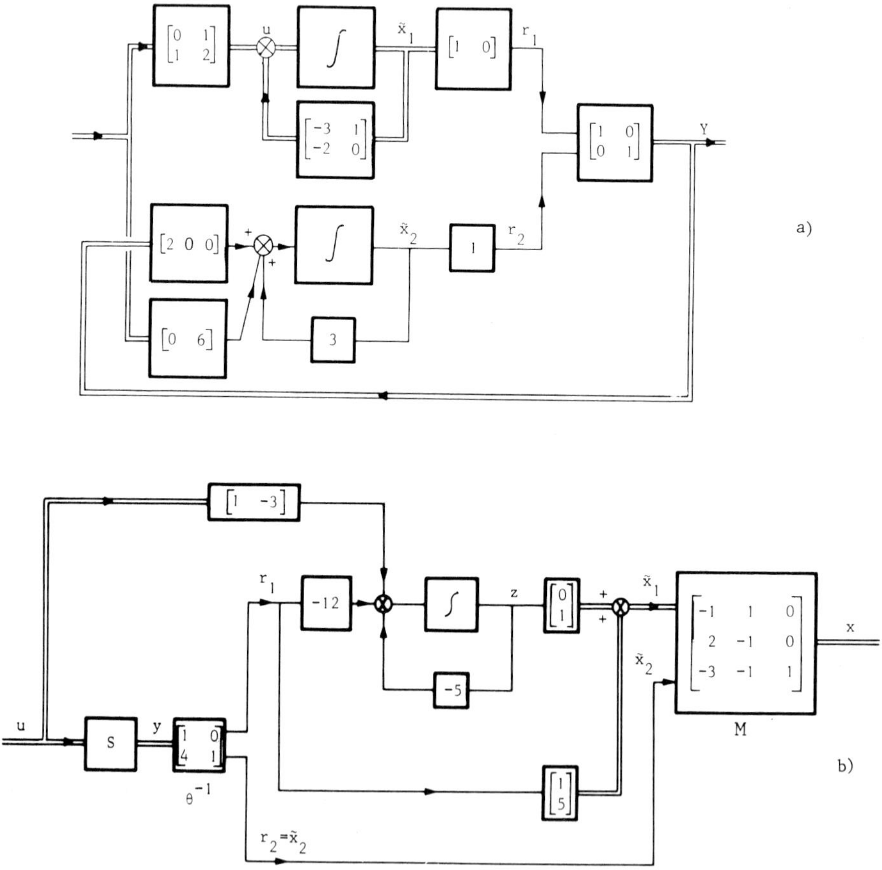

Fig. 4.16 Example of construction of an observer for a multi-output system;

(a) decomposition,
(b) realisation

In this form, we see that, in order to reconstruct the states $\tilde{x}$, it will be sufficient to apply an observer of degree 1 to subsystem 1. Since the dynamics of the observed system are characterised by the modes s = -1, s = -2, we will take for the observer dynamics s = -5; this will cause the transients due to the initial conditions to die away sufficiently fast. Under these conditions, equations (4.61) to (4.64) give

$$\alpha = [-5] \qquad \beta(1) = -12 \qquad \beta(2) = [-5 \quad 1]$$

$$T = [-5 \quad 1] \qquad \beta(2)\,\tilde{B}(1) = [1 \quad -3]$$

$$\gamma = \begin{bmatrix} 0 \\ 1 \end{bmatrix} \qquad \delta = \begin{bmatrix} 1 \\ 5 \end{bmatrix} \qquad \beta_2\,\phi(1) = [0 \quad 0]$$

The reconstruction of all the states can be made as indicated in figure 4.16b.

4.3.5 Case where the system is unobservable

In the above paragraphs, we made the fundamental assumption that the system whose state we wished to reconstruct was completely observable. In the case that this condition is not satisfied, it is quite clear, from the very definition of observability, that observation of the outputs will not enable us to reconstruct the entire state vector.

By way of introduction, let us consider the very simple case of a single-variable unobservable system defined by the equations

$$\dot{x} = \begin{bmatrix} -1 & 0 \\ 0 & -2 \end{bmatrix} x + \begin{bmatrix} 1 \\ 3 \end{bmatrix} u \qquad y = [1 \quad 0]\, x$$

where the state x_2, associated with the mode s = -2, is unobservable. Let us try to reconstruct the states x_1, x_2, by an observer of order 1, with eigenvalue α. We have then

$$z = [t_1 \quad t_2] \begin{bmatrix} x_1 \\ x_2 \end{bmatrix}$$

and the condition $TA - \alpha T = \beta_1 C$ gives

$$[-t_1 \quad -2\,t_2] - \alpha [t_1 \quad t_2] = [\beta_1 \quad 0],$$

so that

$$t_1(1+\alpha) = -\beta_1$$
$$t_2(2+\alpha) = 0$$

A priori, the problem can be solved by

$$t_2 = 0, \quad \forall\alpha$$

or

$$\alpha = -2, \quad \forall t_2 .$$

In the first case,

$$\text{rank}\begin{bmatrix} T \\ C \end{bmatrix} = \text{rank}\begin{bmatrix} t_1 & 0 \\ 1 & 0 \end{bmatrix} = 1 < n.$$

This case must therefore be rejected. In the second case, we have

$$\beta_1 = t_1 \qquad \beta_2 = \begin{bmatrix} t_1 & t_2 \end{bmatrix}\begin{bmatrix} 1 \\ 3 \end{bmatrix} = t_1 + 3\,t_2$$
$$\gamma = \begin{bmatrix} 0 \\ 1/t_2 \end{bmatrix} \qquad \delta = \begin{bmatrix} 1 \\ -\,t_1/t_2 \end{bmatrix}$$

If we take t_1 and t_2 to be connected by the relation $t_1 + 3t_2 = 0$, we have

$$t_1 = \theta_1 \qquad t_2 = -\,\theta_1/3 ,$$

whence

$$\dot z = -\,2\,z + \theta_1\,y$$
$$x_1 = y$$
$$x_2 = -\,\frac{3}{\theta_1}\,z + 3\,y .$$

The realisation is shown in figure 4.17. We note that the observer dynamics are no longer arbitrary, but are imposed by the unobservable modes of the system.

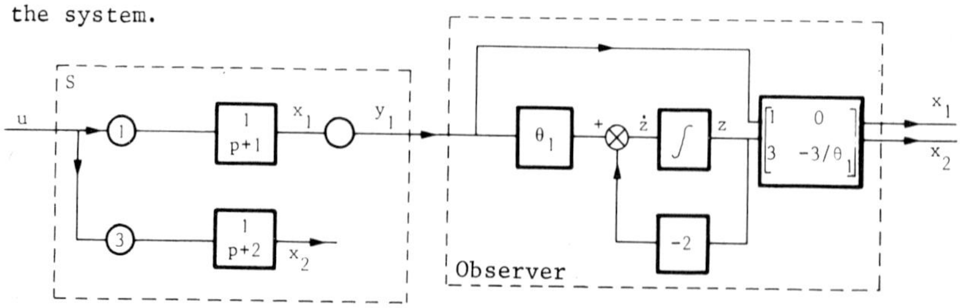

Fig. 4.17 Observer for an unobservable system

Let us now consider the general case. Without loss of generality, we can suppose the equations of the system S to be written in a basis where A appears in Jordan form (which facilitates the application of observability criteria), and let us assume for simplicity that only one Jordan block is associated with each eigenvalue λ_i. Let

$$\dot{x} = \begin{bmatrix} A_{\lambda_1 n_1} & & & \\ & A_{\lambda_2 n_2} & & \\ & & \ddots & \\ & & & A_{\lambda_r n_r} \end{bmatrix} x + Bu$$

$$y = \left[C(1) \;\ldots\ldots\ldots\; C(r)\right] x$$

be the equations of the system, where $A_{\lambda_i n_i}$ represents a Jordan block of order n_i associated with the eigenvalue λ_i, and C(i) is a matrix of dimensions (p, n_i) associated with this block. The equation

$$TA - \alpha T = \beta_1 C$$

can be written out in the form:

$$\left[T(1) \;\; T(2) \;\ldots\; T(r)\right] \begin{bmatrix} A_{\lambda_1 n_1} & & \\ & \ddots & \\ & & A_{\lambda_r n_r} \end{bmatrix} = \alpha\left[T(1) \;\ldots\; T(r)\right] + \beta_1\left[C(1) \;\ldots\; C(r)\right]$$

so that

$$T(i)\, A_{\lambda_i n_i} = \alpha T(i) + \beta_1 C(i)$$

If a state associated with the eigenvalue λ_i is unobservable, we know that the first column of the corresponding block C(i) is null.

The last equation, written out in terms of the columns of C(i), gives

$$\left[{}^{1}T(i) \;\ldots\; {}^{n_i}T(i)\right] \begin{bmatrix} \lambda_i & 1 & & & \\ & \lambda_i & 1 & & \\ & & \ddots & \ddots & \\ & & & \lambda_i & 1 \\ & & & & \lambda_i \end{bmatrix} = \alpha\left[{}^{1}T(i) \;\ldots\; {}^{n_i}T(i)\right] + \beta_1\left[{}^{1}C(i) \;\ldots\; {}^{n_i}C(i)\right]$$

so that

$$^{1}T(i)\ \lambda_i = \alpha\ ^{1}T(i) + \beta_1\ ^{1}C(i)$$

$$^{k-1}T(i) + \lambda_i\ ^{k}T(i) = \alpha\ ^{k}T(i) + \beta_1\ ^{k}C(i) \qquad k = 2, \ldots, n_i\ .$$

Taking account of the fact that $^{1}C(i)$ is null (the unobservability assumption), it is clear that we shall have to take: λ_i = eigenvalue of α.

If, associated with an eigenvalue λ_i, there are k_i unobservable states, it is clear that λ_i, as an eigenvalue of α, has multiplicity k_i. (1) Consequently, in the general case of an unobservable system, we shall only be able to choose n - p - q of the observer modes, where q is the number of unobservable states of the system.

4.3.6 Introduction of the observer into the control loop

Suppose that, instead of using the true states x of the system, we implement feedback control by means of the states x^* reconstructed by an appropriate observer. Let us treat this situation by using, first, state equations, and then, transfer function matrices. The system equations can now be written (cf. figure 4.18)

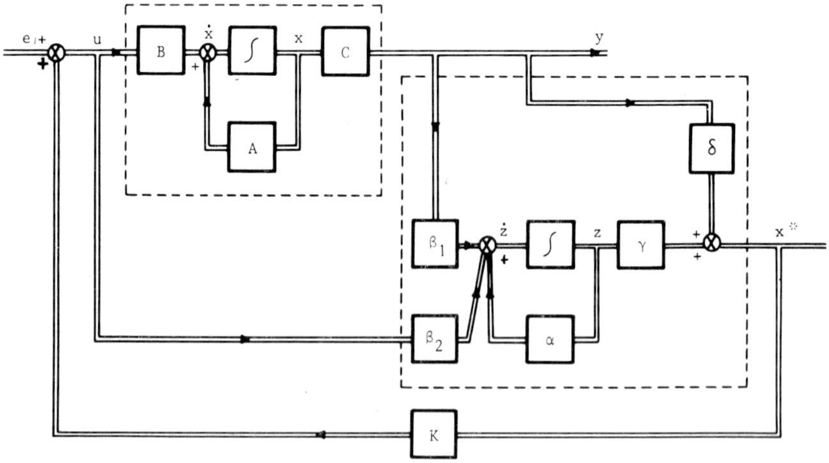

Fig. 4.18 State feedback using reconstructed states. Structure of the loop.

(1) By taking linear combinations, it is in fact possible to generate k_i equations of the type $\lambda_i\,^{1}T(\lambda_i, j) = \alpha\,^{1}T(\lambda_i, j)$, where $T(\lambda_i, j)$ corresponds to the jth Jordan block associated with λ_i.

$$\dot{x} = Ax + Bu \qquad y = Cx$$
$$u = e + Kx^*$$
$$\dot{z} = \alpha z + \beta_1 y + \beta_2 u \qquad (4.66)$$
$$x^* = \gamma z + \delta y \ .$$

The complete system (system being controlled, observer, control loop) can then be written in the form:

$$\begin{bmatrix} \dot{x} \\ \dot{z} \end{bmatrix} = \begin{bmatrix} A + BK\delta C & BK\gamma \\ \beta_1 C + \beta_2 K\delta C & \alpha + \beta_2 K\gamma \end{bmatrix} \begin{bmatrix} x \\ z \end{bmatrix} + \begin{bmatrix} B \\ \beta_2 \end{bmatrix} e \qquad (4.67)$$

$$= \mathcal{A} \begin{bmatrix} x \\ z \end{bmatrix} + \mathcal{B} e \ .$$

Taking account of the relations defining the observer (subsection 4.3.3), the matrix $\mathcal{A}$ can be rewritten in the form:

$$\mathcal{A} = \begin{bmatrix} A + BK - BK\gamma T & BK\gamma \\ T(A+BK) - (\alpha + TBK\gamma)T & \alpha + TBK\gamma \end{bmatrix} \qquad (4.68)$$

where T is the matrix giving $z \to Tx$, and so $\mathcal{A}$ is similar to $\mathcal{A}^*$ (and hence has the same determinant) defined by

$$\mathcal{A}^* = \begin{bmatrix} A + BK & BK\gamma \\ 0 & \alpha \end{bmatrix}$$

In fact, we have

$$\begin{bmatrix} I & 0 \\ -T & I \end{bmatrix} \begin{bmatrix} I & 0 \\ T & I \end{bmatrix} = \mathcal{A}^* \mathcal{A} = M_1 \ M_2$$

and

$$|\mathcal{A}^*| = |\mathcal{A}||M_1| \ |M_2| = |\mathcal{A}| \ .$$

It follows that the characteristic polynomial, of the system compensated using the reconstructed states x^*, is the product of the theoretical characteristic polynomial $|sI - A - BK|$ by that of the observer $|sI - \alpha|$.

Otherwise expressed, the observer does nothing but adjoin its own dynamics.

Let us now discuss the problem with the help of transfer function matrices. We have:

$$Y(s) = [C,\ 0]\ (sI - \mathcal{A})^{-1} \begin{bmatrix} B \\ \beta_2 \end{bmatrix} E(s)$$

with

$$(sI - \mathcal{A})^{-1} = M_2(sI - \mathcal{A}^*)^{-1}M_1$$

i.e.

$$Y(s) = [C,\ 0] \begin{bmatrix} I & 0 \\ T & I \end{bmatrix} \begin{bmatrix} (sI-A-BK)^{-1} & \times \\ 0 & (sI-\alpha)^{-1} \end{bmatrix} \begin{bmatrix} T & 0 \\ -T & I \end{bmatrix} \begin{bmatrix} B \\ \beta_2 \end{bmatrix} E(s)$$

$$= [C,\ 0] \begin{bmatrix} (sI-A-BK)^{-1} & \times \\ 0 & (sI-\alpha)^{-1} \end{bmatrix} \begin{bmatrix} B \\ 0 \end{bmatrix} E(s)$$

since $TB = \beta_2$,

i.e. $$Y(s) = C(sI - A - BK)^{-1}BE(s).$$

The dynamics of the observer do not appear in the expression for the transfer function matrix, its own modes being unobservable for the closed-loop system. This was in fact to be expected, seeing that the expression in terms of transfer function matrices presupposes null initial conditions, and that then the dynamical expression:

$$z - Tx = e^{-\alpha t}(z_0 - Tx_0)$$

reduces to

$$z = Tx \quad .$$

Example:

Let the system be

$$\dot{x} = \begin{bmatrix} 0 & 1 & 0 \\ 0 & 0 & -1 \\ 0 & 0 & 0 \end{bmatrix} x + \begin{bmatrix} 0 & 0 \\ 1 & 0 \\ 0 & 1 \end{bmatrix} u \qquad y = \begin{bmatrix} 1 & 0 & 0 \\ 0 & 0 & 1 \end{bmatrix} x$$

already decomposed into two single-output observable subsystems, and suppose we wish to compensate it in such a way that the closed-loop characteristic equation has one of the following two forms (corresponding respectively to decompositions on the basis ${}^1B$, ${}^2B$, and ${}^2B$, ${}^1B$):

$$s^3 - \beta_3 s^2 - \beta_2 s - \beta_1 = 0 \quad \text{or} \quad (s^2 - \rho_2 s - \rho_1)(s - \rho_3) = 0.$$

We have found that the state feedback matrix is, depending on which case we take:

$$K = \begin{bmatrix} 0 & 0 & 0 \\ -\beta_1 & -\beta_2 & \beta_3 \end{bmatrix} \quad \text{ou} \quad K = \begin{bmatrix} \rho_1 & \rho_2 & 0 \\ 0 & 0 & \rho_3 \end{bmatrix}.$$

We note, however, that, although x_1 and x_3 are accessible and measurable (directly from the outputs), the state x_2 has to be reconstructed by an observer of order 1 ($n - p = 1$), which may be expressed in the form (cf. figure 4.19):

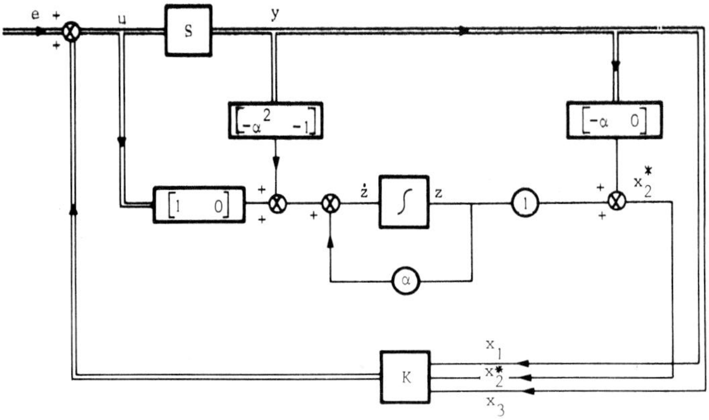

Fig. 4.19 Example.

$$\dot{z} = \alpha z - \alpha^2 y_1 + [\alpha \quad 1]v = \alpha z + [-\alpha^2 \quad -1]y + [1 \quad 0]u$$
$$x_2^* = -\alpha y_1 + z = [-\alpha \quad 0]y + z \quad .$$

Writing the columns in $K = [{}^1K, {}^2K, {}^3K]$ explicitly, the control law becomes:

$$u = e + [{}^1K \quad {}^2K \quad {}^3K]\begin{bmatrix} x_1 \\ x_2^* \\ x_3 \end{bmatrix} = e + [{}^1K - \alpha\,{}^2K \quad {}^3K]y + {}^2Kz \; .$$

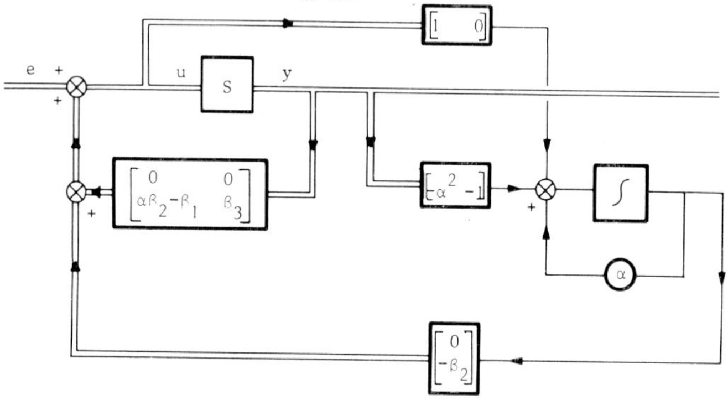

Fig. 4.20 One form of structure for the control loop in the example of subsection 4.3.6 (observer method).

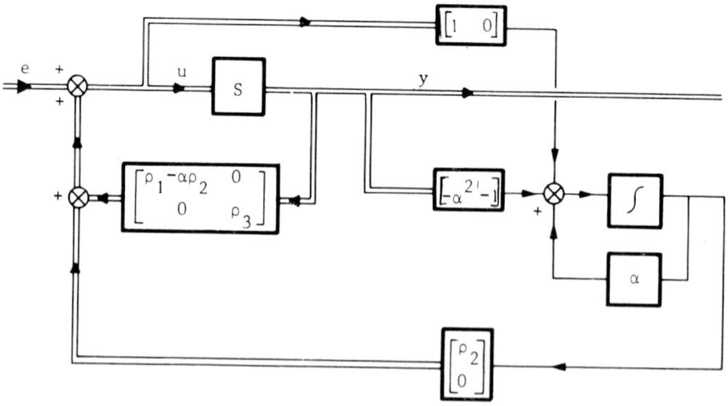

Fig. 4.21 Another form of structure for the control loop in the example of subsection 4.3.6 (observer method).

For the two types of compensator, the structures of the closed loop system are illustrated in figures 4.20 and 4.21. In the first case, the transfer function matrix is given by:

$$Y(s) = \frac{1}{s^3-\beta_3 s^2-\beta_2 s-\beta_1} \begin{bmatrix} s-\beta_3 & -1 \\ -(\beta_2 s+\beta_1) & s^2 \end{bmatrix} \begin{bmatrix} E_1(s) \\ E_2(s) \end{bmatrix}$$

In the second case:

$$Y(s) = \begin{bmatrix} \dfrac{1}{s^2-\rho_2 s-\rho_1} & \dfrac{-1}{(s-\rho_3)(s^2-\rho_2 s-\rho_1)} \\ 0 & \dfrac{1}{s-\rho_3} \end{bmatrix} \begin{bmatrix} E_1(s) \\ E_2(s) \end{bmatrix}$$

If we use a precompensator at the input to the system, we can do without the feedforward element: the structure of the system is then as represented in figure 4.22, with the transfer function matrix given by

$$\begin{bmatrix} Y_1(s) \\ Y_2(s) \end{bmatrix} = \frac{1}{s^3-\beta_3 s^2-\beta_2 s-\beta_1} \begin{bmatrix} -1 \\ s^2 \end{bmatrix} E(s)$$

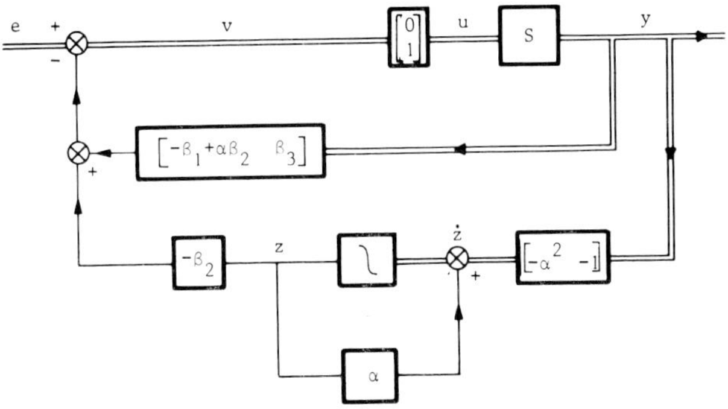

Fig. 4.22 A third form of structure for the control loop in the example of subsection 4.3.6 (observer method).

4.4 DIRECT OUTPUT FEEDBACK BY AUGMENTATION OF THE ORIGINAL SYSTEM

4.4.1 The principle of the method (1)

To the system S being controlled, which is defined by its state equations (4.1), where we assume D = 0, we adjoin a system S´ defined by

$$(S') : \quad \begin{aligned} \dot{\mathcal{U}} &= F\,\mathcal{U} + Gw \\ u &= H\,\mathcal{U} \end{aligned} \qquad (4.69)$$

where $\mathcal{U}$ is a vector of dimension $m\rho$ defined by:

$$\mathcal{U}^T = \left[u^T, \dot{u}^T, \ldots u^{(\rho-1)T}\right]$$

and

$$G = \begin{bmatrix} 0 \\ 0 \\ \vdots \\ I \end{bmatrix} \qquad F = \begin{bmatrix} 0 & I & 0 & \ldots & 0 \\ 0 & 0 & I & \ldots & \\ \vdots & & & & I \\ 0 & & & & 0 \end{bmatrix} \qquad H = \begin{bmatrix} I & 0 & \ldots & 0 \end{bmatrix} \qquad (4.70)$$

The matrices G, F, H, made up from elementary blocks of dimensions (m, m), are respectively of dimensions $(m\rho,\ m)$, $(m\rho,\ m\rho)$, $(m,\ m\rho)$.

If the system S´ is put in cascade ahead of S, the resulting system S^* of dimension $n + m\rho$ is defined by (cf. figure 4.23):

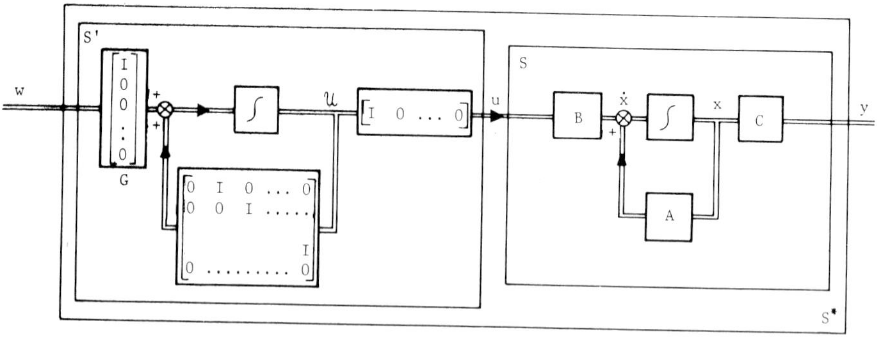

Fig. 4.23 Principle of the augmentation of a system with a view to output feedback compensation.

(1) This section is based on an article by J.B. Pearson, ref. 4.10.

$$\begin{bmatrix} \dot{x} \\ \dot{\mathcal{U}} \end{bmatrix} = \begin{bmatrix} A & BH \\ 0 & F \end{bmatrix} \begin{bmatrix} x \\ \mathcal{U} \end{bmatrix} + \begin{bmatrix} 0 \\ G \end{bmatrix} w = A^* x^* + B^* w \tag{4.71}$$

$$y = \begin{bmatrix} C & 0 \end{bmatrix} \begin{bmatrix} x \\ \mathcal{U} \end{bmatrix} = C^* x^* .$$

It is also clear, in view of the forms chosen for F, G and H, that, if (A, B) is controllable, so is (A^*, B^*). Indeed, the controllability matrix of S^* can be written:

$$\begin{bmatrix} 0 \ldots 0 & 0 & B & AB \ldots A^{n-1} & B \\ \vdots & 0 & I & 0 \ldots\ldots & 0 \\ \vdots & I & 0 & \ldots\ldots\ldots & \vdots \\ I & \ldots\ldots & 0 & \ldots\ldots & 0 \end{bmatrix}$$

which is of rank $m\rho + n$ if $(B, AB, \ldots A^{n-1} B)$ is of rank n.

In view of the results of subsection 4.2.2., it is then possible to find a feedback matrix K^*, of dimensions $(m, m\rho + n)$, which enables us to fix arbitrarily the $m\rho + n$ modes of the system S^* and can be written in the form:

$$K^* = \begin{bmatrix} K(\rho + 1) & K(1) & K(2) \ldots K(\rho) \end{bmatrix} \tag{4.72}$$

where $K(\rho + 1)$ is of dimensions (m, n) and K(i), for $i = 1, \ldots \rho$, are of dimensions (m, m). The corresponding feedback loops are displayed in figure 4.24, and the control law is defined by:

$$w = e + K(\rho + 1)\, x + \sum_{i=1}^{\rho} K(i)\, u^{(i-1)} . \tag{4.73}$$

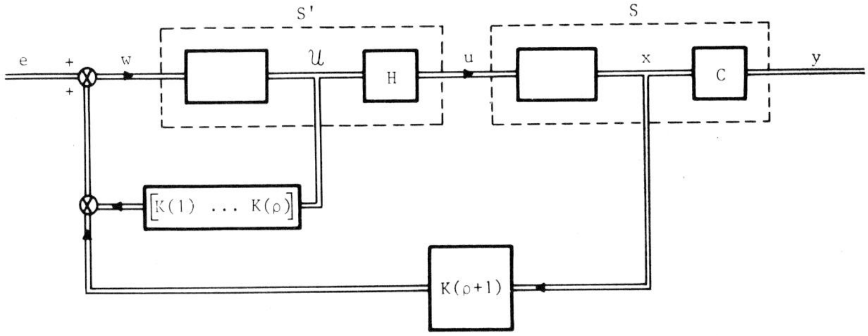

Fig. 4.24 Principle of compensation of the modified system

This being the case, Pearson has shown that the structure of the compensator can be specified, by identifying it with a system which acts on S only by connecting its inputs and outputs in accordance with:

$$u^{(\rho)} - \sum_{i=0}^{\rho-1} N(i)\, u^{(i)} = \sum_{i=0}^{\rho} D(i)\, y^{(i)} \tag{4.74}$$

the matrices N(i) and D(i) being of respective dimensions (m, m) and (m, p). From the defining equations of S, we have, in fact:

$$\begin{aligned} u^{(\rho)} &= \dot{u}^{(\rho-1)} = w = \sum_{i=0}^{\rho-1} N(i)\, u^{(i)} + \sum_{i=0}^{\rho} D(i)\, y^{(i)} = \\ &= e + K(\rho + 1)\, x + \sum_{i=1}^{\rho} K(i)\, u^{(i-1)} \end{aligned} \tag{4.75}$$

and, by using the equations of the system S to express $y^{(i)}$ and demanding that the relation obtained,

$$\begin{aligned} K(\rho + 1)\, x + \sum_{i=1}^{\rho} K(i)\, u^{(i-1)} &= \sum_{i=0}^{\rho-1} N(i)\, u^{(i)} + D(0)\, Cx + D(1)\left[CAx + CBu\right] \\ &\quad + \ldots + D(\rho)\left[CA^{\rho}\, x + \ldots + CBu^{(\rho-1)}\right] \end{aligned} \tag{4.76}$$

should hold for all x and u, we have the relations

$$\begin{aligned} K(1) &= N(0) + D(1)\, CB + \ldots + D(\rho)\, CA^{\rho-1}\, B \\ K(2) &= N(1) + D(2)\, CB + \ldots + D(\rho)\, CA^{\rho-2}\, B \\ &\vdots \\ K(\rho) &= N(\rho - 1) + D(\rho)\, CB \end{aligned} \tag{4.77}$$

and

$$K(\rho + 1) = D(0)\, C + D(1)\, CA + \ldots + D(\rho)\, CA^{\rho} .$$

The last relation can be rewritten as (cf. ref. 4.10)

$$\begin{aligned} K(\rho + 1)^T &= \left[C^T \quad (CA)^T \ldots (CA^{\rho})^T\right] \begin{bmatrix} D(0)^T \\ \vdots \\ D(\rho)^T \end{bmatrix} \\ &= \left[Q_{0\rho}\right] \begin{bmatrix} D(0)^T \\ \vdots \\ D(\rho)^T \end{bmatrix} \end{aligned} \tag{4.78}$$

where we recognise in $Q_{0\rho}$ a matrix constructed in an analogous manner to the observability matrix of the system S.

The matrices K(i), for i = 1,...ρ, and K(ρ + 1) being determined as in subsection 4.2.2, and the matrices D(0),... D(ρ) by equation (4.78), we then deduce the matrices N(0),... N(ρ - 1) from relations (4.77) rewritten in the form:

$$\begin{aligned} N(\rho - 1) &= K(\rho) - D(\rho)\ CB \\ &\vdots \\ N(0) &= K(1) - D(1)\ CB - \ldots - D(\rho)\ CA^{\rho-1}\ B\ . \end{aligned} \qquad (4.79)$$

4.4.2 The order of the compensator

A) We have not yet specified the order of the compensator, which is connected with the index ρ. Equation (4.78) shows however that, in order to determine the D(i), it is necessary for the matrix $Q_{0\rho}$ to be of full rank. Taking account of the analogy between $Q_{0\rho}$ and the observability matrix of the system, it is thus sufficient that $\rho \geqslant \nu - 1$, where ν is the observability index of S, defined by

$$\text{rank}\ \left[C^T,\ (CA)^T, \ldots\ (CA^{\nu-1})^T\right] = n.$$

We can thus take $\rho = \nu - 1$, and the compensator will be of order $m(\nu - 1)$.

B) Reduction of the order. It follows from the above, that the order of the compensator is a function of ν and of the number of inputs m by which we are controlling the system. Now, we have seen that the system may in general be completely controllable by a number m^* (less than m) of inputs.

There then exists a matrix R^*, of dimensions (m, m^*), such that the system (A, BR^*) is controllable from m^* inputs; R^* plays the role of a precompensator, as indicated in figure 4.25: the use of such a pre-compensator enables the compensator order to be reduced from $m(\nu - 1)$ to $m^*(\nu - 1)$.

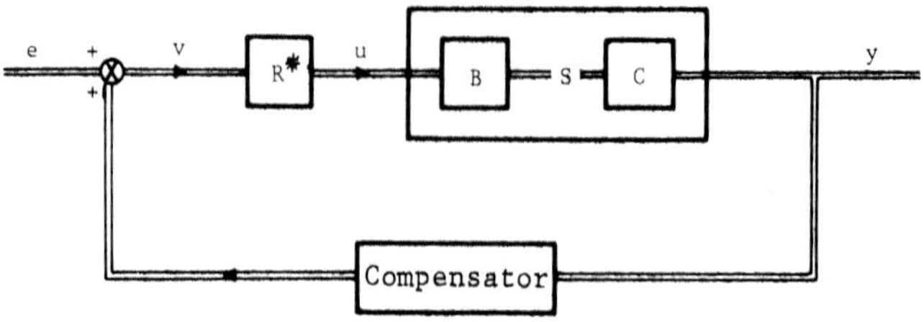

Fig. 4.25 Reduction of the compensator order by precompensation.

4.4.3 Example

Let the system be

$$\dot{x} = \begin{bmatrix} 0 & 1 & 0 \\ 0 & 0 & -1 \\ 0 & 0 & 0 \end{bmatrix} x + \begin{bmatrix} 0 & 0 \\ 1 & 0 \\ 0 & 1 \end{bmatrix} u \qquad y = \begin{bmatrix} 1 & 0 & 0 \\ 0 & 0 & 1 \end{bmatrix} x$$

a) The observability index of the system being 2, the order of the compensator can be taken as equal to the dimension of u, i.e. 2. If we take a compensator of order 2, the system S′ is defined by

$$S' : \quad F = \begin{bmatrix} 0 & 0 \\ 0 & 0 \end{bmatrix} \quad G = \begin{bmatrix} 1 & 0 \\ 0 & 1 \end{bmatrix} \quad H = \begin{bmatrix} 1 & 0 \\ 0 & 1 \end{bmatrix}$$

and the system S^* by

$$A^* = \left[\begin{array}{ccc|cc} 0 & 1 & 0 & 0 & 0 \\ 0 & 0 & -1 & 1 & 0 \\ 0 & 0 & 0 & 0 & 1 \\ \hline 0 & 0 & 0 & 0 & 0 \\ 0 & 0 & 0 & 0 & 0 \end{array}\right] \qquad B^* = \left[\begin{array}{cc} 0 & 0 \\ 0 & 0 \\ 0 & 0 \\ \hline 1 & 0 \\ 0 & 1 \end{array}\right]$$

whose characteristic equation is $s^5 = 0$.

b) To determine the feedback matrix K^*, we put A^* and B^* into the canonical forms

$$\tilde{A}^* = \left[\begin{array}{cc|ccc} 0 & 1 & & & \\ 0 & 0 & & & \\ \hline 0 & 0 & 0 & 1 & 0 \\ -1 & 0 & 0 & 0 & 1 \\ 0 & 0 & 0 & 0 & 0 \end{array}\right] \qquad \tilde{B}^* = \left[\begin{array}{cc} 0 & 0 \\ 0 & 1 \\ \hline 0 & 0 \\ 0 & 0 \\ 1 & 0 \end{array}\right]$$

by using the transformation matrix

$$M = \left[\begin{array}{cc|ccc} 0 & 0 & 1 & 0 & 0 \\ 0 & 0 & 0 & 1 & 0 \\ 1 & 0 & 0 & 0 & 0 \\ 0 & 0 & 0 & 0 & 1 \\ 0 & 1 & 0 & 0 & 0 \end{array}\right]$$

If the desired characteristic equation of the system is

$$(s^3 - \gamma_3 s^2 - \gamma_2 s - \gamma_1)(s^2 - \beta_2 s - \beta_1) = 0$$

we shall take $\tilde{K}^*$ in the form

$$\tilde{K}^* = \begin{bmatrix} 0 & 0 & \gamma_1 & \gamma_2 & \gamma_3 \\ \beta_1 & \beta_2 & 0 & 0 & 0 \end{bmatrix}$$

We deduce that:

$$K^* = \tilde{K}^* M^{-1} = \begin{bmatrix} \gamma_1 & \gamma_2 & 0 & \gamma_3 & 0 \\ 0 & 0 & \beta_1 & 0 & \beta_2 \end{bmatrix} = \begin{bmatrix} K(2) & K(1) \end{bmatrix} .$$

c) The control law is then of the form

$$w = K(2)\ x + K(1)\ u = \dot{u} = N(0)\ u + D(0)\ y + D(1)\ \dot{y}$$

with

$$\begin{bmatrix} C^T & A^T C^T \end{bmatrix} \begin{bmatrix} D^T(0) \\ D^T(1) \end{bmatrix} = K^T(2) = \begin{bmatrix} \gamma_1 & 0 \\ +\gamma_2 & 0 \\ 0 & \beta_1 \end{bmatrix}$$

whence

$$D(0) = \begin{bmatrix} \gamma_1 & 0 \\ 0 & \beta_1 \end{bmatrix}, \; D(1) = \begin{bmatrix} \gamma_2 & 0 \\ 0 & 0 \end{bmatrix}, \; N(0) = K(1) - D(1)CB = \begin{bmatrix} \gamma_3 & 0 \\ 0 & \beta_2 \end{bmatrix}$$

This is illustrated in figure 4.26 .

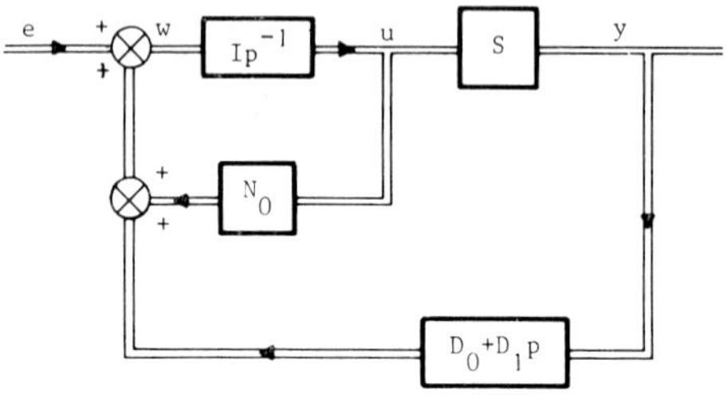

Fig. 4.26 One type of structure for the control loop in the example of subsections 4.3.6 and 4.4.3 (augmentation method).

This compensator can be made easier to realise in the following form, by rewriting

$$w = \dot{u} = N(0)\ u + D(0)\ y + D(1)\ \dot{y}$$

as

$$u = z + D(1)\ y$$

$$\dot{z} = N(0)\ z + [D(0) + N(0)\ D(1)]\ y$$

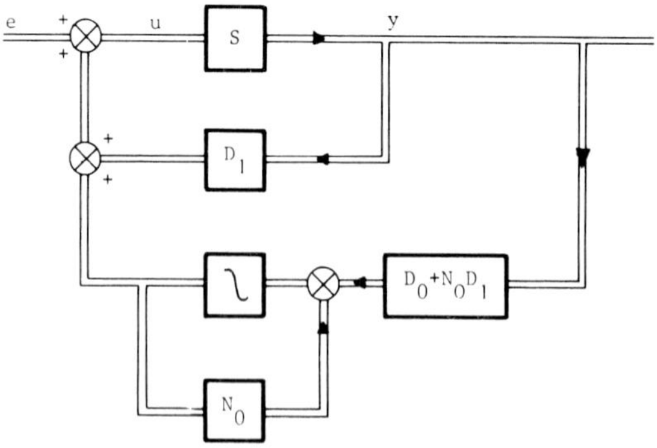

Fig. 4.27 Another type of structure for the control loop in the example of subsections 4.3.6 and 4.4.3 (augmentation method).

which leads to the configuration of figure 4.27 with

$$D(0) + N(0)\, D(1) = \begin{bmatrix} \gamma_1 + \gamma_3\gamma_2 & 0 \\ 0 & \beta_1 \end{bmatrix}$$

The response of the system is then of the form

$$\begin{bmatrix} Y_1(s) \\ \\ Y_2(s) \end{bmatrix} = \begin{bmatrix} \dfrac{s-\gamma_3}{s^3-\gamma_3 s^2-\gamma_2 s-\gamma_1} & \dfrac{-(s-\gamma_3)(s-\beta_2)}{(s^3-\gamma_3 s^2-\gamma_2 s-\gamma_1)(s^2-\beta_2 s-\beta_1)} \\ \\ 0 & \dfrac{s-\beta_2}{s^2-\beta_2 s-\beta_1} \end{bmatrix} \begin{bmatrix} E_1(s) \\ \\ E_2(s) \end{bmatrix}$$

d) Reduction of order of the compensator. Noticing that the system is completely controllable from u_2 alone, we can take

$$u = R^* v \qquad R^* = \begin{bmatrix} 0 \\ 1 \end{bmatrix} \qquad BR^* = \begin{bmatrix} 0 \\ 0 \\ 1 \end{bmatrix}$$

and a compensator of order 1 will be sufficient. Following the same steps as before, we have

$$A^* = \begin{bmatrix} 0 & 1 & 0 & 0 \\ 0 & 0 & -1 & 0 \\ 0 & 0 & 0 & 1 \\ 0 & 0 & 0 & 0 \end{bmatrix}, \ \tilde{A}^* = \begin{bmatrix} 0 & 1 & 0 & 0 \\ 0 & 0 & 1 & 0 \\ 0 & 0 & 0 & 1 \\ 0 & 0 & 0 & 0 \end{bmatrix}, \ M^* = \begin{bmatrix} -1 & 0 & 0 & 0 \\ 0 & -1 & 0 & 0 \\ 0 & 0 & 1 & 0 \\ 0 & 0 & 0 & 1 \end{bmatrix}$$

If the desired characteristic equation is

$$s^4 - \alpha_4 s^3 - \alpha_3 s^2 - \alpha_2 s - \alpha_1 = 0$$

we have

$$\tilde{K}^* = \begin{bmatrix} \alpha_1 & \alpha_2 & \alpha_3 & \alpha_4 \end{bmatrix}$$

and

$$K^* = \tilde{K}^* M^{-1} = \begin{bmatrix} -\alpha_1 & -\alpha_2 & +\alpha_3 & \alpha_4 \end{bmatrix} = \begin{bmatrix} K(2) & K(1) \end{bmatrix},$$

whence

$$D_0 = \begin{bmatrix} -\alpha_1 & \alpha_3 \end{bmatrix} \qquad D(1) = \begin{bmatrix} -\alpha_2 & 0 \end{bmatrix} \qquad N(0) = \alpha_4 .$$

The control law becomes

$$w = \dot{v} = N_0 \, v + D_0 \, y + D(1) \, \dot{y}$$

or, again,

$$v = z + D(1)\, y = z + \begin{bmatrix} -\alpha_2 & 0 \end{bmatrix} y$$

$$\dot{z} = \alpha_0 \, z + \begin{bmatrix} -(\alpha_1 + \alpha_2\alpha_4) & \alpha_3 \end{bmatrix} y \, .$$

The response of the system is then:

$$\begin{bmatrix} Y_1(s) \\ Y_2(s) \end{bmatrix} = \frac{s-\alpha_4}{s^4-\alpha_4 s^3-\alpha_3 s^2-\alpha_2 s-\alpha_1} \begin{bmatrix} -1 \\ s^2 \end{bmatrix} E(s)$$

The control configuration is given in figure 4.28.

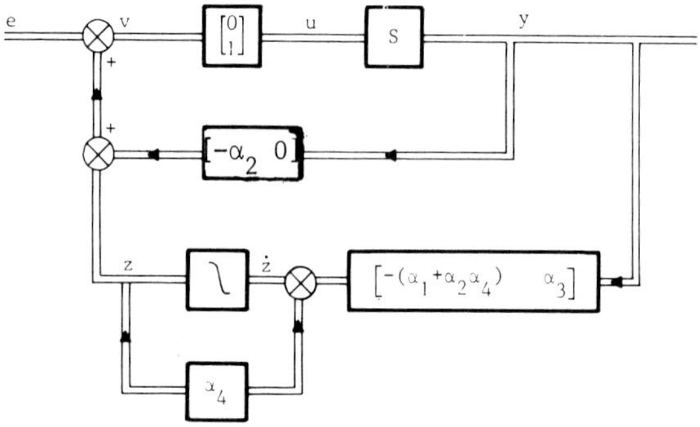

Fig. 4.28 A third type of structure for the control loop in the example of subsections 4.3.6 and 4.4.3 (augmentation method).

The procedures described in this chapter, in particular those involving a state feedback compensator and an observer, form the basis of the techniques most often utilised at the present time for the control of multivariable systems by state-space methods. It is however important to point out, without questioning their advantages, that in some circumstances such systems turn out to be rather sensitive, in the sense that there can be a significant degradation of performance when the parameters of the system or the compensator are perturbed from their nominal values.

We thus appreciate the benefits of methods which will take into account, a priori, such sensitivity problems, in the design of the control system. This is one of the new avenues which are opening up for the control of multivariable systems, in which, very recently, significant results have been obtained, in particular by J.B. Pearson (1) and by E.J. Davison (2).

(1) P.W. Staats and J.B. Pearson: "Robust solution of the linear servomechanism problem", 6th IFAC World Congress (Boston 1975) paper 43.4.

(2) E.J. Davison and A. Goldenberg: "Robust control of a general servomechanism problem", 6th IFAC World Congress (Boston 1975) paper 9.5.

Chapter 5

Non-interactive control

In the general framework of the control of multivariable systems, considerable attention has been given in the past to the idea of non-interaction. From this viewpoint, we usually seek to compensate a system so that: each input affects only one output, and a perturbation of one output, with the inputs fixed at zero, affects no other output.

Insistence on the property of non-interaction sometimes arises from a desire to simplify the control loop. Indeed, when non-interaction is obtained, the multivariable system is transformed into a set of single-variable systems, completely decoupled, which can then be treated by classical methods. Furthermore, control can be imposed loop by loop, without having to modify the other loops. These are, no doubt, the reasons why the study of such systems began, historically, by considering these ideas (cf. ref. 5.1).

Such a design, in which the principle of non-interaction appears as a means for reducing the number of degrees of freedom of the system, may seem unduly restrictive. It is not, however, solely due to the laziness of control engineers. In practice, an obvious advantage of non-interaction is the ability to maintain certain system outputs constant, while others are deliberately altered. Such ideas have been applied, e.g., to the control of boilers, where the temperature and pressure are held fixed while the flow-rate can be varied at will (cf. refs. 5.4, 5.6).

There are numerous real processes where this type of control is desirable.

However, although the concept of a non-interactive system appears theoretically very simple, its design is not without many problems, as much in respect of stability as of physical realisation. One of the purposes of this chapter will be precisely to clarify a number of points regarding these problems.

In the case where non-interaction corresponds to no practical motivation and is introduced only as a way of artificially reducing the number of degrees of freedom of the system, the choice between an interactive and a non-interactive solution will depend on other criteria, connected, e.g., with the notion of sensitivity. One may thus be led to define domains of interaction (or non-interaction) inside which an interactive (or non-interactive) solution will be preferable (cf. ref. 5.24).

5.1 CONCEPTS AND DEFINITIONS OF INTERACTION

In a multivariable system, each input in general affects several outputs. In the case that the system can be validly represented by a linear time-invariant model, characterised by a transfer function matrix N, and the number of inputs m is equal to the number of outputs ($p = m$), the elements of N away from the principal diagonal represent coupling terms between the inputs.

Let us consider, e.g., the control system represented in figure 5.1, where the outputs y_i are required to follow the reference inputs e_i. We shall distinguish various types of non-interaction.

a) A controlled variable y_j is affected only by the reference input e_j and not by the other reference inputs e_k.

b) A reference input e_k affects only the corresponding output y_k, to the exclusion of all the others.

c) Each output is affected only by the corresponding reference input, and each reference input affects only the corresponding output.

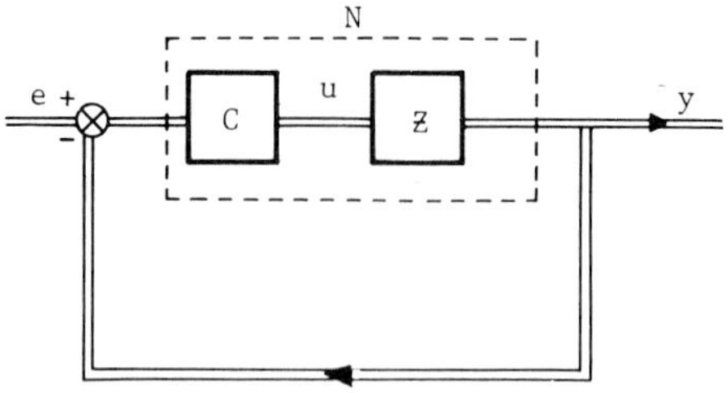

Fig. 5.1 Signal-flow diagram of a multivariable servo-system with unity feedback.

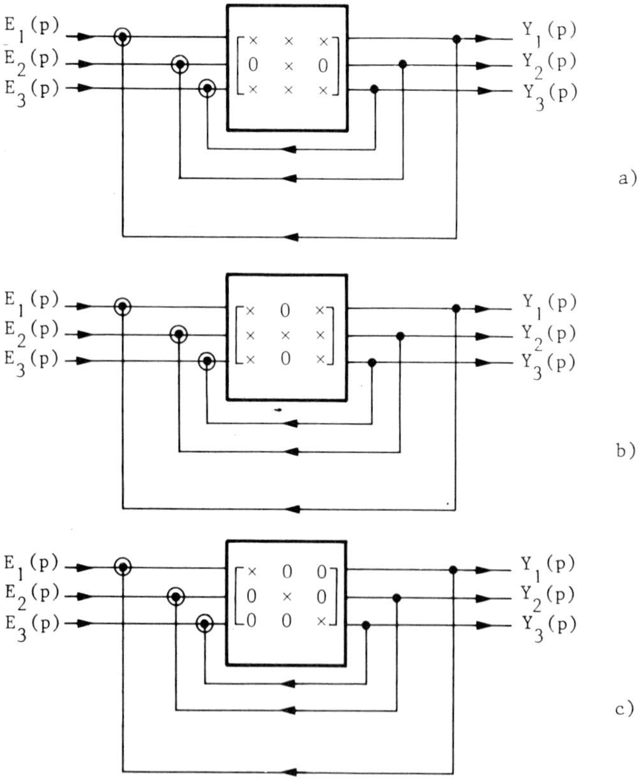

Fig. 5.2 Concepts of non-interaction:
a) y_2 is affected only by e_2, to the exclusion of the other reference inputs;
b) the reference input e_2 affects only the output y_2;
c) complete input-output non-interaction.

For the system represented by figure 5.1, where the feedback is assumed to be unity, conditions a) to c) imply respectively:-

a) The jth row of N contains only the diagonal element, subject of course to a suitable ordering of the components of the vectors e and y. Such a situation is illustrated in figure 5.2a for the case of a three-variable system.

b) The kth column of N contains only the diagonal element (cf.figure 5.2b, with the same assumptions as before).

c) The matrix N reduces to its principal diagonal (cf. figure 5.2c).

Only the last case will be considered in the remainder of this chapter, and we shall agree to call a system non-interactive if it has property c).

Whether or not this property holds, it is important to notice that a perturbation on one output can, by propagating through the system, affect the other outputs. This will be, e.g., the effect on the outputs of an initial condition for one of them, y_i, the inputs being identically zero. In the case that an initial condition for y_i affects only the output y_i itself, we shall say that the system is non-interactive with respect to y_i. If this property holds regardless of which output is considered, we shall say that the system is non-interactive with respect to the outputs (sometimes known as "independent output restoration").

5.2 GENERAL PROBLEMS POSED BY NON-INTERACTIVE CONTROL SYSTEMS

The synthesis of non-interactive control systems can be carried out in two ways:

- either, by using operational methods, i.e. the concept of transfer function matrices;
- or, by using the state-space techniques studied in the preceding chapters.

Initially, operational methods were almost exclusively used, for obvious reasons moreover. More often than not, in fact, the only justification for non-interaction was to reduce a multivariable control problem to a set of single-variable problems, for which classical techniques, using all the tools associated with the idea of a transfer function, were directly applicable.

This kind of approach, which ignored the ideas of controllability and observability, was the cause of much disillusionment, and it would be easy to cite many examples, even from the most reliable sources, where such an approach, conducted without due care, led to the design of control systems which were stable only on paper.

We shall endeavour here to set forth the reasons for these errors and to show that certain constraints must, in general, be imposed on the transfer function matrices, open-loop as well as closed-loop. Such methods very often end up by producing compensators of high order, whose realisation poses delicate problems.

Later on, under the impetus of Morgan (cf. ref. 5.26) and then of Falb and Wolovich (cf. refs. 5.7 and 5.8), the problem of non-interactive control was attacked on the basis of state-space representations. This approach, although sometimes more laborious than the former, has the great merit of allowing an exact knowledge of the limitations imposed on the control system by the supplementary constraints due to the character of the non-interaction sought for, and of showing up the fundamental relations which exist between non-interaction, reproducibility and invertibility. It is thus by exhibiting these that this chapter will begin; in fact, we shall show later that, having once understood the implications of each of these methods, it will sometimes be necessary, and often desirable, to combine them.

Let us consider, as an introductory example, the case of a process having the same number of inputs as of outputs ($p = m$), characterised by its transfer function matrix $Z(s)$. We try to find a compensator $C(s)$ enabling us to decouple the system using unity feedback (cf. figure 5.3).

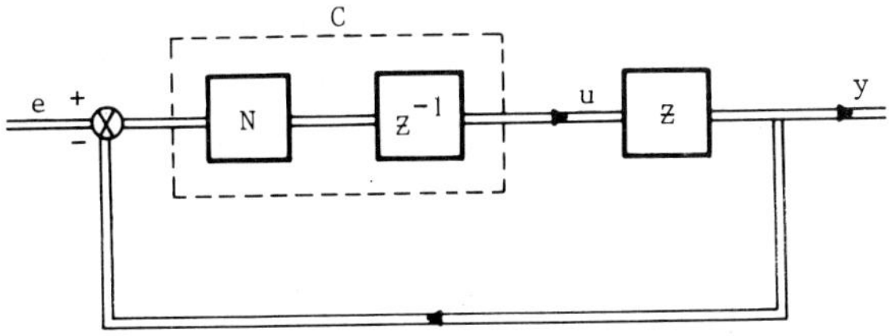

Fig. 5.3 Theoretical structure of forward-path compensator

If N and H are, respectively, the open-loop and closed-loop transfer function matrices, we have:

$$N = ZC \qquad (5.1)$$
$$H = (I + N)^{-1} N .$$

The condition of input-output non-interaction requires that, subject to a suitable ordering of the components of the input vector e and the output vector y, the matrix H should be diagonal. Since

$$H = (I + N)^{-1} N = (I + N^{-1})^{-1} .$$

the diagonalisation of H implies that of N, and we must have:

$$ZC = N = \text{diagonal} \begin{bmatrix} n_{11} & n_{22} & \cdots & n_{mm} \end{bmatrix}$$

and hence, if the matrix Z is invertible:

$$C = Z^{-1} N . \qquad (5.2)$$

It then follows that, conceptually, the compensator is constituted by the cascade of a diagonal matrix N with Z^{-1} (cf. figure 5.3).

This structure of C raises three kinds of problem.

a) First, the realisation depends on the existence of Z^{-1}, i.e. on the invertibility properties of the system to be controlled, which are thus closely linked to the problems of non-interactive control.

b) The presence of dynamical elements in Z implies the presence of predictive elements in Z^{-1}. Thus, the elements of C may not be physically realisable if N does not contain sufficient dynamics. In connection with point a), this will involve the construction of a "minimal-degree inverse".

c) Finally, it follows from the very structure of C that the compensator and the process are closely linked . Cancellations can thus occur in the cascade of NZ^{-1} and Z, giving rise to uncontrollable or unobservable modes. Particular attention must be paid to this problem, at any rate for modes situated in the right half-plane. This gives rise, in general, to constraints on the structures of the open-loop and closed-loop matrices (cf. subsection 5.2.2).

5.2.1 Right-inverse of a linear reproducible process

A - Properties

Let the linear dynamical system S be defined by the set of matrices (A, B, C, D),

$$\begin{aligned} \dot{x} &= Ax + Bu \\ y &= Cx + Du \end{aligned} \tag{5.3}$$

or by its transfer function matrix Z(s). Following our usual notation, the dimensions of the state, input and output vectors are, respectively, n, m, p. We shall say that the system S has a generalised right-inverse if there exists a dynamical system S^*, with transfer function matrix Z^*, such that:

$$Z(s)Z^*(s) = Is^{-\rho} \tag{5.4}$$

where I is the unit matrix of dimensions (p, p) and ρ is an index to be determined.

Referring to figure 5.4, it is quite clear that although, for this relation to be satisfied, we must have $p^* = m$, $m^* = p$, the number of states n^* of S^* is, on the other hand, not specified. A necessary condition for the existence of a right-inverse of Z is that m should be greater than or equal to p, i.e. that the number of inputs of the process should be at least as large as the number of outputs. As for the smallest index ρ for which it is possible to satisfy equation (5.4), it represents the minimum number of differentiations which must be made on the output vector y of the process in order to produce the input vector of the generalised inverse system which, in cascade ahead of S, will generate the input vector u of the process (cf. figure 5.4).

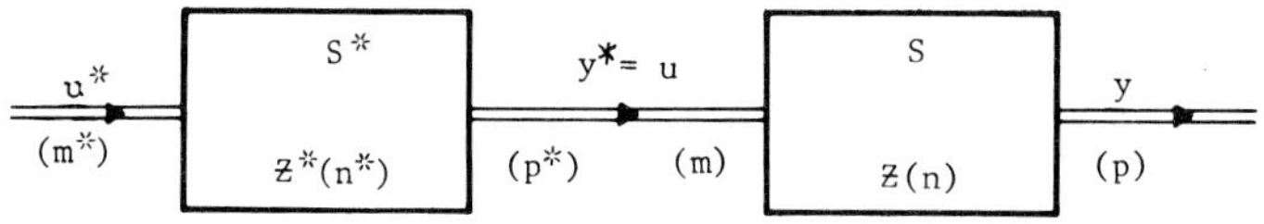

Fig. 5.4 The idea of a generalised right-inverse.

Also from equation (5.4), it is clear that a necessary and sufficient condition for S to have a right-inverse is that

$$\text{rank } Z(s) = p, \text{ for all } s.$$

Now, this condition is none other than the necessary and sufficient condition that any vector y(t) can occur as the output of S, i.e. that, for every y(t), there is a u(t) such that

$$Y(s) = Z(s)U(s).$$

We recognise the condition for reproducibility in the sense of Brockett and Mesarovic, which was presented in chapter 1.

Following a procedure indicated by Massey and Sain (cf. refs. 5.22, 5.23) in the case of left-inverses, suppose that we associate with the continuous-time system S defined by equations (5.3), the discrete-time system S′ defined by:

$$\begin{aligned} x_{k+1} &= Ax_k + Bu_k \\ y_k &= Cx_k + Du_k \end{aligned} \qquad (5.5)$$

where the matrices A, B, C, D are the same as in equations (5.3). The analogy between the systems S and S′ is evident; in particular, the transfer function matrix in z of S′ will be obtained from the transfer function matrix in s of S by the simple substitution of z for s. Hence S has a generalised right-inverse if and only if S′ has one. As it is easier to discuss the discrete-time system than the continuous one, the use of this analogy will enable us to obtain a number of results more simply.

Suppose $x_0 = 0$. With an input sequence $u_0, u_1, \dots u_k$, the system (5.5) associates an output sequence $y_0, y_1, \dots y_k$, such that

$$\begin{bmatrix} y_k \\ y_{k-1} \\ \vdots \\ y_1 \\ y_0 \end{bmatrix} = \begin{bmatrix} D & CB & \cdots\cdots & CA^{k-1}B \\ 0 & D & CB & CA^{k-2}B \\ & & & \vdots \\ & & D & CB \\ \cdots & \cdots & \cdots & D \end{bmatrix} \begin{bmatrix} u_k \\ u_{k-1} \\ \vdots \\ u_1 \\ u_0 \end{bmatrix}$$

or, in matrix form,

$$y(k, 0) = M_k\, u(k, 0) \qquad (5.6)$$

where M_k is a matrix of dimensions $[p(k+1), m(k+1)]$. We can then show that (cf. refs. 5.22, 5.23):-

a) For any integer ρ:

$$\text{rank}\,(M_\rho) - \text{rank}\,(M_{\rho-1}) \leq p$$

the equality holding only if S has a right-inverse of index ρ. (This implies in particular that, if S has an inverse of index ρ, it has one for every index greater than ρ). From the structure of the matrix M_ρ, it is also clear that the equality is satisfied if and only if the first p rows of M_ρ are linearly independent of each other and of the remaining $p\rho$ rows. We can determine ρ by constructing successively:

$$M_0, M_1, \dots M_{k-1}, M_k$$

until the relation

$$\text{rank}\,(M_\rho) - \text{rank}\,(M_{\rho-1}) = p$$

holds.

b) The system S has a right-inverse if and only if

$$\text{rank}\,(M_n) - \text{rank}\,(M_{n-1}) = p.$$

c) The system S has a right-inverse if and only if the matrix N defined by:

$$N = \begin{bmatrix} D & CB & \dots\dots & CA^{n-1}B & \dots\dots & CA^{2n-1}B \\ 0 & D & CB \dots & & & \\ \vdots & & & & & \\ 0 & & & D & & CA^{n-1}B \end{bmatrix}$$

is of rank $(n+1)p$.

This last criterion is precisely that proposed by Brockett and Mesarovic. The complexity of the expression for N shows the advantage of criterion a) over that of Brockett and Mesarovic or even that of Rosenbrock (cf. ref. 5.31).

B - Construction of a right-inverse

We have seen that the discrete system S′ associates with an input sequence $u(\rho, 0)$, an output sequence $y(\rho, 0)$ such that:

$$y(\rho, 0) = M_\rho\, u(\rho, 0)\ . \tag{5.6}$$

The existence of a right-inverse being connected with the property of reproducibility of the system, it follows that, for any arbitrarily given y_ρ, we can find an input sequence $u_0, \dots u_\rho$, which will generate y_ρ, i.e. that there exists a matrix K such that

$$u(\rho, 0) = Ky_\rho \tag{5.7}$$

K being a matrix of dimensions $[m(\rho+1), p]$. If we decompose this matrix into $\rho+1$ matrices K_i of dimensions (m, p) in the form

$$\begin{bmatrix} u_\rho \\ u_{\rho-1} \\ \vdots \\ u_1 \\ u_0 \end{bmatrix} = \begin{bmatrix} K_\rho \\ K_{\rho-1} \\ \vdots \\ K_1 \\ K_0 \end{bmatrix} y_\rho \tag{5.8}$$

it follows from equations (5.6), (5.7), (5.8) that:

$$\begin{bmatrix} y_\rho \\ y_{\rho-1} \\ \vdots \\ y_1 \\ y_0 \end{bmatrix} = \begin{bmatrix} D & CB & & CA^{\rho-1}B \\ 0 & D & CB & \cdots \\ \vdots & & & \\ & & D & CB \\ 0 & \cdots & & D \end{bmatrix} \begin{bmatrix} K_\rho \\ K_{\rho-1} \\ \vdots \\ K_1 \\ K_0 \end{bmatrix} y_\rho$$

We deduce, by identifying the various components y_i, that

$$\begin{aligned} DK_0 &= 0 \\ DK_1 + CBK_0 &= 0 \\ &\cdots \\ DK_{\rho-1} + CBK_{\rho-2} + \ldots + CA^{\rho-2}BK_0 &= 0 \\ DK_\rho + CBK_{\rho-1} + \ldots + CA^{\rho-1}BK_0 &= Ip\ . \end{aligned} \tag{5.9}$$

Equations (5.9) allow us to determine, though in a non-unique fashion, the matrices K_i. It is then possible to give a structure for the minimal right-inverse, in the form indicated in figure 5.5.

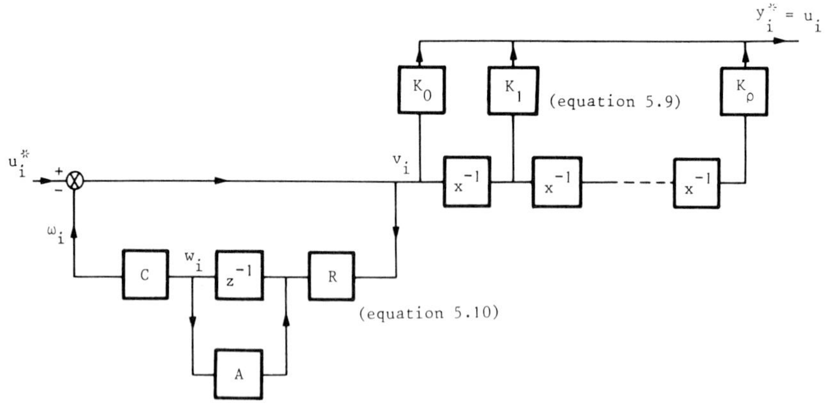

Fig. 5.5 Structure of the generalised right-inverse S^*.

To establish that the proposed system really is an inverse, it is sufficient to verify that: if

$$Y(z) = Z'\,Z'^{*}\,U^{*} = z^{-\rho}\,U^{*}$$

then, to the input sequence

$$u^* = \begin{bmatrix} e_0 & 0 & 0 \ldots 0 \end{bmatrix}$$

there must correspond the output sequence

$$y = \begin{bmatrix} 0 & 0 \ldots e_0 \ldots 0 \end{bmatrix}$$

whose only non-null component occurs at time ρ.

The various signals in the system being as indicated in figure 5.5, the flow of information takes place according to the following recurrence relations:

a) $w_i = (A - RC)w_{i-1} + Ru^*_{i-1}$, with $w_0 = 0$, $w_1 = Re_0$

and consequently $w_i = (A - RC)^{i-1} Re_0$.

b) $v_i = u^*_i - Cw_i$

so that $\quad v_0 = e_0 , \quad v_1 = - CRe_0 , \ldots, v_i = C(A - RC)^{i-1} Re_0$

c) $$y^*_i = u_i = \sum_{j=0}^{\rho} K_j v_{i-j} = K_i e_0 - \sum_{j=0}^{i-1} K_j C(A - RC)^{i-j-1} Re_0$$

d) Also, we have

$$\mathcal{Z}'(z) = C\left[Iz - A\right]^{-1} B + D = D + Cz^{-1}\left[I - Az^{-1}\right]^{-1} B$$

$$\mathcal{Z}'(z) = D + CBz^{-1} + CABz^{-2} + \ldots + CA^{n-1} Bz^{-n} ,$$

which can be rewritten in the form:

$$\mathcal{Z}'(z) = \mathfrak{J}_0 + \mathfrak{J}_1 z^{-1} + \mathfrak{J}_2 z^{-2} + \ldots + \mathfrak{J}_n z^{-n}$$

with

$$\mathfrak{J}_0 = D, \quad \mathfrak{J}_1 = CB, \ldots, \mathfrak{J}_i = CA^{i-1} B$$

and we then have

$$y_i = \sum_{\nu=0}^{i} \mathfrak{J}_\nu y^*_{i-\nu} = \sum_{\nu=0}^{i} \mathfrak{J}_\nu \left[K_{i-\nu} e_0 - \sum_{j=0}^{i-\nu-1} K_j C(A - RC)^{i-\nu-1-j} Re_0\right].$$

It follows that

$$y_0 = \mathcal{G}_0 K_0 e_0 = 0 ;$$

from the first of equations (5.9);

$$y_1 = (\mathcal{G}_0 K_1 + \mathcal{G}_1 K_0) e_0 = 0 ;$$

from the first two of equations (5.9);

$$y_{\rho-1} = \left[\mathcal{G}_0 K_{\rho-1} + \mathcal{G}_1 K_{\rho-2} + \dots + \mathcal{G}_{\rho-1} K_0\right] e_0 = 0 ,$$

$$y_\rho = \left[\mathcal{G}_0 K_\rho + \dots + \mathcal{G}_\rho K_0\right] e_0 = e_0 .$$

It remains to demonstrate that, for all i greater than ρ, the components of the output sequence are null. Let us consider component ρ+1:

$$y_{\rho+1} = \left[\mathcal{G}_{\rho+1} K_0 + \dots + \mathcal{G}_1 K_\rho\right] e_0 - \left[\mathcal{G}_\rho K_0 + \dots + \mathcal{G}_0 K_\rho\right] CRe_0 .$$

$y_{\rho+1}$ will be null if

$$\mathcal{G}_{\rho+1} K_0 + \dots + \mathcal{G}_1 K_\rho = (\mathcal{G}_\rho K_0 + \dots + \mathcal{G}_0 K_\rho) CR$$

which holds if we take R in the form:

$$R = A^\rho BK_0 + A^{\rho-1} BK_1 + \dots + BK_\rho \qquad (5.10).$$

It would be easy to show similarly, making use of equations (5.9) and (5.10), that every $y_{\rho+k}$ is null and that, consequently, the system of figure 5.5 is indeed a right-inverse. As a result of the analogy between the systems S and S´, we easily deduce the structure of the continuous right-inverse; it is as indicated in figure 5.5, with the delay operators z^{-1} replaced by integrators.

Remark: it has been indicated that equations (5.9) do not define the matrices K_i uniquely. On the one hand, it may be possible to take a certain number of them to be null, since, from the equations, if D, CB, ... $CA^{k-2}B$ are null, while $CA^{k-1}B$ is not, we can take $K_\rho = K_{\rho-1} = \dots = K_{\rho-k+1} = 0$. On the other hand, for the non-null matrices, there remains a certain possibility of choice, which we can make use of in particular applications, as the occasion arises.

C - Example

Let the system S be:

$$\dot{x} = \begin{bmatrix} -1 & 0 & 0 \\ 0 & -2 & 0 \\ 1 & 1 & -3 \end{bmatrix} x + \begin{bmatrix} 1 & 0 & 1 \\ 1 & 1 & 1 \\ 0 & 0 & 1 \end{bmatrix} u , \quad y = \begin{bmatrix} 1 & -1 & 0 \\ 2 & 0 & 1 \end{bmatrix} x .$$

We have:

$$M_0 = \begin{bmatrix} 0 & 0 & 0 \\ 0 & 0 & 0 \end{bmatrix}, \qquad M_1 = \begin{bmatrix} 0 & 0 & 0 & 0 & -1 & 0 \\ . & . & . & 2 & 0 & 3 \\ . & . & . & 0 & 0 & 0 \\ . & . & . & 0 & 0 & 0 \end{bmatrix}$$

$$\text{rank } (M_0) = 0, \quad \text{rank } (M_1) = 2$$

and it follows that $\rho = 1$. The matrices K_0 and K_1 are then defined by the equations

$$0 \; K_0 = 0$$

$$0 \; K_1 + \begin{bmatrix} 0 & -1 & 0 \\ 2 & 0 & 3 \end{bmatrix} K_0 = I_2 .$$

We can take, for example,

$$K_1 = 0_{3,2}$$

$$K_0 = \begin{bmatrix} 0 & -1 & 0 \\ 1/2 & 0 & 0 \end{bmatrix}^T$$

whence we get the expression for R:

$$R = ABK_0 = \begin{bmatrix} 0 & 2 & -1 \\ -1/2 & -1 & 1 \end{bmatrix}^T .$$

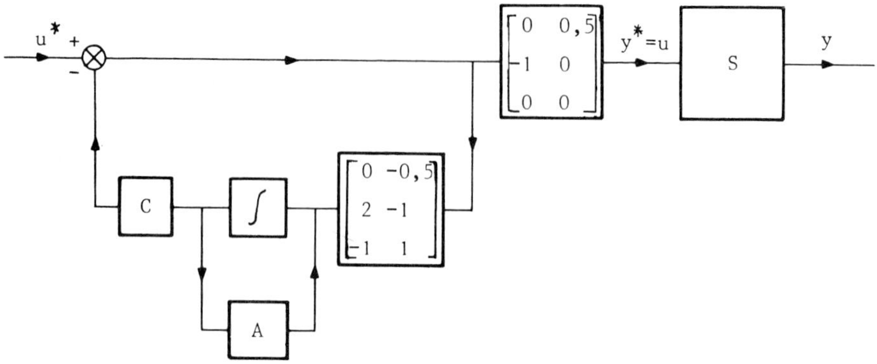

Fig. 5.6 Right-inverse for the example of paragraph 5.2.1.C.

The inverse system can be represented as in figure 5.6. We verify that:

$$Z^* = K_0\left[I+C(I-As^{-1})^{-1}s^{-1}R\right]^{-1}$$

$$= \frac{1}{s(s+4)}\begin{bmatrix} 0.5(s+1) & 0.5(s+1)(s+3) \\ -(s^2+6s+7.5) & 0.5(s+3) \\ 0 & 0 \end{bmatrix}$$

and that

$$ZZ^* = \begin{bmatrix} 1 & 0 \\ 0 & 1 \end{bmatrix} s^{-1}.$$

Remark: The general forms of the matrices K_0 and R are:

$$K_0 = \begin{bmatrix} \alpha & \frac{1-3\beta}{2} \\ -1 & 0 \\ -2\alpha/3 & \beta \end{bmatrix} \qquad R_R = \begin{bmatrix} -\alpha/3 & \frac{\beta-1}{2} \\ 2(1-\alpha/3) & \beta-1 \\ 8\alpha/3-1 & 1-4\beta \end{bmatrix}.$$

We can, for example, choose:

$$\beta = 1, \quad \alpha = 0 \text{ or } 3,$$

if we wish to reduce the importance of the feedback loop in the first part of the inverse system.

5.2.2 Appearance of uncontrollable and unobservable modes (1)

The other important problem [cf. section 5.2, paragraph (c)] associated with the design of non-interactive systems, is connected with the appearance of uncontrollable and unobservable modes, when the cascade of process and compensator is formed.

We know, indeed, that for a closed-loop system to be stable, it is necessary that:

- firstly, the controllable and observable part of the system (that represented by the transfer function matrix) should be closed-loop stable,

- secondly, the uncontrollable and/or unobservable modes of the system (which appear when the compensator is cascaded with the process, and are not represented by the transfer function matrix) should be open-loop stable.

Although the appearance of such modes is hardly probable when the two cascaded subsystems are independent, it is obviously not the same where a process and its compensator are concerned, since their transfer function matrices Z and C are connected by the relation:

$$ZC = N.$$

A - "Transfer function matrix" approach

Although only a state-space approach is capable of completely elucidating the problem (cf. paragraph 5.2.2B and section 5.3), it is undoubtedly worthwhile to try to understand physically, by a "transfer function matrix" approach, the origin of the difficulties which arise.

The strict cancellation of two factors normally remains a purely mathematical concept, never encountered in practice. It is important,

(1) As before, we shall suppose that the process has the same number of inputs as outputs, and that Z(s) is invertible.

indeed, to make an essential distinction between the assumed (or measured) transfer function matrix of the process and the true one. This difference arises from measurement errors, faulty identification, possible linearisations made in the model, and small temporal variations in the process. In view of the numerous reasons why the matrix used in the calculation of the compensator and the true one are not identical, we shall distinguish them by denoting them respectively by $Z(s)$ and $Z^*(s)$. Of course, we shall assume that $Z(s)$ and $Z^*(s)$ are nevertheless sufficiently close to one another.

Let N be the desired open-loop matrix. The compensator C is determined from N and Z by the relation:

$$C = Z^{-1}N$$

i.e. the elements c_{ij} of C are defined by

$$c_{ij} = \sum_{k=1}^{m} \frac{Z_{ki}}{|Z|} n_{kj}$$

where Z_{ik} is the cofactor of the element z_{ik} of Z, so that, in view of the form of N,

$$c_{ij} = Z_{ji}\, n_{jj}/|Z|.$$

The compensator thus determined is utilised to control the actual process with transfer function matrix Z^*, leading to an effective open-loop matrix:

$$N^* = Z^*C$$

an element of which may be expressed in the form

$$n^*_{ij} = \sum_{k=1}^{m} z^*_{ik}\, c_{kj} = \sum_{k=1}^{m} z^*_{ik} \frac{Z_{jk}}{|Z|} n_{jj}$$

$$n^*_{ij} = n_{jj} \sum_{k=1}^{m} z^*_{ik}\, Z_{jk}/|Z|. \tag{5.11}$$

From the assumption of proximity between Z^* and Z, we have: (1)

(1) cf. ref. 5.20 of R.J. Kavanagh.

$$\lim_{Z \to Z^*} \sum_{k=1}^{m} z^*_{ik} Z_{jk} = 0$$

$$\lim_{Z \to Z^*} \sum_{k=1}^{m} z^*_{jk} Z_{jk} = |Z^*| .$$

The true open-loop transfer function matrix N^* will now be written in the form:

$$N^* = \begin{bmatrix} n^*_{11} & \varepsilon_{12} & \cdots & \varepsilon_{1m} \\ \varepsilon_{21} & n^*_{22} & & \vdots \\ \vdots & & \ddots & \\ \varepsilon_{m1} & & & n^*_{mm} \end{bmatrix}$$

where the n^*_{ij} are given by the expressions (5.11).

In particular, we see that the diagonal element n^*_{jj} contains all the poles and zeroes originally in n_{jj}, which is a factor of the expression for n^*_{ij}, together with a number of other factors arising from $\sum z^*_{jk} Z_{jk}/|Z|$. As a result of the limiting conditions, to each denominator factor there corresponds a numerator factor, of such a form that the supplementary poles and zeroes introduced appear as doublets. If they are all in the left half-plane, their effects will be negligible, since the associated residues will be very small. This is obviously not the case if one or more of these dipoles appears in the right half-plane, as it will then result in an instability of the system, even if the corresponding residue is very small.

If we write the matrices Z, C, N in the form:(1)

$$Z = \frac{\mathcal{Z}}{z} \qquad C = \frac{\mathcal{C}}{c} \qquad N = \frac{\mathcal{N}}{n}$$

where $\mathcal{Z}$, $\mathcal{C}$, $\mathcal{N}$ are polynomial matrices and z, c, n are polynomials in s, we have seen that a sufficient condition, in general, is that:

n includes the unstable factors of z

$(\operatorname{adj} \mathcal{Z})\,\mathcal{N}$ includes the unstable factors of $|\mathcal{Z}|$

$\mathcal{N}$ being a diagonal matrix.

(1) cf. E.G. Gilbert, ref. 5.11.

We thus have the following sufficient conditions:

1)

$$1)\ \left[\text{adj}\ \mathcal{Z}\right]\mathcal{N} = \left[\text{adj}\ \mathcal{Z}\right]\begin{bmatrix} n_{11} & & & \\ & n_{22} & & \\ & & \ddots & \\ & & & n_{mm} \end{bmatrix} = \left[{}^{1}\mathcal{Z}n_{11}\ \ldots\ {}^{m}\mathcal{Z}n_{mm}\right]$$

(${}^{i}\mathcal{Z}$ being the ith column of adj $\mathcal{Z}$), and consequently, the unstable zeroes of $|\mathcal{Z}|$ are zeroes of elements of $\mathcal{N}$ (unless they are already contained in a column ${}^{k}\mathcal{Z}$).

2) If all the elements of N are reduced to a common denominator, this denominator must include every unstable factor of z.

Remark: Let us emphasise the importance of these general conditions, and the fact that they are only sufficient conditions which are, in general, unnecessarily strong. It will be shown later that it is possible to sharpen them. In particular, condition 2) does not require that all the denominators of diagonal elements of N should necessarily contain all the unstable factors of z.

We emphasise above all that, from condition 1), if the determinant of the process contains an unstable zero, then any non-interactive control scheme which can be realised will normally have a zero situated in the right half-plane, i.e. will be non-minimum phase.

B - Approach by state-space methods (1)

Let the system be defined by its state equations, put in the form

$$\dot{x} = Ax + Bu$$

$$y = x^{1} \tag{5.12}$$

where m of the n state variables are taken as the components of the output. Let the control vector u take the form (state feedback)

$$u = Le + Rx \tag{5.13}$$

(1) This approach is that proposed by B.S. Morgan in ref. 5.26.

where L and R are constant matrices of respective dimensions (m, m) and (m, n) (cf. figure 5.7). The equations of the closed-loop system can then be written:

$$\dot{x} = (A + BR)\, x + BLe = \alpha e + \beta e$$

$$y = x^1 = Mx \; .$$

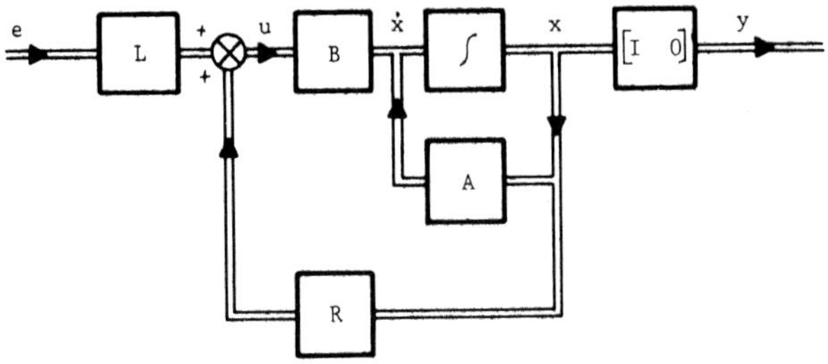

Fig. 5.7 Illustration of the principle of state feedback compensation.

If we decompose the matrices so as to display square blocks of dimensions (m, m), in the forms, e.g.

$$A = \begin{bmatrix} A_{11} & A_{12} \\ A_{21} & A_{22} \end{bmatrix} \qquad B = \begin{bmatrix} B_{11} \\ B_{21} \end{bmatrix} \qquad R = \begin{bmatrix} R_{11} & R_{12} \end{bmatrix}$$

where the matrices A_{11}, B_{11}, R_{11} are of dimensions (m, m), sufficient conditions for non-interaction can be written, with the matrices α and β decomposed like A and B,

$$A_{11} + B_{11} R_{11} = \text{diagonal } [\alpha_{11}]$$

$$\alpha_{12} = 0 = A_{12} + B_{11} R_{12} \tag{5.14}$$

$$\beta_{11} = \text{diagonal}$$

and we shall have, further:

$$\alpha_{21} = A_{21} + B_{21} R_{11} \,, \qquad \alpha_{22} = A_{22} + B_{21} R_{12}$$

$$\beta_{11} = B_{11} L \qquad , \qquad \beta_{21} = B_{21} L \; .$$

If B_{11} is nonsingular [i.e. in fact, if CB is nonsingular (1)] , we can express R and L in the form:

$$L = B_{11}^{-1} \beta_{11}$$

$$R_{11} = B_{11}^{-1} (\alpha_{11} - A_{11}), \qquad R_{12} = - B_{11}^{-1} A_{12}$$

and it follows that α_{21} and α_{22} then have the expressions:

$$\alpha_{21} = A_{21} + B_{21} B_{11}^{-1}(\alpha_{11} - A_{11}) ,$$

$$\alpha_{22} = A_{22} - B_{21} B_{11}^{-1} A_{12} ;$$

this last expression shows that the matrix α_{22} is, under these conditions, uniquely determined by the original system when we impose the non-interaction constraints (5.14).

The system equations being defined by:

$$\begin{bmatrix} \dot{x}^1 \\ \dot{x}^2 \end{bmatrix} = \begin{bmatrix} \alpha_{11} & 0 \\ \alpha_{21} & \alpha_{22} \end{bmatrix} \begin{bmatrix} x^1 \\ x^2 \end{bmatrix} + \begin{bmatrix} \beta_{11} \\ \beta_{21} \end{bmatrix} e \qquad \text{where } x = \begin{bmatrix} x^1 \\ x^2 \end{bmatrix}$$

the dynamics are characterised by the zeros of the product:

$$|sI - \alpha_{11}| \; |sI - \alpha_{22}| .$$

The first factor presents no problem, since α_{11} can be chosen arbitrarily. On the other hand, if the system is such that $|sI - \alpha_{22}|$ contains unstable factors, we cannot, at least within the framework of the compensation structure chosen here, achieve both decoupling and stability simultaneously.

It is important to notice, also, that the equations of the compensated system, in terms of transfer function matrices, become:

$$\begin{aligned} X^1(s) &= (sI-\alpha_{11})^{-1} \beta_{11}E(s) = K_1(s)E(s) \\ X^2(s) &= (sI-\alpha_{22})^{-1} [\alpha_{21}(sI-\alpha_{11})^{-1} \beta_{11} + \beta_{21}]E(s) \\ &= (sI-\alpha_{22})^{-1} [A_{21}+B_{21}B_{11}^{-1} (sI-A_{11})] (sI-\alpha_{11})^{-1}\beta_{11}E(s) \\ &= K_2(s)E(s) \end{aligned}$$

(1) cf. also paragraph 5.3.1.A

and that the closed-loop transfer function matrix is given by:

$$Y(s) = X^1(s) = K_1(s)E(s).$$

The modes of $(sI - \alpha_{22})$ are thus unobservable.

Examples

1) Let the process be defined by the transfer function matrix:

$$Z(s) = \frac{1}{(s-1)(s+1)(s+2)} \begin{bmatrix} s^2+6 & s^2+s+4 \\ 2s^2+7s-9 & s^2+4s-5 \end{bmatrix}$$

which has a representation of the form:

$$\dot{x} = \begin{bmatrix} 1 & 1 & -0,5 \\ 0 & -1 & 0,5 \\ 0 & 0 & -2 \end{bmatrix} x + \begin{bmatrix} 1 & 1 \\ 2 & 1 \\ 10 & 6 \end{bmatrix} u, \qquad y = \begin{bmatrix} 1 & 0 & 0 \\ 0 & 1 & 0 \end{bmatrix} x .$$

With the previous notation, we have:

$$B_{11} = \begin{bmatrix} 1 & 1 \\ 2 & 1 \end{bmatrix}$$

$$\alpha_{22} = A_{22} - B_{21} B_{11}^{-1} A_{12} = [-2] - [10 \quad 6] \begin{bmatrix} -1 & 1 \\ 2 & -1 \end{bmatrix} \begin{bmatrix} -0,5 \\ 0,5 \end{bmatrix} = -3$$

and, consequently, $sI - \alpha_{22}$ has only one factor, which is stable. It will thus be possible to achieve stable non-interactive control and the above method is directly applicable. We have, taking, for example

α_{11} = diagonal (a_i), β_{11} = diagonal (b_i),

$$K_1(s) = \text{diagonal} \left[\frac{b_i}{s-a_i}\right]$$

and

$$K_2(s) = \frac{1}{s+3} \left[\frac{2b_1(s-1)}{s-a_1}, \quad \frac{2b_2(2s+1)}{s-a_2}\right] .$$

Taking, e.g.,

$$a_1 = a_2 = -1, \qquad b_1 = b_2 = 1$$

we shall have:

$$\begin{bmatrix} Y_1(s) \\ Y_2(s) \\ X_3(s) \end{bmatrix} = \begin{bmatrix} \frac{1}{s+1} & 0 \\ 0 & \frac{1}{s+1} \\ \frac{2(s-1)}{(s+1)(s+3)} & \frac{2(2s+1)}{(s+1)(s+3)} \end{bmatrix} E(s)$$

We deduce also:

$$L = B_{11}^{-1}\,\beta_{11} = \begin{bmatrix} -1 & 1 \\ 2 & -1 \end{bmatrix}$$

$$R = \begin{bmatrix} R_{11} & R_{12} \end{bmatrix} = \begin{bmatrix} 2 & 1 & -1 \\ -4 & -2 & 1{,}5 \end{bmatrix}$$

whence we have the structure of a possible control system (cf. figure 5.8).

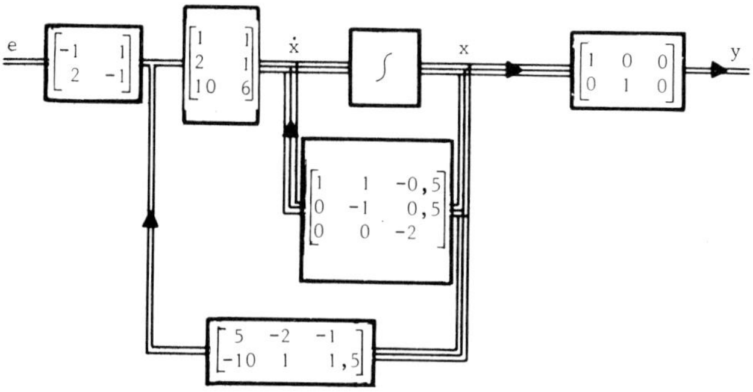

Fig. 5.8 Non-interaction by state feedback, for the example of paragraph 5.2.2.B.1).

2) Consider now the example proposed by Morgan (cf. ref. 5.26):

$$Z(s) = \frac{1}{(s-1)(s+1)(s+2)} \begin{bmatrix} s^2+6 & s^2+s+4 \\ 2s^2-7s-2 & s^2-5s-2 \end{bmatrix}$$

which in state-space form, can be represented by:

$$\dot{x} = \begin{bmatrix} 0 & -1 & 0 \\ -1 & -2 & -2 \\ 1 & 0 & 0 \end{bmatrix} x + \begin{bmatrix} 1 & 1 \\ 2 & 1 \\ 3 & 2 \end{bmatrix} u \,, \qquad y = \begin{bmatrix} 1 & 0 & 0 \\ 0 & 1 & 0 \end{bmatrix} x \,.$$

This time, we have:

$$\alpha_{22} = 0 - \begin{bmatrix} 3 & 2 \end{bmatrix} \begin{bmatrix} -1 & 1 \\ 2 & -1 \end{bmatrix} \begin{bmatrix} 0 \\ -2 \end{bmatrix} = +2$$

and $sI - \alpha_{22}$ has an unstable factor with $s = 2$. It is certainly still possible, since CB is invertible, to produce a non-interactive control scheme, but it will not be stable.

Taking, e.g.

$$\alpha_{11} = \text{diagonal} \begin{bmatrix} a_i \end{bmatrix}, \qquad \beta_{11} = \text{diagonal} \begin{bmatrix} b_i \end{bmatrix}$$

$$a_i = -4, \qquad b_i = 1$$

we shall indeed have

$$Y(s) = \frac{1}{s+4} E(s)$$

but the state x_3 of the system, which is unobservable, has dynamics characterised by $K_2(s)$, i.e.

$$X_3(s) = \frac{1}{(s-2)(s+4)} \begin{bmatrix} s+2, & s+3 \end{bmatrix} E(s)$$

There is thus an instability.

C - Remarks and provisional conclusions

1) We note, in both the above examples, that the unobservable mode which we cannot influence corresponds to a zero of the determinant of Z. In fact, in the first case, we have:

$$|Z| = \frac{-(s+3)}{(s-1)(s+1)(s+2)}$$

and, in the second case:

$$|Z| = \frac{-(s-2)}{(s-1)(s+1)(s+2)}$$

This observation, which reveals the same phenomenon as was indicated in paragraph 5.2.2.A, is moreover general. Indeed, the transfer function matrix of the process is, using equations (5.12) and taking account of the partitioning of the matrices, equal to:

$$Z(s) = \left[I-(sI-A_{11})^{-1}A_{12}(sI-A_{22})^{-1}A_{21}\right]^{-1}(sI-A_{11})^{-1}\left[A_{12}(sI-A_{22})^{-1}B_{21}+B_{11}\right]$$

so that, in another form,

$$Z(s) = \left[(sI-A_{11})-A_{12}(sI-A_{22})^{-1}A_{21}\right]^{-1}\left[A_{12}(sI-A_{22})^{-1}B_{21}+B_{11}\right],$$

$$|Z| = |sI-A_{22}|\,|A_{12}(sI-A_{22})^{-1}B_{21}+B_{11}|\Big/|sI-A|$$

whence

$$|Z| = |B_{11}|\,|sI-A_{22}+B_{21}B_{11}^{-1}A_{12}|\Big/|sI-A| = |B_{11}|\,|sI-\alpha_{22}|\Big/|sI-A|.$$

The zeroes of $|Z|$ are thus those of $|sI - \alpha_{22}|$ and conversely; we see the essential role which these zeroes play in all problems of non-interaction, whatever may be the method finally adopted, whether state feedback or dynamic precompensation.

2) It should also be quite clear that what has just been said does not constitute a full treatment of the problem of non-interaction by state feedback, since the conditions (5.14) would in general represent only sufficient conditions. Our sole intention was to show that the imposition of non-interaction on a system can incur the expense of being unable to control some of the modes (cf. example 2). In anticipation of the developments which will be made in the following subsection, let us also emphasise that the condition "CB invertible" found here is a sufficient, but not necessary, condition for a system (A, B, C) to be capable of non-interactive control by state feedback.

Attention having now been drawn to these fundamental problems, seen from an overall viewpoint, the object of the following paragraphs will be to specify in more detail the design of a non-interactive control system. For the reasons already indicated, we shall begin by treating the problem of state feedback.

5.3 NON-INTERACTION BY STATE-SPACE TECHNIQUES (1)

5.3.1 Necessary and sufficient condition for decoupling by state feedback

Let a system be defined by its state equations

$$\dot{x} = Ax + Bu \qquad\qquad y = Cx$$

the number of outputs being assumed equal to the number of inputs (p = m), and let $\mathcal{G}$ be the matrix defined by:

$$\mathcal{G} = \begin{bmatrix} {}_1CA^{d_1}B \\ {}_2CA^{d_2}B \\ \vdots \\ {}_mCA^{d_m}B \end{bmatrix} \tag{5.15}$$

In this expression, ${}_iC$ represents, in accordance with the notation used in this book, the ith row of the matrix C, and the index d_i is defined as the smallest power of A such that ${}_iCA^{d_i}B$ is non-null, i.e.

$${}_iCB = 0 , \qquad {}_iCAB = 0 , \ \ldots , \qquad {}_iCA^{d_i-1}B = 0 , \qquad {}_iCA^{d_i}B \neq 0 .$$

A - Gilbert's fundamental theorem

Given a control law of the form

$$u = Rx + Le$$

there exist two matrices R and L which diagonalise the system, if and only if the matrix $\mathcal{G}$ is invertible. For these matrices R and L, the closed-loop response in each channel reduces to a multiple integration.

1) Proof of sufficiency

Suppose in fact that $|\mathcal{G}|$ is different from zero: let the matrices R and L be defined by:

$$R = -\mathcal{G}^{-1}A^* \qquad L = \mathcal{G}^{-1} \tag{5.16}$$

(1) The whole of this section, concerning non-interaction by state feedback methods, makes use of the results of E.G. Gilbert, ref. 5.12, and of P.L. Falb and W.A. Wolovich, refs. 5.7 and 5.8.

with

$$A^* = \begin{bmatrix} {}_1CA^{d_1+1} \\ \vdots \\ {}_mCA^{d_m+1} \end{bmatrix} \tag{5.17}$$

The equations of the compensated system $\bar{S}$, whose structure is indicated in figure 5.7, may be written, taking account of equations (5.16) and (5.17):

$$\dot{x} = (A - B\mathcal{G}^{-1} A^*)\, x + B\mathcal{G}^{-1} e = \bar{A}x + \bar{B}e$$

$$y = Cx = \bar{C}x \ .$$

It is then easy to verify that the matrices $\mathcal{G}$ and $\bar{A}^*$, associated with the system $\bar{S}$ thus defined, have very special properties. We have, in fact,

$${}_i\overline{CA}^k\, \bar{B} = {}_iC(A - B\mathcal{G}^{-1} A^*)^k\, B\mathcal{G}^{-1} = {}_iCA^k\, B\mathcal{G}^{-1} + \Sigma\, {}_iCA^{\nu}\, BM + \Sigma\, {}_iCB(B^{\rho}\, A^{\mu})\, N \ .$$

All the terms on the right-hand side of this equation vanish if $k \leq d_i - 1$. For $k = d_i$, the first term alone is non-null and may be written

$${}_iCA^{d_i}\, B\mathcal{G}^{-1} = {}_i\mathcal{G}\mathcal{G}^{-1} = (0 \quad 0 \ \ldots \ 1 \ \ldots \ 0)$$

It follows that: $\bar{d}_i = d_i$ and that $\mathcal{G} = I_m$, the unit matrix of dimensions (m, m).

As for the expression for $\bar{A}^*$, it is easily determined by calculating its ith row

$${}_i\bar{A}^* = {}_i\overline{CA}^{d_i+1} = {}_iC(A - B\mathcal{G}^{-1} A^*)^{d_i+1}$$

$${}_i\bar{A}^* = {}_iCA^{d_i+1} - {}_iCA^{d_i}\, B\mathcal{G}^{-1} A^* + \sum_{k=1}^{d_i-1} {}_iCA^k\, BX + {}_iCBY$$

Only the first two terms on the right-hand side are non-null, and

$${}_i\bar{A}^* = {}_iCA^{d_i+1} - {}_i\mathcal{G}\mathcal{G}^{-1} A^* = {}_iA^* - {}_iA^* = 0 \ .$$

The matrix $\bar{A}^*$ associated with the system $\bar{S}$ is thus null.

The calculation of the transfer function matrix H of the compensated system can then be performed by using Leverrier's algorithm

$$ {}_i\bar{H} = {}_iC(sI-\bar{A})^{-1}\bar{B} = {}_iC(s^{n-1}I + \bar{R}_1 s^{n-2} +\ldots + \bar{R}_{n-1})\bar{B}\phi^{-1}(s) $$

with

$$ \phi(s) = s^n + \bar{a}_{n-1}s^{n-1} +\ldots + \bar{a}_1 s + \bar{a}_0 $$

$$ \bar{A}_1 = \bar{A}, \quad \bar{a}_{n-1} = -\operatorname{trace}\bar{A}_1, \quad \bar{R}_1 = \bar{A} + \bar{a}_{n-1} $$

$$ \bar{A}_2 = \overline{AR}_1, \quad \bar{a}_{n-2} = -1/2\operatorname{trace}A_2, \quad \bar{R}_2 = \bar{A}^2 + \bar{a}_{n-1}\bar{A} + \bar{a}_{n-2}. $$

We then have:

$$ {}_i\bar{H} = \phi^{-1}(s)\left[{}_iC\bar{B}s^{n-1} + \ldots + {}_iC(\bar{A}^{n-1} +\ldots + \bar{a}_1 I)\bar{B}\right] $$

and, taking account of the form $\bar{A}^*$,

$$ {}_i\bar{H} = \phi^{-1}(s)\,{}_iC\bar{A}^{d_i}\bar{B}\left[s^{n-d_i-1} +\ldots + \bar{a}_{d_i+1}\right] $$

$$ {}_i\bar{H} = \left[0 \ldots h_{ii} \ldots 0\right] $$

with

$$ h_{ii} = \phi^{-1}(s)\left[s^{n-d_i-1} +\ldots + \bar{a}_{d_i+1}\right] \tag{5.18} $$

The closed-loop system is thus indeed non-interactive, which proves the sufficiency of the theorem.

2) <u>Proof of necessity</u>

Suppose that there are indeed two matrices R and L which decouple the system, i.e.

$$ {}_iH = {}_iC(sI-A-BR)^{-1}BL = \lambda_i\,{}_iE $$

${}_iE$ being a row-vector whose only nonzero element is in the ith position. Again using Leverrier's algorithm, we have, in view of the fact that ${}_iCA^kB = 0$ for $k < d_i$:

$$ {}_iH = {}_iCA^{d_i}BLs^{-d_i-1} + {}_iCA^{d_i}(A+BR)BLs^{-d_i-2} + \ldots\ldots $$

In particular, considering the terms of highest degree, it follows that:

$$ {}_iCA^{d_i}\ BL = {}_i\mathcal{J}\ L = \lambda_i\ {}_iE $$

where λ_i is a nonzero scalar.

It thus follows that

$$ \mathcal{J}L = \begin{bmatrix} \lambda_1\ {}_1E \\ \vdots \\ \lambda_m\ {}_mE \end{bmatrix} = \Lambda $$

with Λ being a nonsingular diagonal matrix. Since L is also nonsingular, this last equation has a solution only if $\mathcal{J}$ is invertible. This proves not only the necessity of the theorem but also the fact that all the possible matrices L are given by:

$$ L = \mathcal{J}^{-1}\ \Lambda\ . $$

B - Expression for the transfer function matrix

Let us consider equation (5.18) again. In Leverrier's algorithm, we know that the last matrix $\bar{R}_n$ of the sequence $\bar{R}_i$ is null, because of the Cayley-Hamilton theorem:

$$ \bar{R}_n = \bar{A}^n + \bar{a}_{n-1}\ \bar{A}^{n-1} + \ldots + \bar{a}_0\ I = 0\ . $$

In particular, it follows that:

$$ {}_iC\bar{R}_n\ \bar{B} = 0\ ; $$

now,

$$ {}_iC\bar{R}_n\ \bar{B} = {}_iC\bar{A}^{d_i}\ \bar{B}\ \bar{a}_{d_i} + \sum_{\substack{k=0 \\ k\neq d_i}}^{n} \bar{a}_k\ {}_iC\bar{A}^k\ \bar{B}\ . $$

In the expression under the summation sign, all the terms are null, the first d_i-1 by definition of d_i, the rest because of the special properties of $\bar{A}^*$. It follows that

$$_i C\bar{R}_n \bar{B} = {}_i E\bar{a}_{d_i} = 0$$

whence

$$\bar{a}_{d_i} = 0 .$$

We could show similarly that $a_{d_i-1}, \ldots, a_0$ are null. As a result,

$$h_{ii} = \frac{s^{n-d_i-1} + \ldots + \bar{a}_{d_i+1}}{s^n + \ldots + \bar{a}_{d_i+1} s^{d_i+1}} = \frac{1}{s^{d_i+1}} \qquad (5.19)$$

The compensated system is thus characterised not only by input-output decoupling. Each output is the $(d_i+1)^{th}$ integral of the corresponding input, as illustrated in figure 5.9.

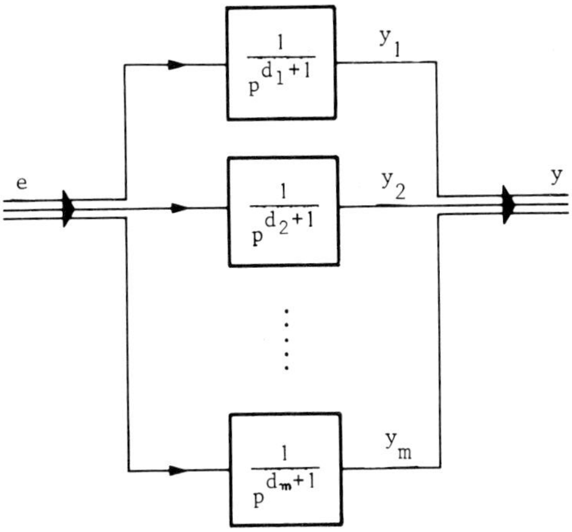

Fig. 5.9 Reduction of the system of figure 5.7 for $L = \mathcal{G}^{-1}$ and $R = -\mathcal{G}^{-1}\bar{A}^*$.

5.3.2 The idea of a non-linear transformation

We demonstrated in paragraph 5.3.1.A, the existence of special systems $\bar{S}$ whose matrices $\mathcal{G}$ and $\bar{A}^*$ reduce to I_m and 0_m, respectively.

We also saw that, if the original system is capable of being controlled non-interactively (i.e. $\mathcal{G}$ is invertible), the relations connecting the evolution and control matrices of S and $\bar{S}$ can be written

$$\bar{A} = A - B\mathcal{G}^{-1} A^* \tag{5.20}$$

$$\bar{B} = B\mathcal{G}^{-1} . \tag{5.21}$$

This transformation, which is not a linear similarity transformation, can be interpreted as resulting from the use of the particular compensation matrices defined by equation (5.16).

The control corresponding to system $\bar{S}$ is in general unacceptable (we have seen that it reduces to a multiple integration). Nevertheless, the structural simplicity of this system leads us to expect that it will be easier to compensate than the original system S. Moreover, if the matrices $\bar{R}$ and $\bar{L}$, associated with $\bar{S}$, are related to the matrices R and L, associated with S, by the equations

$$\bar{L} = \mathcal{G} L \tag{5.22}$$

$$\bar{R} = \mathcal{G} R + A^* \tag{5.23}$$

it is easy to see that the two systems thus compensated have the same overall transfer function matrix. In fact, it can be written

$$\begin{aligned} \bar{H} &= C(sI - \bar{A} - \bar{B}\bar{R})^{-1}\bar{B}\bar{L} \\ &= C(sI - A + B\mathcal{G}^{-1}A^* - B\mathcal{G}^{-1}\mathcal{G}R - B\mathcal{G}^{-1}A^*)^{-1}B\mathcal{G}^{-1}\mathcal{G}L \\ &= C(sI - A - BR)^{-1}BL \\ &= H. \end{aligned}$$

We shall thus begin by compensating the system $\bar{S}$, using techniques which will be detailed in the following subsection. When matrices $\bar{R}$ and $\bar{L}$ giving satisfactory dynamics have been found, we shall deduce the matrices R and L, associated with the original system, from the inverse formulae to (5.22) and (5.23), namely

$$L = \mathcal{G}^{-1} \bar{L} \qquad R = \mathcal{G}^{-1}(\bar{R} - A^*) . \tag{5.24}$$

This set of procedures is summarised in Table 5.1.

TABLE 5.1

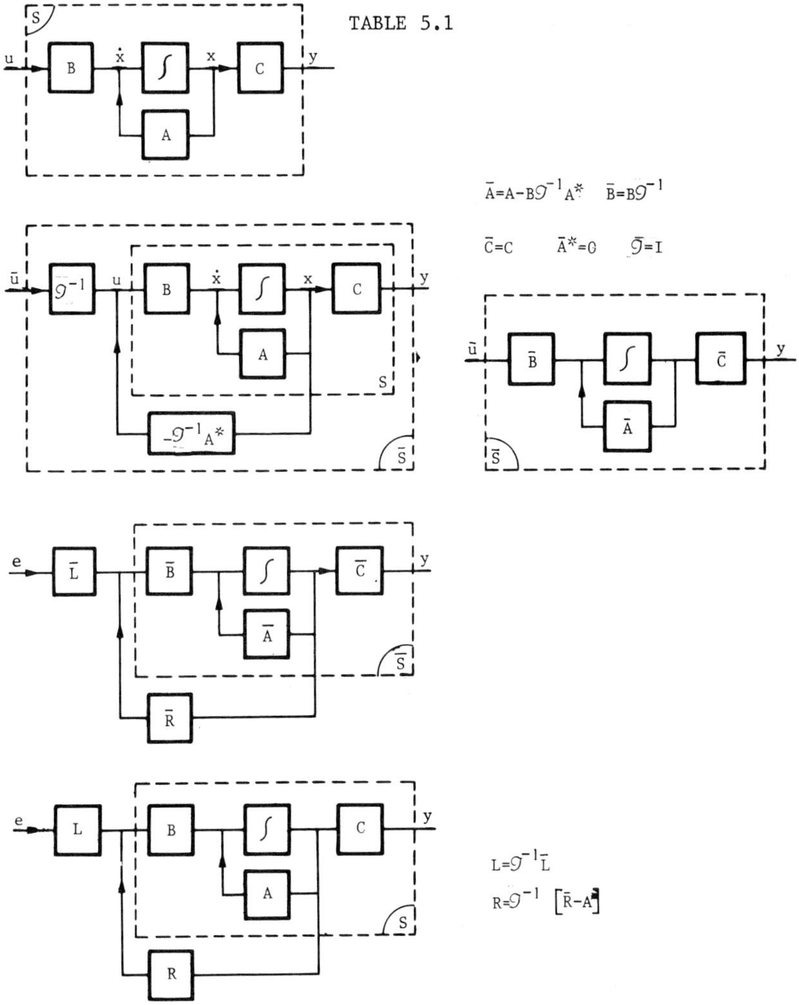

5.3.3 Canonical form of an integrator decoupled system $\bar{S}$

A - Structure

Let the system be defined by its state equations

$$\dot{x} = \bar{A}x + \bar{B}u \ , \qquad \dot{y} = \bar{C}x$$

where we assume that:- the system is controllable; the number of outputs p is equal to the number of inputs m; the system is of the type $\bar{S}$ defined in the previous subsection, i.e. the associated matrices $\mathcal{G}$ and $\bar{A}^*$ are such that

$$\overline{\mathcal{G}} = I_m \qquad \overline{A}^* = 0 \,. \tag{5.25}$$

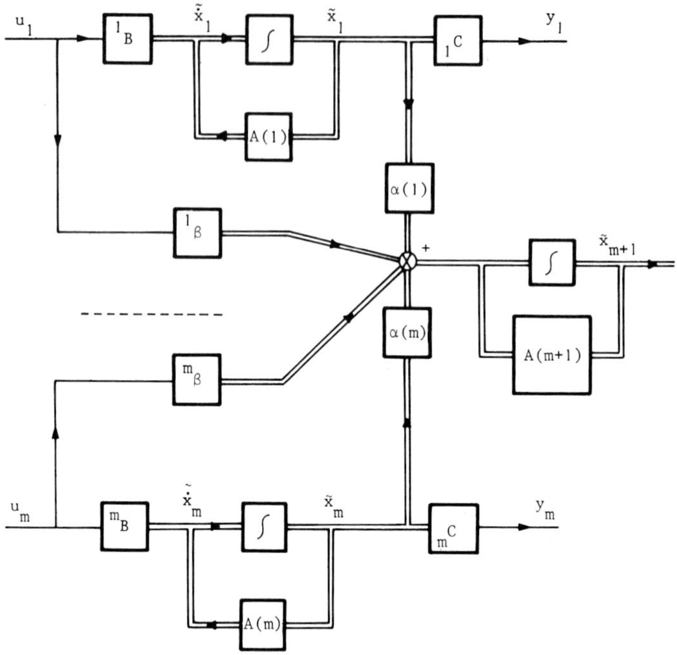

Fig. 5.10 Canonical structure of $\tilde{S}$.

These assumptions enable us to find a linear transformation defined by the matrix $T(\tilde{x} = Tx)$ (1), such that the matrices A, B, C, of the resultant system, have the structures (2) indicated in equations (5.26):

$$\tilde{A} = \begin{bmatrix} A(1) & 0 & 0 & \vdots & 0 & 0 \\ 0 & A(2) & 0 & \vdots & 0 & 0 \\ \cdots & \cdots & \cdots & \cdots & \cdots & \cdots \\ 0 & 0 & \cdots & \cdots & A(m) & \\ \alpha(1) & \alpha(2) & \cdots & \cdots & \alpha(m) & A(m+1) \end{bmatrix}, \quad \tilde{B} = \begin{bmatrix} {}^1B & & & \\ & {}^2B & & \\ & & \ddots & \\ & & & {}^mB \\ {}^1\beta & {}^2\beta & \cdots & {}^m\beta \end{bmatrix}$$

(1) We have therefore the relations

(2) These structures were, to our knowledge, first pointed out by E.G. Gilbert, ref. 5.12.

$$\tilde{A} = \begin{bmatrix} A(1) & 0 & 0 & \vdots & 0 & 0 \\ 0 & A(2) & 0 & \vdots & 0 & 0 \\ \cdots & \cdots & \cdots & \cdots & \cdots & \cdots \\ 0 & 0 & \cdots & \cdots & A(m) & \\ \alpha(1) & \alpha(2) & \cdots & \cdots & \alpha(m) & A(m+1) \end{bmatrix}, \quad \tilde{B} = \begin{bmatrix} {}^1B & & & \\ & {}^2B & & \\ & & \ddots & \\ & & & {}^mB \\ {}^1\beta & {}^2\beta & \cdots & {}^m\beta \end{bmatrix}$$

$$\tilde{C} = \begin{bmatrix} {}_1C & 0 & \cdots & \cdots & 0 \\ 0 & {}_2C & \ddots & & \vdots \\ 0 & \cdots & 0 \quad {}_mC & 0 & \cdots \; 0 \end{bmatrix}$$

where A(i) is a square matrix of dimensions (n_i, n_i), of the form

$$A(i) = \left[\begin{array}{cc|c} \begin{matrix} 0 \\ \vdots \\ 0 \end{matrix} & \begin{matrix} I_{d_i} \\ \cdots\cdots\; 0 \end{matrix} & 0 \\ \hline \multicolumn{2}{c|}{\Gamma(i)} & \phi(i) \end{array}\right], \qquad (5.27)$$

while ${}^iB$, ${}_iC$ are column and row vectors

$${}^iB = \begin{bmatrix} 0 & 0 & \cdots & 1 & {}^ib^T \end{bmatrix}^T \qquad (5.28)$$

$${}_iC = \begin{bmatrix} 1 & 0 & \cdots & 0 \end{bmatrix} . \qquad (5.29)$$

In equation (5.27), $\phi(i)$ is a square matrix of dimensions (r_i, r_i), the dimensions of $\Gamma(i)$ being (r_i, d_i+1). In the expression for ${}^iB$, the first d_i elements are null, and we have:

$$n_i = r_i + d_i + 1 . \qquad (5.30)$$

B - Finding the transformation matrix

Let us decompose the transformation matrix T into row blocks, of dimensions n_i, corresponding to the dimensions of the submatrices A(i). The equation $\tilde{A}T = T\bar{A}$ becomes, in view of the structure of A,

$$\begin{bmatrix} T(1) \\ T(2) \\ \vdots \\ T(m+1) \end{bmatrix} \bar{A} = \begin{bmatrix} A(1) & & & 0 \\ 0 & & & \vdots \\ \vdots & & & \\ 0 & & A(m) & 0 \\ \alpha(1) & \dots & \alpha(m) & A(m+1) \end{bmatrix} \begin{bmatrix} T(1) \\ T(2) \\ \vdots \\ T(m+1) \end{bmatrix}$$

i.e. m matrix equations of the form:

$$T(i)\ \bar{A} = A(i)\ T(i) \qquad \text{for} \qquad i = 1, 2, \dots, m \qquad (5.30)$$

and one equation of the form:

$$T(m+1)\ \bar{A} = \sum_{i=1}^{m} \alpha(i)\ T(i) + A(m+1)\ T(m+1)\ . \qquad (5.31)$$

If, in any of the first m equations (5.30), we display the n_i rows of T(i) in the form

$$\begin{bmatrix} {}_1T(i) \\ {}_2T(i) \\ \vdots \\ {}_{n_i}T(i) \end{bmatrix} \bar{A} = \left[\begin{array}{c|cccc|c} 0 & 1 & & & & \\ 0 & 0 & 1 & & & \\ \vdots & & & \ddots & & 0 \\ & & & 0 & 1 & \\ 0 & 0 & \dots & & 0 & \\ \hline & & \Gamma(i) & & & \phi(i) \end{array}\right] \begin{bmatrix} {}_1T(i) \\ {}_2T(i) \\ \vdots \\ {}_{n_i}T(i) \end{bmatrix} \qquad (5.32)$$

identifying the first d_i+1 rows gives:

$${}_jT(i)\ \bar{A} = {}_{j+1}T(i) \qquad j = 1, 2, \dots, d_i \qquad (5.33)$$

and

$${}_{d_i+1}T(i)\ \bar{A} = 0\ . \qquad (5.34)$$

The equation $\bar{C} = \tilde{C}T$, in its turn, becomes, displaying the m rows ${}_i\bar{C}$ of $\bar{C}$,

$${}_i\bar{C} = {}_iCT(i)$$

i.e.

$${}_i\bar{C} = \begin{bmatrix} 1 & 0 & \dots & 0 \end{bmatrix} \begin{bmatrix} {}_1T(i)^T & {}_2T(i)^T \dots & {}_{n_i}T(i)^T \end{bmatrix}^T = {}_1T(i)\ . \qquad (5.35)$$

Equations (5.33) can, in view of (5.35), be rewritten to give:

$$
\begin{aligned}
{}_1T(i) &= {}_i\bar{C} \\
{}_2T(i) &= {}_i\bar{C}\,\bar{A} \qquad\qquad (5.36) \\
&\ldots\ldots\ldots\ldots
\end{aligned}
$$

which is to say

$$
\begin{aligned}
{}_{d_i+1}T(i) &= {}_i\bar{C}\,\bar{A}^{d_i}\,, \\
{}_kT(i) &= {}_i\bar{C}\bar{A}^{k-1}\,, \qquad k = 1, 2, \ldots, d_i + 1\,;
\end{aligned}
$$

as for equation (5.34), it can be rewritten as

$$
{}_i\bar{C}\,\bar{A}^{d_i+1} = 0
$$

which holds because of the assumption made about $\bar{A}^*$.

Equations (5.36) thus give us expressions for the first d_i+1 rows among the n_i contained in each block T(i). It also follows from equations (5.36) and the special form of , that these d_i+1 row vectors are orthogonal to all the column vectors of $\bar{B}$, $\bar{A}\bar{B}$, ... $\bar{A}^{n-1}\bar{B}$, with the exception of the ith column of each of these matrices.

As for the equation $\tilde{B} = T\bar{B}$, it becomes

$$
\begin{aligned}
T(i)\,\bar{B} &= \begin{bmatrix} 0 & 0 \ldots & {}^iB & 0 \ldots 0 \end{bmatrix} \qquad i = 1, 2, \ldots, m \\
T(m+1)\,\bar{B} &= \begin{bmatrix} {}^1\beta & {}^2\beta \ldots {}^m\beta \end{bmatrix} \qquad\qquad (5.37)
\end{aligned}
$$

where ${}^iB$ and ${}^i\beta$ are column vectors of respective dimensions n_i and $n - \sum n_i$. Writing out one of the first m of the matrix equations (5.37) in the form

$$
\begin{bmatrix} {}_1T(i) \\ {}_2T(i) \\ \vdots \\ {}_{d_i}T(i) \\ {}_{d_i+1}T(i) \\ \vdots \\ {}_{n_i}T(i) \end{bmatrix} \bar{B} = \left[\begin{array}{c|c|c} 0 \ldots 0 & 0 & 0 \ldots 0 \\ \cdot & & \cdot \\ \cdot & & \cdot \\ \cdot & & \cdot \\ \cdot & & \cdot \\ \cdot & 0 & \cdot \\ \cdot & & \cdot \\ \cdot & 1 & \cdot \\ & \vdots & \\ 0 & {}^ib & 0 \end{array}\right]
$$

gives

$$
{}_kT(i)\,\bar{B} = 0 \qquad\qquad k = 1, 2, \ldots, d_i
$$

and

$${}_{d_i+1}T(i)\ \bar{B} = {}_iE$$

i.e., in view of equations (5.36)

$${}_i\bar{C}\ \bar{A}^{k-1}\ \bar{B} = 0 \qquad k = 1, 2, \ldots, d_i$$

$${}_i\bar{C}\ \bar{A}^{d_i}\ \bar{B} = E_i$$

which are automatically satisfied as a result of the form assumed for in the original hypotheses. If we now consider the equations relating to the last r_i rows of each block T(i), we find that they are of two types: those arising from (5.32), which can be written

$${}_{d_i+1+k}T(i)\ \bar{A} = \left[{}_k\Gamma(i) \quad {}_k\phi(i)\right] \begin{bmatrix} {}_1T(i) \\ \vdots \\ {}_{d_i+1}T(i) \\ \vdots \\ {}_{d_i+1+r_i}T(i) \end{bmatrix} \qquad k = 1, 2, \ldots, r_i \tag{5.38}$$

where ${}_k\Gamma(i)$, ${}_k\phi(i)$ denote the kth rows of $\Gamma(i)$, $\phi(i)$, and those arising from (5.37), which we can write as

$${}_{d_i+1+k}T(i)\ \bar{B} = \left[0 \ldots 0 \quad b_{ik} \quad 0 \ldots 0\right] \tag{5.39}$$

where b_{ik} is the kth element of the column vector ${}^ib$. If, in the last equation, we display the columns of $\bar{B}$ in the form:

$$\bar{B} = \left[{}^1\bar{B} \quad {}^2\bar{B}, \ldots, {}^m\bar{B}\right]$$

it follows that

$${}_{d_i+1+k}T(i)\ {}^j\bar{B} = 0 \qquad k = 1, \ldots, r_i\ ; \quad j \neq i$$

$${}_{d_i+1+k}T(i)\ {}^i\bar{B} = b_{ik}$$

so that the last r_i rows of T(i) are orthogonal to every column of $\bar{B}$ except the ith.

Writing out equations (5.38), and taking account of (5.36), gives:

$$_{d_i+1+k}T(i)\,\bar{A} = {}_k\Gamma(i)\begin{bmatrix} {}_i\bar{C} \\ {}_i\bar{C}\,\bar{A} \\ \vdots \\ {}_i\bar{C}\bar{A}^{d_i} \end{bmatrix} + {}_k\phi(i)\begin{bmatrix} {}_{d_i+1+1}T(i) \\ \vdots \\ {}_{d_i+1+r_i}T(i) \end{bmatrix} \qquad (5.40)$$

$$_{d_i+1+k}T(i)\,\bar{A} = \sum_{\nu=1}^{d_i+1} \gamma_{k\nu}\;{}_i\bar{C}\,\bar{A}^{\nu-1} + \sum_{\mu=1}^{r_i} \phi_{k\mu}(i)\;{}_{d_i+1+\mu_i}T(i)$$

where $\gamma_{k\nu}(i)$ and $\phi_{k\mu}(i)$ denote the νth and μ^{th} elements of the row vectors $_k\Gamma(i)$ and $_k\phi(i)$. Multiplying equation (5.40) on the right by $\bar{B}$, we have:

$$_{d_i+1+k}T(i)\,\bar{A}\,\bar{B} = \sum_{\nu=1}^{d_i+1} \gamma_{k\nu}(i)\;{}_i\bar{C}\,\bar{A}^{\nu-1}\,\bar{B} + \sum_{\mu=1}^{r_i} \phi_{k\mu}(i)\;{}_{d_i+1+\mu}T(i)\,\bar{B}\,, \quad k = 1, \ldots, r_i$$

i.e., taking account of the form of $\bar{\mathcal{G}}$ and of equations (5.37):

$$_{d_i+1+k}T(i)\,\bar{A}\,\bar{B} = \gamma_{k,d_i+1}(i)\;{}_iE + \sum_{\mu=1}^{r_i} \phi_{k\mu}(i)\,\left[0 \ldots 0 \;\; b_{\mu i} \;\; 0 \ldots 0\right].$$

Displaying as before the m columns of $\bar{B}$, it follows that

$$k = 1, \ldots, r_i \qquad \begin{aligned} &_{d_i+1+k}T(i)\,\bar{A}\;{}^j\bar{B} = 0 \qquad \text{if} \quad j \neq i \\ &_{d_i+1+k}T(i)\,\bar{A}\;{}^i\bar{B} = \gamma_{k,d_i+1}(i) + \sum_{\mu=1}^{r_i} \phi_{k\mu}(i)\,b_{\mu i}\,. \end{aligned} \qquad (5.41)$$

From (5.39) and (5.41), it thus follows that the last r_i rows of each block T(i) are orthogonal to all the columns of $\bar{B}$ and $\bar{A}\bar{B}$ except the ith.

If we multiply (5.40) on the right by $\bar{A}\bar{B}$, we shall obtain, similarly,

$$\begin{aligned} _{d_i+1+k}T(i)\,\bar{A}^2\,B &= \sum_{\nu=1}^{d_i+1} \gamma_{k\nu}(i)\;{}_i\bar{C}\,\bar{A}^{\nu}\,\bar{B} + \sum_{\mu=1}^{r_i} \phi_{k\mu}(i)\;{}_{d_i+1+\mu}T(i)\,\bar{A}\,\bar{B} \\ &= \gamma_{k,d_i}(i)\;{}_iE + \sum_{\mu=1}^{r_i} \phi_{k\mu}(i)\;{}_{d_i+1+\mu}T(i)\,\bar{A}\,\bar{B} \end{aligned} \qquad k = 1, 2, \ldots, r_i$$

and, in consequence of equations (5.41),

$$k = 1, 2, \ldots, r_i \qquad _{d_i+1+k}T(i)\,\bar{A}^2\;{}^jB = 0 \quad \text{if } j \neq i$$

$$_{d_i+1+k}T(i)\,\bar{A}^2\;{}^iB = \gamma_{kd_i}(i) + \sum_{\mu=1}^{r_i} \phi_{k\mu}(i)\left[\gamma_{\mu,d_i+1}(i) + \sum_{\nu=1}^{r_i} \phi_{\mu\nu}(i)\,b_{\nu i}\right].$$

Proceeding thus successively, we establish that all the rows of T(i), the last r_i as well as the first d_i+1, are orthogonal to the sequence of vectors

$$({}^1\bar{B} \ldots {}^{i-1}\bar{B} \quad {}^{i+1}\bar{B} \ldots {}^{m}\bar{B} \quad \bar{A}\,{}^1B \ldots \bar{A}\,{}^{i-1}\bar{B} \quad \bar{A}\,{}^{i+1}\bar{B} \ldots \bar{A}^{m-1}\,{}^{m}\bar{B}) \; .$$

The subspace spanned by the independent vectors orthogonal to this sequence will be called $\mathcal{E}_i$, and n_i will be taken to be its dimension. It can be shown that (cf. e.g. ref. 5.12):

1) the subspaces $\mathcal{E}_i$ and $\mathcal{E}_j$, with $i \neq j$, are disjoint;

2) the first d_i+1 rows of each block T(i) are independent.

Remarks: In the case that r_i is nonzero, it is quite clear that the above conditions do not define the rows ${}_{d_i+1+k}T(i)$ uniquely. This means that the decomposition is not unique but can lead to different forms of T, and hence of the $\Gamma(i)$ and ${}^ib$.

If $\sum n_i$ is equal to n, the transformation matrix is completely determined. If $n - \sum n_i > 0$, the rows of T(m+1) will be taken so as to complete the whole space.

It remains to show that this choice, which ensures that T has full rank, is compatible with the structures chosen at the beginning, i.e. that we can satisfy the equations:

$$T(m+1)\,\bar{B} = \left[{}^1\beta \quad {}^2\beta \ldots {}^m\beta\right]$$

and

$$T(m+1)\,\bar{A} = \sum_{i=1}^{m} \alpha(i)\,T(i) + A(m+1)\,T(m+1) \; .$$

The first equation is satisfied by taking

$${}^k\beta = T(m+1)\,{}^k\bar{B}$$

and the second by determining the $\alpha(i)$ from:

$$T(m+1)\,\bar{A} - A(m+1)\,T(m+1) = \sum_{i=1}^{m} \alpha(i)\,T(i) \; .$$

This matrix equation represents a set of $n(n - \sum n_i)$ scalar equations, to determine the coefficients of the matrices $\alpha(i)$. It follows that, even though the dimension n_{m+1} is uniquely determined, the matrices associated with the space $\mathcal{E}_{m+1}$ are not.

C - Examples

1) Let the system S be defined by the state equations

$$\dot{x} = \begin{bmatrix} 1 & 1 & 0 \\ 0 & 2 & 0 \\ 0 & 1 & 3 \end{bmatrix} x + \begin{bmatrix} 1 & 1 \\ -1 & 1 \\ 0 & 0 \end{bmatrix} u \qquad y = \begin{bmatrix} 1 & 0 & 0 \\ 0 & 0 & 1 \end{bmatrix} x$$

a) Since the matrix takes the form

$$\mathfrak{J} = \begin{bmatrix} {}_1CB \\ {}_2CAB \end{bmatrix} = \begin{bmatrix} 1 & 1 \\ -1 & 1 \end{bmatrix}$$

the system is decouplable (R, L) and $d_1 = 0$, $d_2 = 1$.

b) The matrices $\bar{A}$, $\bar{B}$, $\bar{C}$ of the system $\bar{S}$ associated with S are

$$\bar{A} = A - B\mathfrak{J}^{-1} A^* = \begin{bmatrix} 0 & 0 & 0 \\ 0 & -3 & -9 \\ 0 & 1 & 3 \end{bmatrix}, \quad \bar{B} = B\mathfrak{J}^{-1} = \begin{bmatrix} 1 & 0 \\ 0 & 1 \\ 0 & 0 \end{bmatrix}, \quad \bar{C} = C$$

c) We next put the system $\bar{S}$ into the canonical form $\tilde{S}(\tilde{A}, \tilde{B}, \tilde{C})$ by means of the transformation T. To do this, we determine the spaces $\mathcal{E}_1$, $\mathcal{E}_2$. $\mathcal{E}_1$ is generated by the row vectors orthogonal to ${}^2\bar{B}$, $\bar{A}\,{}^2\bar{B}$,... . Only the first two vectors of this sequence are independent, and hence $\mathcal{E}_1$ is of dimension 1. It follows that $n_1 = 1$, $r_1 = 0$ and $T(1) = {}_1\bar{C} = (1,0,0)$. $\mathcal{E}_2$, generated by the single vector ${}^1\bar{B}$, is of dimension 2. It follows that $n_2 = 2$, $r_2 = 0$ and

$$T(2) = \begin{bmatrix} {}_2\bar{C} \\ {}_2\bar{C}\,\bar{A} \end{bmatrix} = \begin{bmatrix} 0 & 0 & 1 \\ 0 & 1 & 3 \end{bmatrix}.$$

We have then:

$$T = \begin{bmatrix} 1 & 0 & 0 \\ 0 & 0 & 1 \\ 0 & 1 & 3 \end{bmatrix} \qquad T^{-1} = \begin{bmatrix} 1 & 0 & 0 \\ 0 & -3 & 1 \\ 0 & 1 & 0 \end{bmatrix}$$

and

$$\tilde{A} = T\bar{A}T^{-1} = \begin{bmatrix} 0 & 0 & 0 \\ 0 & 0 & 1 \\ 0 & 0 & 0 \end{bmatrix}, \quad \tilde{B} = T\bar{B} = \begin{bmatrix} 1 & 0 \\ 0 & 0 \\ 0 & 1 \end{bmatrix}, \quad \tilde{C} = CT^{-1} = \begin{bmatrix} 1 & 0 & 0 \\ 0 & 1 & 0 \end{bmatrix}$$

We notice that, in this case, the space $\mathcal{E}_{m+1}$ is empty ($n_1 + n_2 = n$) and so are the matrices $\Gamma(1)$ and $\phi(1)$ ($r_1 = r_2 = 0$).

2) Let us now consider the system, also controllable and observable, defined by the matrices

$$A = \begin{bmatrix} 0 & 1 & 0 \\ 2 & 3 & 0 \\ 1 & 1 & 1 \end{bmatrix} \qquad B = \begin{bmatrix} 0 & 0 \\ 1 & 0 \\ 0 & 1 \end{bmatrix} \qquad C = \begin{bmatrix} 1 & 1 & 0 \\ 0 & 0 & 1 \end{bmatrix}$$

In this case we have, following the same steps as before:

a) $d_1 = d_2 = 0$:

$$\mathfrak{J} = \begin{bmatrix} {}_1CB \\ {}_2CB \end{bmatrix} = I_2 \qquad A^* = \begin{bmatrix} 2 & 4 & 0 \\ 1 & 1 & 1 \end{bmatrix}$$

b)

$$\bar{A} = \begin{bmatrix} 0 & 1 & 0 \\ 0 & -1 & 0 \\ 0 & 0 & 0 \end{bmatrix} \qquad \bar{B} = \begin{bmatrix} 0 & 0 \\ 1 & 0 \\ 0 & 1 \end{bmatrix} \qquad \bar{C} = \begin{bmatrix} 1 & 1 & 0 \\ 0 & 0 & 1 \end{bmatrix}$$

c) $\mathcal{E}_1$, defined by the row vectors orthogonal to ${}^2\bar{B}$, is of dimension 2, while $\mathcal{E}_2$, orthogonal to ${}^1\bar{B}$ and $\bar{A}\,{}^1\bar{B}$, is of dimension 1. It follows that

$$n_1 = 2 \qquad r_1 = 1 \qquad n_2 = 1 \qquad r_2 = 0$$

The first and last rows of T are then given, respectively, by:

$$_1T(1) = {}_1\bar{C} = \begin{bmatrix} 1 & 1 & 0 \end{bmatrix}$$

$$_1T(2) = {}_2\bar{C} = \begin{bmatrix} 0 & 0 & 1 \end{bmatrix} .$$

The second row $_2T(1)$ of $T(1)$ should be independent of $_1\bar{C}$ and orthogonal to $_2\bar{B}$, hence of the form $(\alpha, \beta, 0)$ with $\alpha \neq \beta$. We can take, for example:

$$_2T(1) = \begin{bmatrix} 0 & 1 & 0 \end{bmatrix} .$$

In this case,

$$\tilde{A} = T\bar{A}T^{-1} = \begin{bmatrix} 0 & 0 & 0 \\ 0 & -1 & 0 \\ 0 & 0 & 0 \end{bmatrix}, \quad \tilde{B} = \begin{bmatrix} 1 & 0 \\ 1 & 0 \\ 0 & 1 \end{bmatrix}, \quad \tilde{C} = \begin{bmatrix} 1 & 0 & 0 \\ 0 & 0 & 1 \end{bmatrix}.$$

We may note that we could equally well have taken for $_2T(1)$ the row vector $(1, 0, 0)$. We should have had

$$\tilde{A} = \begin{bmatrix} 0 & 0 & 0 \\ 1 & -1 & 0 \\ 0 & 0 & 0 \end{bmatrix}, \quad \tilde{B} = \begin{bmatrix} 1 & 0 \\ 0 & 0 \\ 0 & 1 \end{bmatrix}, \quad \tilde{C} = \begin{bmatrix} 1 & 0 & 0 \\ 0 & 0 & 1 \end{bmatrix}.$$

This variety of decompositions is by no means an embarassment, since it will always lead, as we shall see, to the same feedback matrix R.

We note that, in this case, the space $\mathcal{E}_{m+1}$ is still empty but that, since r_1 is nonzero, we have the matrices $\Gamma(1)$ and $\phi(1)$, namely, $\Gamma(1) = 0$, $\phi(1) = -1$, $^1b = 1$, in the first decomposition, and $\Gamma(1) = 1$, $\phi(1) = -1$, $^1b = 0$, in the second.

3) Let us consider, as a final example, the system defined by

$$A = \begin{bmatrix} 0 & 1 & 2 \\ 1 & 2 & 0 \\ 1 & 3 & 1 \end{bmatrix} \qquad B = \begin{bmatrix} 1 & 1 \\ 0 & 1 \\ 1 & 0 \end{bmatrix} \qquad C = \begin{bmatrix} 1 & 0 & 0 \\ 1 & 0 & 1 \end{bmatrix}.$$

We have successively

$$d_1 = 0 \qquad \mathcal{G} = \begin{bmatrix} 1 & 1 \\ 2 & 1 \end{bmatrix} \qquad A = \begin{bmatrix} 0 & 1 & 2 \\ 1 & 4 & 3 \end{bmatrix}$$
$$d_2 = 0$$

and then

$$\bar{A} = \begin{bmatrix} 0 & 0 & 0 \\ 2 & 4 & -1 \\ 0 & 0 & 0 \end{bmatrix} \qquad \bar{B} = \begin{bmatrix} 1 & 0 \\ 2 & -1 \\ -1 & 1 \end{bmatrix} \qquad \bar{C} = \begin{bmatrix} 1 & 0 & 0 \\ 1 & 0 & 1 \end{bmatrix}$$

The space $\mathcal{E}_1$ is of dimension $1(n_1 = 1,\ r_1 = 0)$ and the space $\mathcal{E}_2$ is also of dimension $1(n_2 = 1,\ r_2 = 0)$. It follows, since $n_1 + n_2 < n$, that the spaces $\mathcal{E}_1$ and $\mathcal{E}_2$ must be complemented by $\mathcal{E}_{m+1}$ in order to generate the complete space. Hence

$$T = \begin{bmatrix} T(1) \\ T(2) \\ T(3) \end{bmatrix} = \begin{bmatrix} {}_1\bar{C} \\ {}_2\bar{C} \\ T(3) \end{bmatrix} = \begin{bmatrix} 1 & 0 & 0 \\ 1 & 0 & 1 \\ & T(3) & \end{bmatrix}$$

The last row T(3) is chosen so that the matrix T should be of full rank; if $T(3) = (\alpha, \beta, \gamma)$, this entails $\beta \neq 0$ and we can take

$$T = \begin{bmatrix} 1 & 0 & 0 \\ 1 & 0 & 1 \\ 0 & 1 & 0 \end{bmatrix}$$

whence

$$\tilde{A} = \begin{bmatrix} 0 & 0 & 0 \\ 0 & 0 & 0 \\ 3 & -1 & 4 \end{bmatrix} \qquad \tilde{B} = \begin{bmatrix} 1 & 0 \\ 0 & 1 \\ 2 & -1 \end{bmatrix} \qquad \tilde{C} = \begin{bmatrix} 1 & 0 & 0 \\ 0 & 1 & 0 \end{bmatrix}.$$

We note in this case that the existence of A(m+1) leads to the occurrence of the non-null entries $\alpha(1) = 3$, $\alpha(2) = -1$, $A(m+1) = 4$, and ${}^{1}\beta = 2$, ${}^{2}\beta = -1$.

5.3.4 Derivation of compensation matrices associated with the system S

A - Matrices associated with the system $\tilde{S}$

1) The compensation matrix. The previous transformation allowed us to put the system $\bar{S}$ in a special canonical form $\tilde{S}$. At the same time, the matrices $\bar{R}$ and $\bar{L}$ were transformed into $\tilde{R}$ and $\tilde{L}$, connected with the former by the relations

$$\tilde{R} = \overline{RT}^{-1} \qquad \tilde{L} = \bar{L} \,. \tag{5.42}$$

The problem is thus brought down to finding matrices $\tilde{R}$ and $\tilde{L}$ which enable us to compensate $\tilde{S}$ in such a way that not only does the compensated system remain non-interactive but that, further, we can, at least to some extent, specify the dynamics.

Taking account of the form of $\tilde{A}$ and of what has been said about the characterisation of A(m+1), we shall assume, following E.G. Gilbert (cf. ref. 5.12), that $\tilde{R}$ has a structure of the form (1)

$$\tilde{R} = \begin{bmatrix} {}_{1}\psi & \cdots\cdots\cdots & 0 \\ & {}_{2}\psi \quad \cdots\cdots & 0 \\ & & \vdots \\ \cdots\cdots\cdots & {}_{m}\psi & 0 \end{bmatrix} \tag{5.43}$$

$\tilde{R}$ is a matrix of dimensions (m, n); the ${}_{i}\psi$ are row vectors of dimensions n_i corresponding to those of the blocks A(i); the last block of $n - \sum n_i$ columns obviously appears only if the space $\mathcal{E}_{m+1}$ is not empty.

(1) In fact, it can be shown that this structure is not only sufficient but necessary.

The closed-loop system equations are

$$\dot{\tilde{x}} = (\tilde{A} + \tilde{B}\tilde{R})\,\tilde{x} + \tilde{B}\tilde{L}e$$

$$y = \tilde{C}\tilde{x}$$

with

$$\tilde{A} + \tilde{B}\tilde{R} = \begin{bmatrix} A(1) + {}^{1}B\,{}_{1}\psi & 0 & \vdots & & 0 \\ 0 & A(2) + {}^{2}B\,{}_{2}\psi & \vdots & & 0 \\ 0 & 0 & \vdots & & \vdots \\ \vdots & \vdots & \vdots & A(m) + {}^{m}B\,{}_{m}\psi & \vdots \\ \alpha(1) + {}^{1}\beta\,{}_{1}\psi & \alpha(2) + {}^{2}\beta\,{}_{2}\psi & \vdots & \alpha(m) + {}^{m}\beta\,{}_{m}\psi & A(m+1) \end{bmatrix}$$

Moreover, we saw in paragraph 5.3.1.A that all the possible L matrices are of the form:

$$L = \mathcal{D}^{-1}\,\Lambda$$

where Λ is a nonsingular diagonal matrix. As $\tilde{\mathcal{D}} = I_m$ (1), it follows that:

$$\tilde{L} = \Lambda = \text{diagonal}\,(\lambda_1, \lambda_2, \ldots \lambda_m);$$

the closed-loop transfer function matrix thus becomes:

$$\tilde{H} = \begin{bmatrix} {}_{1}C & \cdots\cdots & 0 \\ & {}_{2}C \;\ddots & \vdots \\ & {}_{m}C & 0 \end{bmatrix} \begin{bmatrix} Ip - A(1) - {}^{1}B\,{}_{1}\psi & & & 0 \\ 0 & \ddots & & \vdots \\ \vdots & & Ip - A(m) - {}^{m}B\,{}_{m}\psi & 0 \\ -\alpha(1) - {}^{1}\beta\,{}_{1}\psi & & -\alpha(m) - {}^{m}\beta\,{}_{m}\psi & Ip - A(m+1) \end{bmatrix}$$

$$\times \begin{bmatrix} \lambda_1\,{}^{1}B & 0 & & \\ 0 & \lambda_2\,{}^{2}B & & \\ \vdots & & \ddots & \\ 0 & & & \lambda_m\,{}^{m}B \\ \lambda_1\,{}^{1}\beta & \cdots\cdots & & \lambda_m\,{}^{m}\beta \end{bmatrix}.$$

(1) In fact, $\bar{\mathcal{D}} = I_m$; in view of relations (5.27a), we have

$${}_{i}\tilde{C}\tilde{A}^{d_i}\tilde{B} = {}_{i}\bar{C}T^{-1}(T\bar{A}T^{-1})^{\tilde{d}_i}\,T\bar{B} = {}_{i}\bar{C}T^{-1}\,T(\bar{A})^{\tilde{d}_i}\,T^{-1}\,T\bar{B} = {}_{i}\bar{C}\,\bar{A}^{\tilde{d}_i}\,\bar{B}\,.$$

It follows that $d_i = \bar{d}_i$ and $\tilde{\mathcal{D}} = \bar{\mathcal{D}} = I_m$.

$$\tilde{H} = \begin{bmatrix} {}_{1}C & \cdots & 0 \\ & {}_{2}C & \ddots & \vdots \\ & & {}_{m}C & 0 \end{bmatrix} \begin{bmatrix} Ip - A(1) - {}^{1}B\,{}_{1}\psi & & & 0 \\ 0 & \ddots & & \vdots \\ \vdots & & Ip - A(m) - {}^{m}B\,{}_{m}\psi & 0 \\ -\alpha(1) - {}^{1}\beta\,{}_{1}\psi & \cdots & -\alpha(m) - {}^{m}\beta\,{}_{m}\psi & Ip - A(m+1) \end{bmatrix}^{-1} \times$$

$$\times \begin{bmatrix} \lambda_1\,{}^{1}B & 0 & & \\ 0 & \lambda_2\,{}^{2}B & & \\ \vdots & & \ddots & \\ 0 & & & \lambda_m\,{}^{m}B \\ \lambda_1\,{}^{1}\beta & \cdots & \cdots & \lambda_m\,{}^{m}\beta \end{bmatrix} .$$

It follows that:

- the characteristic polynomial of $\tilde{H}$ is equal to the product of the characteristic polynomials of the matrices $A(i) + {}^{i}B\,{}_{i}\psi$ $(i = 1,\ldots m)$ and of the matrix $A(m+1)$;
- the ith row ${}_{i}\tilde{H}$ of $\tilde{H}$ contains only null entries, with the exception of the diagonal element:

$$\tilde{h}_{ii} = {}_{i}C\left[Ip - A(i) - {}^{i}B\,{}_{i}\psi\right]^{-1}\,{}^{i}B\,\lambda_i \,.$$

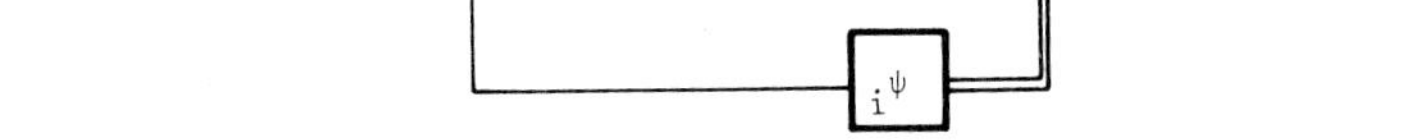

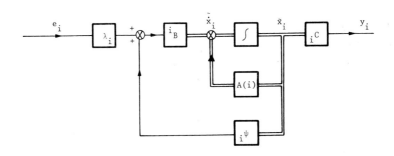

Fig. 5.11 Interpretation of the subsystem with transfer function $\tilde{h}_{ii}$.

This expression shows that $\tilde{h}_{ii}$ can be interpreted as the transfer function of a single-variable system having $A(i)$ as evolution matrix, ${}^{i}B$ as control matrix, and compensation by the matrices ${}_{i}\psi$ and λ_i (cf. figure 5.11) We know that, if this system is controllable (1), it is possible to determine ${}_i\psi$ so that the n_i modes of the closed-loop system can be imposed arbitrarily (cf. chapter 4).

Let us first find the characteristic equation of the single-variable system. From equation (5.27), A(i) appears in the form of a block-triangular matrix: one of its diagonal blocks is a Jordan block of dimension d_i+1 associated with the eigenvalue zero, the other is $\phi(i)$. It follows that

$$|sI-A(i)| = s^{d_i+1} \; |sI-\phi(i)| = s^{n_i} + \nu_{n_i-1}(i)\, s^{n_i-1} + \ldots + \nu_{d_i+1}(i)\, s^{d_i+1} \qquad (5.43a)$$

if

$$s^{r_i} + \nu_{n_i-1}(i)\, s^{r_i-1} + \ldots + \nu_{d_i+1}(i) = \nu(s)$$

denotes the characteristic polynomial associated with $\phi(i)$. The problem is to determine ${}_i\psi$ so that the characteristic polynomial corresponding to $A(i) + {}^iB\,{}_i\psi$ is:

$$\mu(s) = s^{n_i} + \mu_{n_i-1}(i)\, s^{n_i-1} + \ldots + \mu_0(i) = 0$$

where the $\mu_k(i)$ are arbitrarily chosen ($k = 0,\ldots n_i-1$).

This compensation problem was treated in chapter 4. The system $S_i\ [A(i),\ {}^iB]$ being controllable, we know that there exists a linear transformation defined by a matrix M(i) $[\tilde{x}_i = M(i)\ x_i^*]$, allowing us to put the matrices A(i) and ${}^iB$ simultaneously in the forms:

$$A^*(i) = M^{-1}(i)\ A(i)\ M(i) = \begin{bmatrix} 0 & 1 & 0 & \cdots & 0 \\ 0 & 0 & 1 & \cdots & 0 \\ \vdots & \vdots & & & \vdots \\ \vdots & \vdots & & 0 & 1 \\ -\nu_0(i) & \cdots & \cdots & \cdots & -\nu_{n_i-1}(i) \end{bmatrix}$$

(1) The system $[A(i),{}^iB]$ is controllable. Indeed, if it were not, it would follow immediately from the forms of $\tilde{A}$ and $\tilde{B}$ that the columns $\tilde{A}^j\ {}^k\tilde{B}$, for $j=1,\ldots n-1$ and $k = 1,\ldots m$, would not span the space $_n$ and hence that the system S would not be controllable, contrary to the original assumption.

where $\nu_k(i) = 0$ for $k = 0, 1, \ldots d_i$, from equation (5.43a), and

$$ {}^{i}B^{*} = M(i)^{-1}\ {}^{i}B = \begin{bmatrix} 0 \\ 0 \\ \vdots \\ 1 \end{bmatrix} $$

This linear transformation also converts the feedback matrix ${}_{i}\psi$ into

$$ {}_{i}\psi^{*} = {}_{i}\psi\ M(i) $$

and we know (cf. chapter 4) that it will then suffice, in order to achieve our purpose, to take:

$$ {}^{i}B^{*}\ {}_{i}\psi^{*} = \left[A^{*}(i) + {}^{i}B^{*}\ {}_{i}\psi^{*} \right] - A^{*}(i) = $$

$$ = \begin{bmatrix} 0 & 1 & 0 & \cdots\cdots & \\ 0 & 0 & 1 & \cdots\cdots & \\ \vdots & & & 1 \cdots\cdots & \\ \vdots & & & & 1 \\ -\mu_0(i) & \cdots\cdots & & & -\mu_{n_i-1}(i) \end{bmatrix} - \begin{bmatrix} 0 & 1 & 0 & \cdots\cdots \\ 0 & 0 & 1 & \cdots\cdots \\ & & & \\ & & & 1 \\ -\nu_0(i) & \cdots\cdots & & -\nu_{n_i-1}(i) \end{bmatrix} $$

where the first companion matrix has the desired characteristic equation. It follows that

$$ {}_{i}\psi^{*} = \left[{}_{i}\mu - {}_{i}\nu \right] $$

denoting by ${}_{i}\nu$ and ${}_{i}\mu$ the rows:

$$ {}_{i}\nu = \left[-\nu_0(i) \quad -\nu_1(i) \ \ldots - \nu_{n_i-1}(i) \right], \qquad {}_{i}\mu = \left[-\mu_0(i) \ \ldots - \mu_{n_i-1}(i) \right] $$

the first d_i+1 components of ${}_{i}\nu$ being null; and consequently

$$ {}_{i}\psi = {}_{i}\psi^{*} \left[M(i) \right]^{-1} = \left[{}_{i}\mu - {}_{i}\nu \right] \left[M(i) \right]^{-1} . $$

2) <u>Calculation of the transfer function.</u> It is then easy to calculate $\tilde{h}_{ii}$. In fact,

$$\tilde{h}_{ii} = {}_iC\left[sI-A(i) - {}^iB_i\psi\right]^{-1} {}^iB\lambda_i$$

$$= {}_iC\left[sI-A^*(i) - {}^iB^*{}_i\psi^*\right]^{-1} {}^iB^*\lambda_i$$

since the linear transformation does not alter the transfer function. On account of the special forms of

$$sI-A^*(i) - {}^iB^*{}_i\psi^* \quad \text{and} \quad {}^iB^*$$

we have

$$\tilde{h}_{ii} = \frac{[1,0,\ldots 0]\ M(i)}{\mu(s)} \begin{bmatrix} 1 \\ s \\ \vdots \\ s^{n_i-1} \end{bmatrix} \lambda_i$$

The numerator is equal to the product of the first row of M(i) by the column vector $\lambda_i(1, s, \ldots s^{n_i-1})^T$. To simplify the notation, we shall, for the time being, write A(i) as A and M(i) as M, suppressing the indices. We know (cf. chapter 4, paragraph 4.2.3.B) that M is defined by its columns in the form

$$M = \left[{}^1M \quad {}^2M \ldots {}^nM\right] \qquad \text{with} \qquad {}^nM = {}^iB$$

$${}^{n-1}M = \left[A - I\nu_{n-1}\right] B$$

$$\cdots\cdots\cdots\cdots\cdots\cdots$$

$${}^1M = \left[A^{n-1} + \nu_{n-1} A^{n-2} + \ldots + \nu_{d+1}\right] B$$

In view of the forms of A and B, we then have successively

	${}^iB$	$A\,{}^iB$	$A^{d_i}\,{}^iB$	$A^{d_i+1}\,{}^iB$ ...	$A^{n_i-1}\,{}^iB$
first d_i elements	0	0	1	0	0
	⋮	⋮	0	0	0
	⋮	⋮	⋮	⋮	⋮
	0	1	0	0	0
$(d_i+1)^{th}$ elements	1	0	0	.	.
last r_i elements	×	×	×	.	.
	×	×	.	.	.
	×	×	×	.	.

The first elements of the columns $A^{k\,i}B$ are thus all null except for that corresponding to $A^{d_i\,i}B$, which is unity. It follows that

$$\tilde{h}_{ii} = \frac{1}{\mu(s)} \left[\nu_{n_i - r_i}, \ldots \nu_{n_i - 1}, 1, 0, \ldots 0\right] \begin{bmatrix} 1 \\ s \\ \vdots \\ s^{n_i - 1} \end{bmatrix}$$

so that

$$\tilde{h}_{ii} = \frac{s^{r_i} + \nu_{n_i - 1} s^{r_i - 1} + \ldots + \nu_{d_i + 1}}{\mu(s)}$$

B - Matrices associated with the original system

Knowing the matrices $\tilde{R}$ and $\tilde{L}$, we deduce the matrices $\bar{R}$ and $\bar{L}$, associated with the system $\bar{S}$, by using the inverse transformation $\mathcal{C}^{-1}$, i.e. from equations (5.42) rewritten in the form:

$$\bar{R} = \tilde{R}T$$

$$\bar{L} = \tilde{L} \; .$$

Moreover, since the first transformation (from S to $\bar{S}$) converted the matrices R and L into

$$\bar{R} = \mathcal{G}R + A^{*} \qquad \bar{L} = \mathcal{G}L \; ;$$

we deduce, from the inverse transformation:

$$R = \mathcal{G}^{-1}(\bar{R} - A^{*}) = \mathcal{G}^{-1}(\tilde{R}T - A^{*})$$

$$L = \mathcal{G}^{-1}\bar{L} = \mathcal{G}^{-1}\tilde{L} \; .$$

C - Observability and stability

From the form of $\tilde{H}$ given in paragraph 5.3.4.A.1), it is important to notice that the modes associated with $[sI - A(m+1)]^{-1}$ are not connected with the outputs. These modes are thus unobservable for the closed-loop system and do not appear in the expression for the transfer function matrix.

From the point of view of stability, it is nevertheless indispensable that these modes should be stable. As they depend only on the original system and are unaffected by the compensation matrices R and L, it follows that, in practice, a non-interactive solution can be envisaged only if $| \quad | \neq 0$ and if the equation

$$|sI - A(m+1)| = 0$$

has all its roots in the left half-plane. Expressed another way, although the assumption of controllability enabled us, in the framework of an interactive solution, to fix all the modes of the closed-loop system by a suitable choice of feedback matrix, the extra constraints imposed by non-interaction no longer allow us this latitude.

D - Procedure in practice

The complete procedure for decoupling a system by state feedback is summarised in Table 5.2.

E - Examples

1) Take the first example of paragraph 5.3.3.C. We saw that it was possible to put the system into the form

$$\tilde{A} = \begin{bmatrix} 0 & 0 & 0 \\ 0 & 0 & 1 \\ 0 & 0 & 0 \end{bmatrix} \qquad \tilde{B} = \begin{bmatrix} 1 & 0 \\ 0 & 0 \\ 0 & 1 \end{bmatrix} \qquad \tilde{C} = \begin{bmatrix} 1 & 0 & 0 \\ 0 & 1 & 0 \end{bmatrix}$$

Since in this case $n_1 + n_2 = n$, it is possible to specify arbitrarily all the modes of the decoupled system. Suppose we wish to impose the modes corresponding to $s + c$ and $s^2 + as + b$. As the systems $[A(i),\ {}^{i}B]$ are already in canonical form, the ${}_{i}\psi$ can be determined immediately. We have:

$${}_1\psi = -c$$

$${}_2\psi = \begin{bmatrix} -b & -a \end{bmatrix} - \begin{bmatrix} 0 & 0 \end{bmatrix} = \begin{bmatrix} -b & -a \end{bmatrix}$$

whence

TABLE 5.2.

Procedure for decoupling by state feedback

Given: a system S

$$\dot{x} = Ax + Bu \qquad y = Cx \ .$$

i) Determine the matrix $\mathcal{J}$ (5.15).

If $\mathcal{J}$ is invertible, the problem is soluble. Determine A^* (5.17).

ii) Transform the system S into $\bar{S}(\bar{A}, \bar{B}, \bar{C})$.

$$\bar{A} = A - BJ^{-1} A^*, \qquad \bar{B} = BJ^{-1} , \qquad \bar{C} = C \ .$$

Verify that $\mathcal{J} = I_m$ and $\bar{A}^* = 0$.

iii) Transform the system $\bar{S}$ into $\tilde{S}$ by the linear transformation $\tilde{x} = Tx$ (paragraph 5.3.3.B).

Matrices $\tilde{R}$ and $\tilde{L}$ can then be found (paragraph 5.3.4.A) for:

- decoupling the system $\tilde{S}$
- specifying $\sum_{i=1}^{m} n_i$ modes

The matrix $\tilde{R}$ has a structure of the form

$$\tilde{R} = \begin{bmatrix} {}_1\psi & \cdots\cdots & 0 \\ \vdots & {}_2\psi \ \cdots & 0 \\ \vdots & & \vdots \\ \vdots\cdots & {}_m\psi & 0 \end{bmatrix} \qquad \text{with } {}_i\psi = \left[{}_i\nu - {}_i\mu\right] \left[M(i)\right]^{-1} .$$

The matrix $\tilde{L}$ is diagonal.

iv) Making the inverse transformations of those applied in steps iii) and ii), obtain:

$$\bar{R} = \tilde{R}T \qquad \bar{L} = \tilde{L}$$

then

$$R = \mathcal{J}^{-1}(\bar{R} - A^*) = \mathcal{J}^{-1}(\tilde{R}T - A^*)$$

and

$$L = \mathcal{J}^{-1}\,\bar{L} = \mathcal{J}^{-1}\,\Lambda \ .$$

$$\tilde{R} = \begin{bmatrix} -c & 0 & 0 \\ 0 & -b & -a \end{bmatrix}$$

and

$$\bar{R} = \tilde{R}T = \begin{bmatrix} -c & 0 & 0 \\ 0 & -a & -b-3a \end{bmatrix}$$

so that

$$R = \mathfrak{J}^{-1}\left[\bar{R} - A^*\right] = \frac{1}{2}\begin{bmatrix} -(c+1) & a+4 & b+3a+9 \\ -(c+1) & -(a+6) & -(b+3a+9) \end{bmatrix}$$

$$L = \mathfrak{J}^{-1}\Lambda = +\frac{1}{2}\begin{bmatrix} \lambda_1 & -\lambda_2 \\ \lambda_1 & \lambda_2 \end{bmatrix}.$$

We verify that the closed-loop transfer function matrix is given by:

$$\begin{bmatrix} Y_1 \\ Y_2 \end{bmatrix} = \begin{bmatrix} \dfrac{\lambda_1}{s+c} & 0 \\ 0 & \dfrac{\lambda_2}{s^2+as+b} \end{bmatrix} \begin{bmatrix} E_1 \\ E_2 \end{bmatrix}$$

2) Let us now consider the second example. It is still possible to specify all the modes of the decoupled system arbitrarily, since $n_1 + n_2 = n$. However, the fact that r_1 is nonzero forces us to put the subsystem $[A(1),\ {}^1B]$ into canonical form. We have:

$$A(1) = \begin{bmatrix} 0 & 1 \\ 0 & -1 \end{bmatrix} \qquad {}^1B = \begin{bmatrix} 1 \\ 1 \end{bmatrix}.$$

The matrix $M(1) = \begin{bmatrix} 1 & 1 \\ 0 & 1 \end{bmatrix}$ puts it into the form:

$$A^*(1) = M^{-1}(1)\ A(1)\ M(1) = \begin{bmatrix} 0 & 1 \\ 0 & -1 \end{bmatrix} \qquad {}^1B^* = \begin{bmatrix} 0 \\ 1 \end{bmatrix}$$

whence, if we impose the modes corresponding to $s^2 + as + b$, we have:

$$\overset{*}{{}_1\psi} = \mu(1) - \nu(1) = \begin{bmatrix} -b & 1-a \end{bmatrix}$$

$$_1\psi = \overset{*}{{}_1\psi}\, M(1)^{-1} = \begin{bmatrix} -b & -a+b+1 \end{bmatrix}$$

and

$$\tilde{R} = \begin{bmatrix} -b & -a+b+1 & 0 \\ 0 & 0 & -c \end{bmatrix} \qquad \tilde{L} = \begin{bmatrix} \lambda_1 & \\ & \lambda_2 \end{bmatrix}.$$

We deduce from the inverse transformations:

$$R = \mathfrak{J}^{-1}\left[\tilde{R}T - A^*\right] = \begin{bmatrix} -(b+2) & -(a+3) & 0 \\ -1 & -1 & -(1+c) \end{bmatrix}, \qquad L = \begin{bmatrix} \lambda_1 & \\ & \lambda_2 \end{bmatrix}.$$

It may be verified that the closed-loop system transfer function matrix is given by:

$$\begin{bmatrix} Y_1 \\ Y_2 \end{bmatrix} = \begin{bmatrix} \dfrac{\lambda_1(s+1)}{s^2+as+b} & 0 \\ 0 & \dfrac{\lambda_2}{s+c} \end{bmatrix} \begin{bmatrix} E_1 \\ E_2 \end{bmatrix}$$

3) Finally, we consider the third example of paragraph 5.3.3.C. In this case, $n_1 + n_2 < n$: we can impose only two of the three modes of the closed-loop system. If we require the modes $s + a$ and $s + b$, it is sufficient to take:

$$\tilde{R} = \begin{bmatrix} -a & 0 & 0 \\ 0 & -b & 0 \end{bmatrix} \qquad \tilde{L} = \begin{bmatrix} \lambda_1 & \\ & \lambda_2 \end{bmatrix}$$

whence

$$R = \begin{bmatrix} a-b-1 & -3 & -(b+1) \\ 1+b-2a & 2 & b-1 \end{bmatrix}$$

$$L = \mathcal{G}^{-1}\tilde{L} = \begin{bmatrix} -\lambda_1 & \lambda_2 \\ 2\lambda_1 & -\lambda_2 \end{bmatrix}$$

giving a transfer function matrix of the form

$$H = \begin{bmatrix} \frac{\lambda_1}{s+a} & 0 \\ 0 & \frac{\lambda_2}{s+b} \end{bmatrix}$$

In fact, in this case, we must not forget that the modes corresponding to $A(m+1)$ are unobservable for the closed-loop system. It is thus important to be sure that these modes are stable. Now, in the present case:

$$|sI-A(m+1)| = s-4$$

which corresponds to an unstable mode in the right half-plane. The system is thus intrinsically unstable, and it will not be possible in this case to obtain a system simultaneously non-interactive and stable (even though the original system was controllable).

5.3.5 Non-interaction by output feedback

In the preceding paragraphs, the non-interaction of a system was achieved (provided that $\mathcal{G}$ was invertible) under the assumption that all the states of the system were measurable. We saw that, in this case, it is possible not only to ensure input-output decoupling of the closed-loop system but also to specify arbitrarily a number $\sum n_i = n - n_{m+1}$ of the modes.

In practice, where all the system states are not necessarily accessible, it is important to see what further constraints arise from the fact that the utilisable state vector is incomplete, particularly in the case that the only measurable states are constituted by the outputs themselves.

A - Static output feedback

1) General considerations. Suppose that the control law is of the form

$$u = Le + Ky \,. \tag{5.44}$$

where L and K are constant matrices (cf. figure 5.12).

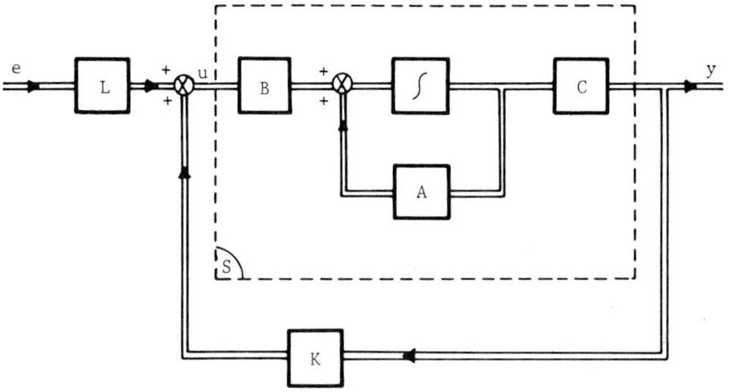

Fig. 5.12 Illustration of the principle of non-interaction by direct output feedback.

The previous study enables us to assert that (L, K) form a diagonalising pair of matrices if:

$$\mathcal{G} \text{ is invertible}$$
$$L = \mathcal{G}^{-1}\Lambda,$$

and if, further, there exists a matrix K of dimensions (m,m) such that

$$KC = R$$

where R is the feedback matrix whose structure was defined above. This extra constraint, as we shall see, has the consequences that:

- the system S may be diagonalisable by (L, R) without being diagonalisable by (L, K);

- the system S may be diagonalisable by (L, K), but at the cost of supplementary constraints on the modes of the decoupled system, i.e. the number of modes that can be arbitrarily fixed may be less than $n - n_{m+1}$. These constraints can, in particular, prevent the realisation of a non-interactive stable system.

2) Examples. Before making these ideas theoretically precise, let us first consider some examples.

a) Consider the first example treated in paragraphs 5.3.3.C and 5.3.4.E. We found that, in this case, the compensator R was of the form:

$$R = \frac{1}{2}\begin{bmatrix} -(c+1) & a+4 & b+3a+9 \\ -(c+1) & -(a+6) & -(b+3a+9) \end{bmatrix}. \quad (5.45)$$

On account of the special form of the matrix C, the structure of the product matrix KC is

$$KC = \begin{bmatrix} k_1 & 0 & k_2 \\ k_3 & 0 & k_4 \end{bmatrix}. \quad (5.46)$$

The existence of a feedback matrix K depends on the possibility of equating the forms (5.45) and (5.46); to do this, we would have to be able to satisfy the equations

$$a + 4 = 0 \qquad\qquad a + 6 = 0$$

which is impossible. It will thus not be possible to decouple the system (L, K), whatever dynamics may be imposed on the closed-loop system.

b) In the case of example 2, on the other hand, it is possible to equate KC and R by taking K in the form

$$K = \begin{bmatrix} -(b+2) & 0 \\ -1 & -(1+c) \end{bmatrix}$$

provided that we further impose the condition $b = a+1$. That is to say, although the use of state feedback enables us to control arbitrarily the three modes of the decoupled system, output feedback will only allow us to specify two of them arbitrarily, in the form:

$$s+c, \; s^2 + as + a + 1 .$$

3) Remark. It is important to notice that the conditions that there should exist an output feedback matrix K, which diagonalises the system, have been stated in terms of KC = R and not $\tilde{K}\tilde{C} = \tilde{R}$. It must not be forgotten, in fact, that between K, $\tilde{K}$, and $\tilde{C}$, C, there is a transformation which is not linear, since

$$\tilde{R} = \tilde{K}\tilde{C} = [\,\mathcal{G}(KC) + A^*]\,T^{-1}\ .$$

In general, we must have:

$$R = KC = -\,\mathcal{G}^{-1}A^* + \mathcal{G}^{-1}\tilde{R}T$$

i.e. taking account of the forms $\tilde{R}$ and T,

$$KC = -\,\mathcal{G}^{-1}A^* + \mathcal{G}^{-1}\begin{bmatrix} {}_1\psi T(1) \\ \vdots \\ {}_m\psi T(m) \end{bmatrix}.$$

This last relation shows that, although the matrix T(m+1) does not affect the possibility of passing from R to K, on the other hand, the output feedback compensation imposes, at least in general (since the identification R = KC may be impossible whatever the ${}_i\psi$ are), supplementary constraints on the ${}_i\psi$ in the form of linear relations between the n_i elements of the row vector ${}_i\psi$, i.e. in fact, on the transfer functions of each decoupled path, since these depend on the elements of the row vector:

$${}_i\mu = {}_i\nu + {}_i\psi M(i)\ .$$

In particular, in the case that all the r_i are null ($d_i + 1 = n_i$ for i = 1,...m), it follows from the form of the matrices T(i) that

$$KC = -\,\mathcal{G}^{-1}A^* + \mathcal{G}^{-1}\begin{bmatrix} \sum_{k=1}^{d_i+1} {}_1^k\psi\ {}_iCA^{k-1} \\ \vdots \\ \sum_{k=1}^{d_i+1} {}_m^k\psi\ {}_mCA^{k-1} \end{bmatrix}.$$

Consider, for example, cases 1 and 2 of paragraphs 5.3.3.C and 5.3.4.E. In the first case, where all the r_i are null, we have:

$$KC = \begin{bmatrix} k_1 & 0 & k_2 \\ k_3 & 0 & k_4 \end{bmatrix}$$

$$KC = \frac{1}{2}\begin{bmatrix} \psi_{11} - 1 & -\psi_{22} + 4 & -(3\psi_{22} + \psi_{21} - 9) \\ \psi_{11} - 1 & \psi_{22} - 6 & 3\psi_{22} + \psi_{21} - 9 \end{bmatrix}$$

and the identification is impossible. In the second case, we should have had:

$$KC = \begin{bmatrix} 2\psi_{12} + \psi_{11} - 2 & 4\psi_{12} + \psi_{11} - 4 & 0 \\ -1 & -1 & -1 + \psi_{21} \end{bmatrix}$$

and hence the condition:

$$\psi_{12} = 1 .$$

It follows that

$$\mu(1) = \begin{bmatrix} -\mu_0(1) & -\mu_1(1) \end{bmatrix} = \begin{bmatrix} -(1 + \psi_{11}) & -\psi_{11} \end{bmatrix}$$

and that the corresponding polynomial is of the form

$$s^2 + \psi_{11}s + (\psi_{11} + 1).$$

B - Dynamic output feedback

In the preceding paragraph, we supposed that the outputs were fed back via a constant matrix K. It may be asked whether the use of dynamic feedback can free us from the constraints pointed out above. We saw in chapter 4 that, in general, the passage from state to output feedback can be made in two different ways, given certain observability assumptions:

i) by augmenting the order of the original system, according to the procedure suggested by Pearson (1);

(1) cf. ref. 4.10.

ii) by using an observer to reconstruct the unknown states of the process.

Application of Pearson's method (1) The method consists of considering, instead of the system S(A, B, C), the "augmented" system S´(A´, B´, C´) defined by:

$$A' = \begin{bmatrix} A & BH \\ 0 & F \end{bmatrix} \qquad B' = \begin{bmatrix} 0 \\ G \end{bmatrix} \qquad C' = \begin{bmatrix} C & 0 \end{bmatrix}$$

where the matrices G, F, H, of respective dimensions $(m\rho, m)$, $(m\rho, m\rho)$, $(m, m\rho)$, $\rho + 1$ being the observability index of the system, are composed of elementary blocks of dimensions (m, m):

$$F = \begin{bmatrix} 0 & I & 0 \dots 0 \\ 0 & 0 & I \dots 0 \\ \dots\dots & \dots\dots & \dots\dots \\ \dots\dots & \dots\dots & I \\ 0 & 0 \dots\dots & 0 \end{bmatrix} \qquad G = \begin{bmatrix} 0 \\ 0 \\ \vdots \\ I \end{bmatrix} \qquad H = \begin{bmatrix} I & 0 \dots 0 \end{bmatrix} .$$

It has been shown, moreover, that, if (A, B) is controllable, so is (A´, B´).

The first problem which arises is to verify that, if S can be diagonalised, so can S´, i.e. to find the connection between $\mathcal{G}$ and $\mathcal{G}'$. We have generally:

$$C' A'^k B' = \sum_{\nu=1}^{k-1} CA^{\nu-1} BHF^{k-\nu} G$$

with, on account of the forms of H, F, G,

$$HF^{\mu} G = 0 \qquad \mu = 1, \dots, \rho - 2 ,$$

$$HF^{\rho-1} G = I_m .$$

It follows that

$$C' A'^k B' = CA^{k-\rho} B$$

and that

(1) For more detail, refer to ref. 4.4.

$$_iC' \, A'^k \, B' = 0 \qquad k = 1, \ldots, \rho + d_i - 1$$

$$_iC' \, A'^{\rho+d_i} \, B' = {}_iCA^{d_i} \, B \, .$$

Consequently, if the original system has an index d_i, relative to $_iC$, the augmented system has the index $\rho + d_i$ relative to $_iC'$, and the matrices $\mathcal{D}$ and $\mathcal{D}'$ are identical (1). Thus, if S is decouplable, so is S´. It follows also that the ith row of the matrix **A´** is equal to:

$$_iA'^{*} = \left[{}_iCA^{\rho+d_i+1} \quad {}_iCA^{\rho+d_i} \, BH \ldots {}_iCA^{d_i+1} \, BHF^{\rho-1} \right] .$$

Quite clearly, if S is of type $\bar{S}$, so is S´.

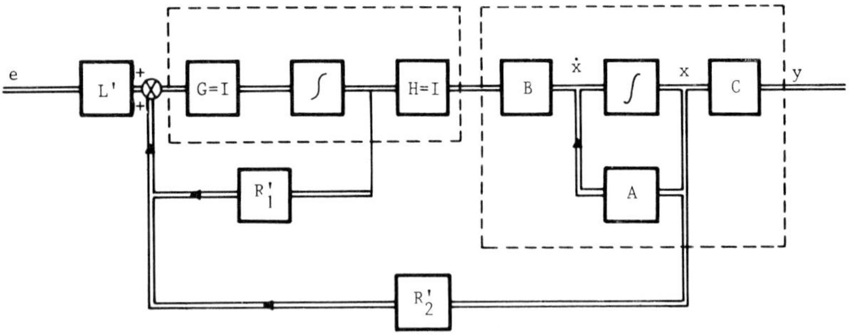

Fig. 5.13 Illustration of the principle of non-interaction by augmentation of the system.

Hence, a possible method of synthesis is the following:

1) Pass from the system S(A, B, C) to the system S´(A´, B´, C´) by the method reviewed above. The system S´ is of order $n + m\rho$, where $\rho + 1$ is the observability index of S.

2) Find, by the method of subsection 5.3.4, matrices R´ and L´ which decouple the system S´. As $\mathcal{D} = \mathcal{D}'$, the precompensation matrices L and L´ are the same.

(1) These two properties, $d_i' = d_i + \rho$, $\mathcal{D}' = \mathcal{D}$, can be deduced directly from the frequency-domain interpretation of d_i and in terms of the transfer function matrix, since the system (F,G,H), placed ahead of the system S, has the effect of multiplying the transfer function matrix by $s^{-\rho}I$.

3) Transform the state feedback R' into a dynamic output feedback K by the method of paragraph 5.3.5.A.

In order to be able to use this method, it is important, however, to verify, as a preliminary, that the number of controllable modes of the system S is unaltered by the augmentation $S \to S'$, and that all the modes of the system (F, G, H) can be controlled arbitrarily. Now, we have seen that the dimensions n_i of the subspaces ${}_i$ are determined by the numbers of independent row vectors orthogonal to the vectors ${}^1B, \ldots {}^{i-1}B, {}^{i+1}B, \ldots A^{n-1}\,{}^mB$, i.e. if this sequence Σ_i of vectors contains k_i independent vectors, we have $n_i = n - k_i$. It is easy to see that the space $\mathcal{E}_i'$ corresponding to $\mathcal{E}_i$ is of dimension $n_i + \rho$. This follows from the structure of the controllability matrix of S', which can be written in the form:

$$\left[\begin{array}{cccc|cccc} 0 & 0 & \ldots\ldots & 0 & B & AB & \ldots & A^{n-1}B \\ 0 & \ldots\ldots & \ldots & I & & & & \\ \ldots & \ldots & \ldots & \ldots & & & & \\ 0 & 0 & I & \ldots & & & & \\ 0 & I & \ldots & \ldots & & & & \\ 0 & 0 & \ldots\ldots & 0 & 0 & & & \end{array}\right]$$

and the fact that the sequence Σ_i' contains $\rho(m-1) + k_i$ independent vectors if Σ_i contains k_i. We thus have

$$n_i' = n + m\rho - \rho(m-1) - k_i = n - k_i + \rho = n_i + \rho.$$

From the relations $d_i' = d_i + \rho$, $n_i' = n_i + \rho$, we deduce that $r_i' = r_i$ and that the dimensions of the matrices $A(m+1)$ and $A'(m+1)$ are the same. Since the modes which cannot be controlled are those corresponding to the matrix $A(m+1)$, the augmentation of the original system does not alter the original possibilities with regard to arbitrary specification of the modes of the compensated system. Furthermore, since the systems S and $\tilde{S}$ have the same transfer function matrix, and the transfer function matrix of S' is obtained from that of S by multiplying by $s^{-\rho}I$, there results the following correspondence between the structures of $A(i)$ and $A'(i)$:

$$A(i) = \left[\begin{array}{c|ccc|c} 0 & & & & \\ 0 & & I_{d_i} & & \\ 0 & & & & 0 \\ \vdots & & & & \\ 0 & 0 & \dots & 0 & \\ \hline & & \Gamma(i) & & \phi(i) \end{array}\right]$$

$$A'(i) = \left[\begin{array}{c|ccc|c} 0 & & & & \\ 0 & & I_{d_i+\rho} & & \\ \vdots & & & & 0 \\ \vdots & & & & \\ 0 & 0 & \dots & 0 & \\ \hline & & \Gamma'(i) & & \phi(i) \end{array}\right] \qquad (5.47)$$

Example: Let us take example 2 of paragraph 5.3.3.C, where the system S is defined by

$$A = \begin{bmatrix} 0 & 1 & 0 \\ 2 & 3 & 0 \\ 1 & 1 & 1 \end{bmatrix} \qquad B = \begin{bmatrix} 0 & 0 \\ 1 & 0 \\ 0 & 1 \end{bmatrix} \qquad C = \begin{bmatrix} 1 & 1 & 0 \\ 0 & 0 & 1 \end{bmatrix}.$$

We have seen that, by state feedback, it is possible to decouple the system and to specify its three modes arbitrarily, since $n_1 + n_2 = n$. Let us try to transform this state feedback into an output feedback.

a) The observability index being $\nu = 2$, we have $\rho = \nu - 1 = 1$, and the matrices A', B', C' become:

$$A' = \left[\begin{array}{ccc|cc} 0 & 1 & 0 & 0 & 0 \\ 2 & 3 & 0 & 1 & 0 \\ 1 & 1 & 1 & 0 & 1 \\ \hline 0 & 0 & 0 & 0 & 0 \\ 0 & 0 & 0 & 0 & 0 \end{array}\right], \quad B' = \left[\begin{array}{cc} 0 & 0 \\ 0 & 0 \\ 0 & 0 \\ \hline 1 & 0 \\ 0 & 1 \end{array}\right], \quad C' = \left[\begin{array}{ccc|cc} 1 & 1 & 0 & 0 & 0 \\ 0 & 0 & 1 & 0 & 0 \end{array}\right]$$

whence

$$\mathfrak{J}' = \begin{bmatrix} 1 & 0 \\ 0 & 1 \end{bmatrix} = \mathfrak{J} \qquad d_1' = 1 = d_1 + \rho \qquad d_2' = 1 = d_2 + \rho$$

$$A'^{*} = \begin{bmatrix} 8 & 14 & 0 & 4 & 0 \\ 3 & 5 & 1 & 1 & 1 \end{bmatrix}.$$

b) We pass from the system S' to $\bar{S}'$ by

$$\bar{A}' = A' - B'\mathcal{G}^{-1} = \begin{bmatrix} 0 & 1 & 0 & 0 & 0 \\ 2 & 3 & 0 & 1 & 0 \\ 1 & 1 & 1 & 0 & 1 \\ -8 & -14 & 0 & -4 & 0 \\ -3 & -5 & -1 & -1 & -1 \end{bmatrix} \qquad \bar{B}' = B'\mathcal{G}^{-1} = \begin{bmatrix} 0 & 0 \\ 0 & 0 \\ 0 & 0 \\ 1 & 0 \\ 0 & 1 \end{bmatrix}$$

$$\bar{C} = C' = \begin{bmatrix} 1 & 1 & 0 & 0 & 0 \\ 0 & 0 & 1 & 0 & 0 \end{bmatrix}.$$

c) We next find the canonical form $\tilde{S}'$ corresponding to $\bar{S}'$. The space $\mathcal{E}_1'$ is defined by the row vectors orthogonal to ${}^2\bar{B}'$, $\bar{A}'\,{}^2\bar{B}'$, whence $n_1' = 3$ (we note that indeed $n_1' = n_1 + \rho$). The space $\mathcal{E}_2'$ defined by the row vectors orthogonal to ${}^1\bar{B}'$, $\bar{A}'\,{}^1\bar{B}'$, $\bar{A}'^2\,{}^1\bar{B}'$, is of dimension 2, and we have

$$n_2' = 2 \qquad n_2' = n_2 + 1 \qquad r_2' = 0 .$$

The transformation matrix T is then

$$T = \begin{bmatrix} 1 & 1 & 0 & 0 & 0 \\ 2 & 4 & 0 & 1 & 0 \\ 0 & 1 & 0 & 0 & 0 \\ 0 & 0 & 1 & 0 & 0 \\ 1 & 1 & 1 & 0 & 1 \end{bmatrix}$$

and so

$$\tilde{A}' = \begin{bmatrix} 0 & 1 & 0 & 0 & 0 \\ 0 & 0 & 0 & 0 & 0 \\ 0 & 1 & -1 & 0 & 0 \\ 0 & 0 & 0 & 0 & 1 \\ 0 & 0 & 0 & 0 & 0 \end{bmatrix}, \quad \tilde{B}' = \begin{bmatrix} 0 & 0 \\ 1 & 0 \\ 0 & 0 \\ 0 & 0 \\ 0 & 1 \end{bmatrix},$$

$$\tilde{C}' = \begin{bmatrix} 1 & 0 & 0 & 0 & 0 \\ 0 & 0 & 0 & 1 & 0 \end{bmatrix}$$

We note that $\tilde{A}'$ is related to $\tilde{A}$ as indicated in (5.47). The system thus appears in the form of two single-variable subsystems, the first characterised by:

$$A'(1) = \begin{bmatrix} 0 & 1 & 0 \\ 0 & 0 & 0 \\ 0 & 1 & -1 \end{bmatrix} \qquad {}^{1}B' = \begin{bmatrix} 0 \\ 1 \\ 0 \end{bmatrix}$$

the second by:

$$A'(2) = \begin{bmatrix} 0 & 1 \\ 0 & 0 \end{bmatrix} \qquad {}^{2}B' = \begin{bmatrix} 0 \\ 1 \end{bmatrix}$$

The latter being already in canonical form, the compensating row vector $_2\psi'$ can be determined directly. If we impose the dynamics corresponding to $(s + c)(s + c')$, we have

$$_2\psi' = \begin{bmatrix} -cc' & -(c + c') \end{bmatrix} .$$

The second subsystem is put into the canonical form

$$A'(1)^* = \begin{bmatrix} 0 & 1 & 0 \\ 0 & 0 & 1 \\ 0 & 0 & -1 \end{bmatrix} \qquad {}^{1}B'^* = \begin{bmatrix} 0 \\ 0 \\ 0 \end{bmatrix}$$

by the transformation defined by

$$M = \begin{bmatrix} 1 & 1 & 0 \\ 0 & 1 & 1 \\ 0 & 1 & 0 \end{bmatrix}$$

whence, if we specify the dynamics $(s^2 + as + b)(s + \alpha)$:

$$_1\psi'^* = \begin{bmatrix} -b\alpha & -(a\alpha + b) & 1 - (a + \alpha) \end{bmatrix}$$

and

$$_1\psi' = {}_1\psi'^* M^{-1} = \begin{bmatrix} -b\alpha & 1 - (a + \alpha) & (b + 1 - a)(\alpha - 1) \end{bmatrix} .$$

It follows that $\tilde{R}'$ is of the form:

$$\tilde{R}' = \begin{bmatrix} -b\alpha & 1 - (a + \alpha) & (b + 1 - a)(\alpha - 1) & 0 & 0 \\ 0 & 0 & 0 & -cc' & -(c + c') \end{bmatrix}$$

The closed-loop transfer function matrix thus takes the form:

$$Z(s) = \begin{bmatrix} \dfrac{\lambda_1'(s+1)}{(s^2+as+b)(s+\alpha)} & 0 \\ 0 & \dfrac{\lambda_2'}{(s+c)(s+c')} \end{bmatrix}$$

We shall get effectively the same results as before by taking, e.g.

$$c' \gg c \qquad \lambda_2' = \lambda_2 c'$$

$$b \ll \alpha^2 \qquad \lambda_1' = \lambda_1 \alpha.$$

Indeed, in this case, we shall have

$$Z(s) = \begin{bmatrix} \dfrac{\lambda_1(s+1)}{(s^2+as+b)(1+\tau_1 s)} & 0 \\ 0 & \dfrac{\lambda_2}{(s+c)(1+\tau_2 s)} \end{bmatrix}$$

The previous dynamics have been modified only by the appearance of small time constants $\tau_1 = 1/\alpha$, $\tau_2 = 1/c'$, negligible in comparison with the dynamics of the closed-loop system itself. We deduce the compensation matrix R':

$$R' = \mathcal{J}'^{-1}(\bar{R}' - A'^{*}) = \begin{bmatrix} R_2' & R_1' \end{bmatrix}.$$

$$R' = \begin{bmatrix} -b\alpha-2(a+\alpha)-6 & -b\alpha-4(a+\alpha)+(b+1-a)(\alpha-1)-10 & 0 & -3-(a+\alpha) & 0 \\ -3-(c+c') & -5-(c+c') & -cc'-(c+c')-1 & -1 & -1-(c+c') \end{bmatrix}$$

and

$$L' = \mathcal{J}^{-1}\,\Lambda' = \begin{bmatrix} \lambda_1' & \\ & \lambda_2' \end{bmatrix}$$

The structure of the compensated system is given in figure 5.13.

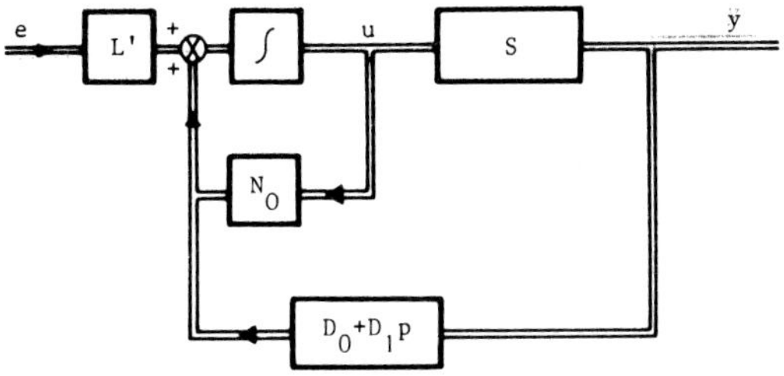

Fig. 5.14 Structure resulting from the augmentation method.

d) It is now easy to pass from state feedback to output feedback by the method described in chapter 4. For this purpose, we replace the control law

$$w = L' e + R_1' u + R_2' x$$

by the control law

$$\dot{w} = \dot{u} = N_0 u + D_0 y + D_1 \dot{y} + L' e$$

with the defining relations

$$\begin{bmatrix} C^T & A^T C^T \end{bmatrix} \begin{bmatrix} D_0^T \\ D_1^T \end{bmatrix} = \begin{bmatrix} R_2' \end{bmatrix}^T \qquad \text{et} \qquad N_0 = R_1' - D_1 CB .$$

In the present case, we have, using the first equation and taking D_i in the form:

$$D_i = \begin{bmatrix} d_{i1} & d_{i2} \\ d_{i3} & d_{i4} \end{bmatrix}$$

the relations:

$$d_{13} = -1 \qquad d_{03} + d_{14} = -1 - (c + c')$$

$$d_{04} + d_{14} = -cc' - (c + c') - 1$$

$$2 d_{11} = -2(a + \alpha) - 4 + (b + 1 - a)(\alpha - 1) = b(\alpha - 1) - a(\alpha - 1) - a - 5$$

$$d_{01} + d_{12} = -b\alpha - 2 - (b + 1 - a)(\alpha - 1) = -2 b\alpha - 1 + b - \alpha + a(\alpha - 1)$$

$$d_{02} + d_{12} = 0 .$$

Otherwise expressed, the structures of the matrices D_0 and D_1 are of the forms:

$$D_1 = \begin{bmatrix} -(a + \alpha) - 2 + \frac{1}{2}(b + 1 - a)(\alpha - 1) & d_{12} \\ -1 & d_{14} \end{bmatrix}$$

$$D_0 = \begin{bmatrix} -d_{12} - b\alpha - 2 - (b + 1 - a)(\alpha - 1) & -d_{12} \\ -d_{14} - 1 - (c + c') & -d_{14} - 1 - cc' - (c + c') \end{bmatrix}$$

the coefficients d_{12}, d_{14} being arbitrary, and N_0:

$$N_0 = \begin{bmatrix} -1 - \frac{1}{2}(b + 1 - a)(\alpha - 1) & -d_{12} \\ 0 & -d_{14} - 1 - (c + c') \end{bmatrix}$$

The structure of the control system is indicated in figure 5.14. We can realise the control law in a simpler form by setting

$$u = z + D_1\, y + L'\, e$$

$$\dot{x} = N_0\, z + (D_0 + N_0\, D_1)\, y \quad .$$

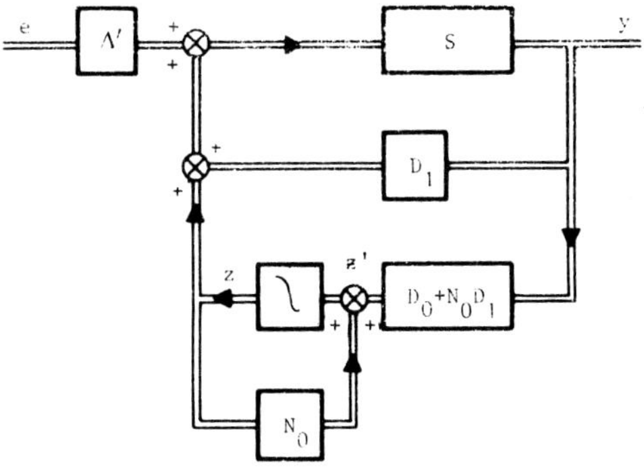

Fig. 5.15 Another possible form of structure

The corresponding structure is represented diagrammatically in figure 5.15.

Suppose we had originally wished to control the system according to the dynamics

$$Y_1(s) = \frac{2E_1(s)}{1 + 2s} \, , \quad Y_2(s) = \frac{E_2(s)}{1+0.5s}$$

We should then have

$$\lambda_1 = 1 \qquad \lambda_2 = 2$$

$$a = 1{,}5 \qquad b = 0{,}5 \qquad 'c = 2 \; .$$

If we required, e.g., that the magnitudes of the extra time constants introduced should not exceed 10% of the original time constants in each channel, we could take:

$$\alpha = 5 \qquad \lambda_1' = 5 \qquad c' = 20 \qquad \lambda_2' = 40 \ .$$

In this case, we should have

$$D_0 = \begin{bmatrix} -d_{12} - 4{,}5 & -d_{12} \\ -d_{14} - 23 & -d_{14} - 63 \end{bmatrix} \qquad N_0 = \begin{bmatrix} -1 & -d_{12} \\ 0 & -d_{14} - 1 - 22 \end{bmatrix}$$

$$D_1 = \begin{bmatrix} -8{,}5 & d_{12} \\ -1 & d_{14} \end{bmatrix} \qquad D_0 + N_0 D_1 = \begin{bmatrix} 4 & -2d_{12} - d_{12}d_{14} \\ 0 & -d_{14} - 63 - d_{14}(d_{14} + 23) \end{bmatrix}.$$

We could take, e.g.

$$d_{14} = 0 \qquad d_{12} = 70$$

so that

$$D_1 = \begin{bmatrix} -8{,}5 & 70 \\ -1 & 0 \end{bmatrix} \quad N_0 = \begin{bmatrix} -1 & -70 \\ 0 & -23 \end{bmatrix} \quad D_0 + N_0 D_1 = \begin{bmatrix} 4 & -140 \\ 0 & -63 \end{bmatrix}.$$

5.4 NON-INTERACTION BY OPERATIONAL METHODS

5.4.1 State-space compensation techniques and operational methods

The indices d_i and the matrix $\mathcal{D}$ have been defined, in the preceding paragraphs, by reference to a state-space representation. It is possible quite easily to find a frequency-domain interpretation for them, in terms of the transfer function matrix.

We have seen, indeed, that the ith row $_iH$ of H can be expressed in the form

$$\begin{aligned} {}_iH &= {}_iC(sI-A)^{-1}B \\ &= \phi^{-1}(s)\left[{}_iCA^{d_i}Bs^{n-d_i-1} + \ldots + ({}_iCA^{n-1}B + \ldots + a_{d_i+1}\,{}_iCA^{d_i}B)\right] \end{aligned}$$

and it follows that: d_i may be defined as the smallest integer k such that

$$\lim_{s \to \infty} s^{k+1} \; {}_i H(s)$$

is finite and nonzero, and ${}_i\mathcal{G}$ is precisely this limit.

Thus, the matrix $\mathcal{G}$ can be found directly by inspection of the transfer function matrix. Expressed in another way: d_i is equal to the difference between the degree of $\psi(s)$ (the common denominator of the transfer function matrix) and the highest degree appearing in the ith row of $H(s)\,\psi(s)$, augmented by 1.

Example:

$$H = \begin{bmatrix} p^2 + p & p + 1 \\ 3 & 5\,p + 4 \end{bmatrix} \frac{1}{p^3 + \alpha p^2 + \beta p + \gamma}$$

$$d_1 = 0 \qquad {}_1\mathcal{G} = \begin{bmatrix} 1 & 0 \end{bmatrix} \qquad d_2 = 1 \qquad {}_2\mathcal{G} = \begin{bmatrix} 0 & 5 \end{bmatrix}.$$

We see also that, in a certain sense, the possibility of compensating the system so as to make it non-interactive depends on its high-frequency properties.

From what has been said, det H = 0 implies det $\mathcal{G}$ = 0, but it can happen that det H is not zero even though det $\mathcal{G}$ is. Although non-interaction cannot then be obtained in the classical manner by state feedback as above, this does not mean that it is impossible to achieve it by another route. We shall see examples in the following paragraphs.

5.4.2 General constraints on the open-loop and closed-loop transfer function matrices

A - Review

Attention has already been drawn, in paragraph 5.2.2.A, to the fact that, when using transfer function matrix methods, the necessity of avoiding the appearance of uncontrollable and/or unobservable modes imposes certain constraints on the structure of the transfer function matrices, open-loop as well as closed-loop.

We thus gave as sufficient conditions:

- the elements of the open-loop matrix N contain, as zeroes, all the unstable zeroes of $|Z|$;

- the elements of N contain, as poles, the unstable poles of the elements of Z. It was pointed out, on the other hand, that these conditions are, in general, too strong and consequently entail unnecessary constraints.

We shall try here to make them sharper. Let us point out, however, that, since the transfer function matrix approach is ill-adapted to the purpose envisaged, we should see what follows rather as practical methods of guidance in the solution of particular problems than as strictly necessary and sufficient conditions.

Let us consider the first example of paragraph 5.2.2.B, where the transfer function matrix is defined by

$$Z(s) = \frac{1}{(s-1)(s+1)(s+2)} \begin{bmatrix} s^2+6 & s^2+s+4 \\ 2s^2+7s-9 & s^2+4s-5 \end{bmatrix} = \frac{\mathcal{Z}}{z}$$

All the elements z_{ij} of Z contain the unstable pole $s = 1$, and the determinant of $\mathcal{Z}$ is equal to $-(s-1)(s+1)(s+2)(s+3)$. If we write the open-loop transfer function matrix

$$N = \frac{\mathcal{N}}{n}$$

Gilbert's constraints give:

n contains the unstable factor $s = 1$,

$$(\mathrm{adj}\,\mathcal{Z})\mathcal{N} = \begin{bmatrix} s^2+4s-5 & -(s^2+s+4) \\ -(2s^2+7s-9) & s^2+6 \end{bmatrix} \begin{bmatrix} \mathcal{N}_{11} & 0 \\ 0 & \mathcal{N}_{22} \end{bmatrix}$$

also contains the unstable pole $s = 1$, i.e. each $\mathcal{N}_{ii}$ contains $s - 1$. As the factor $s - 1$ cannot be common to n and $\mathcal{N}$, the difficulty can only be resolved by taking s = 1 to be a second-order pole of N.

B - Special case. Gilbert's rules

Before going further, we point out some results, also given by Gilbert (cf. ref. 5.12), which apply only to the case where:

- all the poles of the matrices Z, N, C, are simple;
- all the matrices associated with poles of Z are of rank 1.

Remark: We note that both these conditions are satisfied in the example of the preceding paragraph, since

$$Z(s) = \frac{1}{6(s-1)}\begin{bmatrix}7 & 6\\ 0 & 0\end{bmatrix} + \frac{1}{2(s+1)}\begin{bmatrix}-7 & -4\\ 14 & 8\end{bmatrix} + \frac{1}{3(s+2)}\begin{bmatrix}10 & 6\\ -15 & -9\end{bmatrix}$$

under these conditions, if we define

$$G = Z^{-1} = \begin{bmatrix}{}^1g & {}^2g & \dots & {}^mg\end{bmatrix}$$

and the indices l_i by:

$$l_i = 0 \quad \text{if} \quad n_{ii} \text{ is analytic for } s = \lambda,$$
$$l_i = 1 \quad \text{if} \quad n_{ii} \text{ has a simple pole at } s = \lambda,$$

λ being a pole of Z or N, the constraints can be imposed by requiring that:

$$\text{rank}\begin{bmatrix}\lim_{s\to\lambda}(s-\lambda)N_{11}\,{}^1g, \dots \lim_{s\to\lambda}(s-\lambda)N_{mm}\,{}^mg\end{bmatrix}$$

$$= \sum_{i=1}^{m} l_i - 1 \quad \text{for every unstable pole of } Z;$$

$$= \sum_{i=1}^{m} l_i \quad \text{for every unstable pole of } C \neq \text{pole of } Z.$$

These relations ensure, within the framework of the assumptions made, that every unstable mode is controllable and observable.

In the above example:

$$Z^{-1} = \left[{}^1g, \ {}^2g\right] = \begin{bmatrix} \frac{-(s^2+4s-5)}{s+3} & \frac{(s^2+s+4)}{s+3} \\ \frac{(2s^2+7s-9)}{s+3} & \frac{-(s^2+6)}{s+3} \end{bmatrix}$$

For s = 1, we should thus have

$$0 = \sum_{i=1}^{2} l_i - 1$$

whence it follows that $l_1 = 1$, $l_2 = 0$, and that the first element of N contains s = 1 as a pole. Since for every other value of s with positive real part, the columns ${}^1g$ and ${}^2g$ are independent, this is the only constraint which we have to satisfy.

On the other hand, let us consider the second example of paragraph 5.2.2.B,

$$Z(s) = \frac{1}{(s-1)(s+1)(s+2)} \begin{bmatrix} s^2+6 & s^2+s+4 \\ 2s^2-7s-2 & s^2-5s-2 \end{bmatrix}$$

For s = 1, we have $l_1 = l_2 = 1$ and N contains s = 1 as a pole. Since the determinant of Z contains the unstable pole s = 2, the unstable poles of ${}^1g$ and ${}^2g$ (s = 2) must, on the contrary, be included among the zeroes of N, which will then be of the form:

$$N = \begin{bmatrix} \frac{(s-2)}{s-1} N'_{11} & 0 \\ 0 & \frac{(s-2)}{s-1} N'_{22} \end{bmatrix}$$

C - The general case

1) In the general case where not all the poles of Z are of rank 1, the above criteria are not directly applicable; however, by means of rather tedious matrix manipulations (cf. refs. 5.4 and 5.10), the following

practical conditions can be obtained.

As regards the open-loop matrix N:

- each element n_{jj} contains, as poles, the unstable poles of every transfer function in the jth row of Z;
- each element n_{jj} contains, as zeroes, the right half-plane zeroes of $|Z|$;
- each element n_{jj} contains, as zeroes, the right half-plane zeroes common to all the transfer functions in the jth row of Z (even when such a zero cancels with a pole in the factorisation of $|Z|$).

Analogous constraints can be imposed directly on the closed-loop matrix H:

- each h_{jj} must contain, as zeroes, the right half-plane zeroes of the determinant of Z;
- the unstable poles appearing in an element of the jth row of Z must be zeroes of $1 - h_{jj}$;
- each h_{jj} must contain, as zeroes, the unstable zeroes common to all the transfer functions in the jth row of Z.

These constraints also should be regarded as practical rules enabling us to design non-interactive control schemes on sound bases. Although they amount to a substantial improvement on those given in paragraph 5.2.2.A, they are still not necessary: it is possible in special cases to relax them even further. What is important is to recall that they exist and are directly connected with the constraints of non-interaction, as we have seen in the course of the analyses of both the state-space techniques and Gilbert's methods (and, consequently, a weakening of the non-interaction constraints, partial decoupling, may free us from them).

2) It might also have been thought that, by working in terms of system invariants, rather than the transfer function matrix, a definitive solution could be given. This approach has been tried by Kalman (cf. ref. 5.18) who posed the problem in the following terms:-

Given a square, proper, rational, nonsingular matrix Z, is it possible to define matrices such that:

1) $C = Z^{-1}N$ is proper, N being diagonal

2) $\delta(N) = \delta(C) + \delta(Z)$ (irreducibility of representations)

3) $\delta(N)$ is minimal?

The only result obtained, apparently, is the following, which is also a sufficient condition:

- if Z is decomposed into Smith-McMillan canonical form,

$$Z = A \begin{bmatrix} \varepsilon_1/\psi_1 & & \\ & \ddots & \\ & & \varepsilon_m/\psi_m \end{bmatrix} B$$

- and if ε_m and ψ_1 are relatively prime, there exists a rational matrix C satisfying the above conditions. It can be chosen so that

$$ZC = N = \frac{\varepsilon_m}{\theta\psi_1} I$$

where θ is a polynomial relatively prime to ε_m.

Let the system be defined by the transfer function matrix

$$Z(s) = \frac{1}{(s-1)(s+1)(s+2)} \begin{bmatrix} s^2+6 & s^2+s+4 \\ 2s^2-7s-2 & s^2-5s-2 \end{bmatrix}$$

$$\Delta_1 = 1,\ \Delta_2 = -(s^4-5s^2+4) = -(s-2)\psi,$$

$$\varepsilon_1 = 1,\ \psi_1 = \psi = (s-1)(s+1)(s+2),$$

$$\varepsilon_2 = -(s+2),\ \psi_2 = 1.$$

We can then, in view of Kalman's criterion, choose as closed-loop matrix

$$N = \frac{\varepsilon_2}{\theta\psi_1} I = \frac{-(s-2)}{\theta(s-1)(s+1)(s+2)} I$$

where θ is prime to $s - 2$. In particular, if we take $\theta = 1$, the realisation of N requires six state variables although Z needs only three, and we have to introduce three more into the compensator C for non-interaction to be possible.

Two remarks are in order:-

1 - The condition that ε_m and ψ_1 should be relatively prime is not necessary. The theorem thus will not apply in some cases where non-interactive control nevertheless remains possible.

2 - Complete controllability and observability of all the modes have been required, although this is in practice obligatory only for those that are unstable.

D - Example

Let the process be defined by

$$Z(s) = \frac{1}{(s-1)(s+1)(s+2)} \begin{bmatrix} s^2+6 & s^2+s+4 \\ 2s^2-7s-2 & s^2-5s-2 \end{bmatrix}$$

1) Since all the elements of Z contain the unstable pole s = 1, it must be a pole of the elements of the open-loop matrix N.

2) Since $|Z| = -(s-2)/(s-1)(s+1)(s+2)$, the unstable factor s-2 must be a zero of the elements of N.

It may be verified that these constraints do indeed ensure the complete controllability and observability of all the unstable modes of the cascade ZC and hence the effective stability of the closed-loop system.

Suppose that, on the contrary, we had wanted to take an open-loop matrix not satisfying these constraints, e.g.:

$$N = \frac{-(s-2)}{(s+1)(s+2)(s+3)} I$$

which would have led to

$$C = \begin{bmatrix} s^2-5s-2 & -(s^2+s+4) \\ -(2s^2-7s-2) & s^2+6 \end{bmatrix} \frac{1}{(s+1)(s+2)(s+3)}$$

A state representation of the cascade ZC would then have been:

$$\dot{x} = \begin{bmatrix} 1 & 0 & 0 & -7 & 18 & 7 & 3 \\ 0 & -1 & 0 & 0 & -2 & 7 & 2 \\ 0 & 0 & -2 & -1 & 0 & 10 & 3 \\ & & & -1 & & & \\ & & & & -2 & & \\ & & & & & -3 & \\ & & & & & & -3 \end{bmatrix} x + \begin{bmatrix} 0 & 0 \\ 0 & 0 \\ 0 & 0 \\ 1 & -1 \\ 2 & -1 \\ 11 & -5 \\ -37 & 15 \end{bmatrix} \varepsilon$$

$$y = \begin{bmatrix} 1 & -3 & 2 & 0 & 0 & 0 & 0 \\ -1 & -3 & 4 & 0 & 0 & 0 & 0 \end{bmatrix} x$$

which, by the transformation

$$x = M\tilde{x} \qquad \text{with } M = \begin{bmatrix} 1 & 0 & 3{,}5 & 0 & -6 & -0{,}75 & -1{,}75 \\ 0 & 1 & 0 & 0 & 2 & -1 & -3{,}5 \\ 0 & 0 & -1 & 1 & 0 & -3 & -10 \\ 0 & 0 & 1 & 0 & 0 & 0 & 0 \\ 0 & 0 & 0 & 0 & 1 & 0 & 0 \\ 0 & 0 & 0 & 0 & 0 & 0 & 1 \\ 0 & 0 & 0 & 0 & 0 & 1 & 0 \end{bmatrix}$$

can be put in the form

$$\begin{bmatrix} 1 & & & & & & \\ & -1 & & & & & \\ & & -1 & & & & \\ & & & -2 & & & \\ & & & & -2 & & \\ & & & & & -3 & \\ & & & & & & -3 \end{bmatrix} x + \begin{bmatrix} 0 & 0 \\ -2{,}5 & -0{,}5 \\ 1 & -1 \\ 0 & -6 \\ 2 & -1 \\ -37 & 15 \\ 11 & -5 \end{bmatrix} \varepsilon$$

$$y = \begin{bmatrix} 1 & -3 & 1{,}5 & 2 & -12 & -3{,}75 & 11{,}25 \\ -1 & -3 & -7{,}5 & 4 & 0 & -8{,}25 & 34{,}75 \end{bmatrix} x$$

showing the uncontrollability of the mode s = 1. The system is thus unstable, even though the transfer function matrix does not express this fact.

If we had taken:

$$N = \frac{1}{(s+1)(s+2)(s+3)} \; I$$

we should have had:

$$C = \frac{-1}{(s-2)(s+1)(s+2)(s+3)} \begin{bmatrix} s^2-5s-2 & -(s^2+s+4) \\ -(2s^2-7s-2) & s^2+6 \end{bmatrix}$$

and the cascade would have had the state representation (1):

$$\dot{x} = \frac{1}{30} \begin{bmatrix} 30 & & & -70 & 135 & 42 & 18 & -1 \\ & -30 & & 0 & -15 & 42 & 12 & 3 \\ & & -60 & -90 & 0 & 60 & 18 & 4 \\ & & & -30 & & & & \\ & & & & -60 & & & \\ & & & & & -90 & & \\ & & & & & & -90 & \\ & & & & & & & 60 \end{bmatrix} x + \begin{bmatrix} 0 & 0 \\ 0 & 0 \\ 0 & 0 \\ 1 & -1 \\ 2 & -1 \\ 11 & -5 \\ -37 & 15 \\ 4 & 5 \end{bmatrix} \varepsilon$$

$$y = \frac{1}{6} \begin{bmatrix} 1 & -3 & 2 & 0 & \cdots & 0 \\ 1 & -3 & 4 & 0 & \cdots & 0 \end{bmatrix} x$$

so that, in reduced form:

(1) The elements not shown are null.

$$\dot{\tilde{x}} = \begin{bmatrix} 1 & & & & & & & \\ \hline & -1 & & & & & & \\ & & -1 & & & & & \\ & & & -2 & & & & \\ & & & & -2 & & & \\ & & & & & -3 & & \\ & & & & & & -3 & \\ & & & & & & & 2 \end{bmatrix} \tilde{x} + \begin{bmatrix} 0 & 0 \\ \hline \times & \times \\ \times & \times \\ \times & \times \\ \times & \times \\ \times & \times \\ \times & \times \\ 4/30 & 5/30 \end{bmatrix} \varepsilon$$

$$y = \left[\begin{array}{ccccccc|c} 1 & \times & \times & \times & \times & \times & \times & 0 \\ 1 & \times & \times & \times & \times & \times & \times & 0 \end{array}\right] \tilde{x}$$

thus revealing the unobservability of the unstable mode $s = 2$ and the uncontrollability of the unstable mode $s = 1$.

5.5 REALISATION OF NON-INTERACTIVE CONTROL SCHEMES

5.5.1 Design

A - Constraints

It has been shown in the preceding paragraphs that the realisation of a non-interactive control scheme can be envisaged in two principal ways: either from a state-space viewpoint, or by means of an operational representation.

As regards the methods based on state-space, we have seen that it is not always possible, by state feedback, to obtain a system simultaneously decoupled and stable; it all depends, in fact, on the nature of the modes contained in the matrix $\bar{A}$ of subsection 5.3.3 which are unobservable for the closed-loop system and must of necessity be stable.

As for the operational methods, we saw that their utilisation imposes certain constraints on the open-loop (or closed-loop) transfer function matrices, which it is not always easy to take into account. Moreover, the physical realisation of the compensator will often be complicated.

In fact, let us consider again the process defined by the transfer function matrix

$$Z(s) = \frac{1}{(s-1)(s+1)(s+2)} \begin{bmatrix} s^2+6 & s^2+s+4 \\ 2s^2-7s-2 & s^2-5s-2 \end{bmatrix}$$

If we wish to use state feedback, we find the matrix $\tilde{A}$, which takes the form:

$$\tilde{A} = \begin{bmatrix} 0 & 0 & 0 \\ 0 & 0 & 0 \\ 0,66 & 0,66 & 2 \end{bmatrix}$$

Consequently, $A(m+1) = 2$ and $|sI-A(m+1)|$, which is equal to $s - 2$, is not a Hurwitz polynomial. Thus we cannot control this system by the state feedback method.

Operational methods, on the other hand, do enable us to design a stable decoupled system. We recall that we were led to take an open-loop transfer function matrix of the form:

$$N = \begin{bmatrix} \frac{(s-2)}{(s-1)} \frac{n_1}{d_1} & 0 \\ 0 & \frac{(s-2)}{(s-1)} \frac{n_2}{d_2} \end{bmatrix} .$$

where the n_i and d_i are polynomials such that $d_i(2) \neq 0$, $n_i(1) \neq 0$. The corresponding closed-loop transfer function matrix becomes:

$$H = \begin{bmatrix} \frac{P_1(s)}{Q_1(s)} & 0 \\ 0 & \frac{P_2(s)}{Q_2(s)} \end{bmatrix} , \quad \text{with } P_i(s) = (s-2)n_i , \quad Q_i(s) = (s-2)n_i+(s-1)d_i .$$

Let us try to realise a stable closed-loop system by taking:

$$n_i = k, \quad d_i = b_0 + b_1 s .$$

We have then:

$$Q_i(s) = b_1 s^2 + (b_0 - b_1 + k)s - (b_0 + 2k)$$

and the stability conditions become

$$\begin{aligned} & b_1 > 0 \\ & b_0 - b_1 + k > 0 \\ & b_0 + 2\,k < 0\,. \end{aligned}$$

If, e.g., as in the example proposed by Morgan (ref. 5.26), we wish to take, for the closed-loop dynamics, a quadratic factor $s^2 + 3s + 2$, identifying it with $Q_i(s)$ gives:

$$\begin{aligned} & b_1 = 1 \\ & b_0 - b_1 + k = 3 \\ & b_0 + 2\,k = -\,2 \end{aligned}$$

so that $b_1 = 1$, $b_0 = 10$, $k = -6$, and the open-loop matrix is equal to

$$N = \frac{-6(s-2)}{(s+10)(s-1)}\, I \quad .$$

We deduce that

$$C = Z^{-1}N = \frac{6}{(s-1)(s+10)} \begin{bmatrix} s^2-5s-2 & -(s^2+s+4) \\ -(2s^2-7s-2) & s^2+6 \end{bmatrix}$$

The overall factor of 6 can always be suppressed by the use of a precompensator, outside the loop, with coefficient -1/6.

Remark: A state-space representation of the system constituted by the cascade SC takes the form:

$$S: \quad \dot{x} = \begin{bmatrix} 1 & & \\ & -1 & \\ & & -2 \end{bmatrix} x + \begin{bmatrix} 7 & 6 \\ 7 & 4 \\ 10 & 6 \end{bmatrix} u\,, \qquad y = 1/6 \begin{bmatrix} 1 & -3 & 2 \\ -1 & -3 & 4 \end{bmatrix} x$$

$$C: \quad \dot{\xi} = \begin{bmatrix} 1 & & \\ & -10 & \\ & & -10 \end{bmatrix} \xi + \begin{bmatrix} 1 & 1 \\ -148 & 94 \\ 268 & -106 \end{bmatrix} \varepsilon$$

$$u = \begin{bmatrix} 6 & -1 & 0 \\ -7 & 0 & -1 \end{bmatrix} \frac{6}{11}\, \xi + 6 \begin{bmatrix} -1 & 1 \\ 2 & -1 \end{bmatrix} \varepsilon$$

$$\text{SC}: \quad \begin{bmatrix} \dot{x} \\ \dot{\xi} \end{bmatrix} = (1/11)\begin{bmatrix} 11 & 0 & & & -7 & -6 \\ & -11 & & 14 & -7 & -4 \\ & & -22 & 18 & -10 & -6 \\ & & & 11 & 0 & 0 \\ & & & & -110 & 0 \\ & & & & 0 & -110 \end{bmatrix} \begin{bmatrix} x \\ \xi \end{bmatrix} + \begin{bmatrix} 30 & 6 \\ 6 & 18 \\ 12 & 24 \\ 1 & 1 \\ -148 & 94 \\ 268 & -106 \end{bmatrix} \varepsilon$$

$$y = (1/6)\begin{bmatrix} 1 & -3 & 2 & 0 & 0 & 0 \\ -1 & -3 & 4 & 0 & 0 & 0 \end{bmatrix} \begin{vmatrix} x \\ \xi \end{vmatrix} .$$

Putting the system in diagonal form by using the transformation matrix:

$$M = \begin{bmatrix} 0 & 1 & 0 & 0 & 6/11 & 7/11 \\ 7 & 0 & 1 & 0 & 4/9 & 7/9 \\ 6 & 0 & 0 & 1 & 3/4 & 5/4 \\ 11 & 0 & 0 & 0 & 0 & 0 \\ 0 & 0 & 0 & 0 & 0 & 11 \\ 0 & 0 & 0 & 0 & 11 & 0 \end{bmatrix}$$

the cascade takes the form:

$$x = \begin{bmatrix} 1 & & & & & \\ & 1 & & & & \\ & & -1 & & & \\ & & & -2 & & \\ & & & & -10 & \\ & & & & & -10 \end{bmatrix} \tilde{x} + (1/11)\begin{bmatrix} 1 & 1 \\ 278 & 64 \\ \times & \times \\ \times & \times \\ \times & \times \\ \times & \times \end{bmatrix} \varepsilon$$

$$y = (1/6)\begin{bmatrix} -9 & 1 & \times & \times & \times & \times \\ 3 & -1 & \times & \times & \times & \times \end{bmatrix} \varepsilon$$

which shows the complete controllability and observability of the unstable modes s = 1. The compensated system represented by the transfer function matrix H is thus genuinely stable.

B - Combination of state-space and transfer function matrix methods

We can envisage combining the two previous methods: as a first step, state-space techniques would be used to achieve a partial input-state non-interaction. Then, operational methods would be used, in a second stage, to obtain complete input-output non-interaction. This will often allow a more economical realisation.

Let us once more take the second example of paragraph 5.2.2.B, where the system S is represented by its transfer function matrix

$$Z(s) = \frac{1}{(s-1)(s+1)(s+2)} \begin{bmatrix} s^2+6 & s^2+s+4 \\ 2s^2-7s-2 & s^2-5s-2 \end{bmatrix}$$

or by its state equations

$$\dot{x} = \begin{bmatrix} 0 & -1 & 0 \\ -1 & -2 & -2 \\ 1 & 0 & 0 \end{bmatrix} x + \begin{bmatrix} 1 & 1 \\ 2 & 1 \\ 3 & 2 \end{bmatrix} u \qquad y = \begin{bmatrix} 1 & 0 & 0 \\ 0 & 1 & 0 \end{bmatrix} x .$$

We have seen that, in this case, $|sI-\alpha_{22}|$ being equal to s - 2, state-space methods do not allow us directly to achieve stability and decoupling at the same time.

We can, however, make a linear transformation $x = T\tilde{x}$ such that $|sI-\tilde{\alpha}_{22}|$ is a Hurwitz polynomial, where $\tilde{\alpha}_{22}$ denotes the counterpart of α_{22} in the transformed system. For example, if we take:

$$T = \begin{bmatrix} 1 & 0 & 0 \\ 0 & 0 & 1 \\ 0 & 1 & 0 \end{bmatrix}$$

we have

$$|sI-\tilde{\alpha}_{22}| = s+3$$

We then establish non-interaction between the states $\tilde{x}_1$, $\tilde{x}_2$, and the inputs e_1, e_2, by imposing the conditions

$$\tilde{\alpha}_{11} = \text{diag}\left[\tilde{a}_i\right] \qquad \tilde{\alpha}_{12} = 0$$

$$\tilde{\beta}_{11} = \text{diag}\left[\tilde{b}_i\right] .$$

We have then

$$\tilde{K}_1(s) = \frac{\tilde{b}_1}{s-\tilde{a}_1} I, \quad \tilde{K}_2(s) = \left[\frac{-(s+2)\tilde{b}_1}{(s+3)(s-\tilde{a}_1)} \ , \ \frac{(s-2)\tilde{b}_2}{(s+3)(s-\tilde{a}_2)} \right]$$

and

$$\tilde{X}(s) = \begin{bmatrix} \tilde{K}_1(s) \\ \tilde{K}_2(s) \end{bmatrix} E(s), \quad X(s) = \begin{bmatrix} {}_1\tilde{K}_1(s) \\ \tilde{K}_2(s) \\ {}_2\tilde{K}_1(s) \end{bmatrix} E(s) = \Gamma(s)E(s) \qquad (5.48)$$

where ${}_i\tilde{K}_1$ denotes the ith row of $\tilde{K}_1$.

The system defined by equation (5.48) and illustrated in figure 5.16a), is then controlled by a classical servo-loop with unity feedback and a precompensator C_0, as shown in figure 5.16b). The only constraint which will have to be imposed on the closed-loop transfer function matrix is that it should contain, as a zero, the unstable factor s − 2 which is a zero of the determinant of $C\Gamma$ (it will be noticed that, on the other hand, there is no further constraint on the poles, which have already been brought into the left half-plane). If we impose, e.g., as before:

$$K = \frac{-(s-2)}{s^2+3s+2} I$$

it follows that:

$$C_0 = (C\Gamma)^{-1}K(I-K)^{-1} = \begin{bmatrix} \dfrac{-(s-2)(s-\tilde{a}_1)}{\tilde{b}_1 s(s+4)} & 0 \\ \dfrac{-(s-2)(s-\tilde{a}_2)}{\tilde{b}_2 s(s+4)} & \dfrac{-(s+3)(s-\tilde{a}_2)}{\tilde{b}_2 s(s+4)} \end{bmatrix}$$

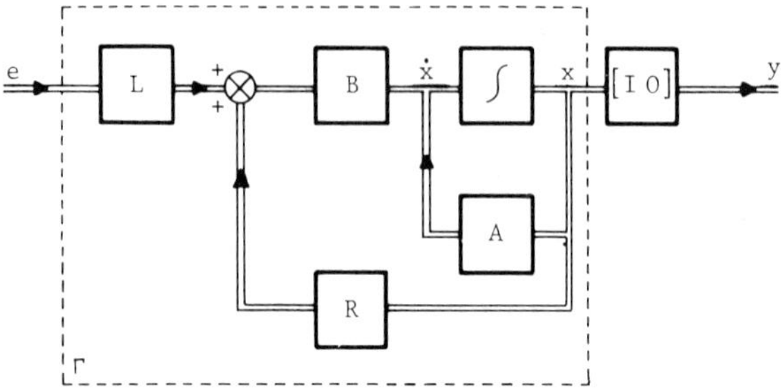

a) First stage: state feedback compensation.
Non-interaction between e and the states x_1, x_3.

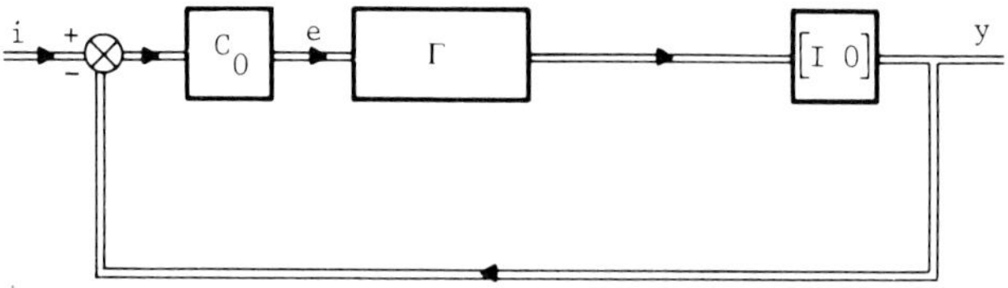

b) Second stage: input-output non-interaction.

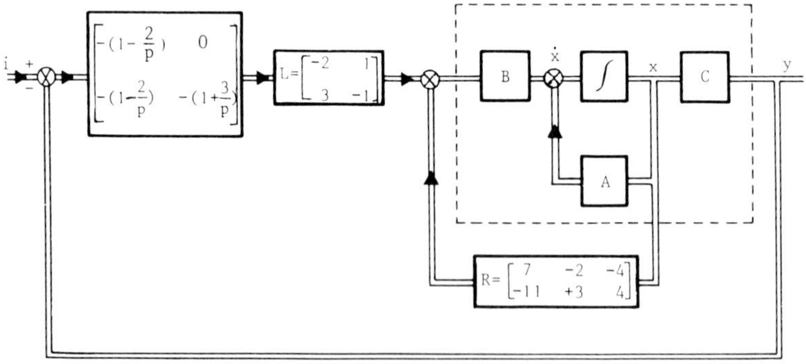

c) Realisation

Fig. 5.16 Input-output decoupling:-

a) Partial input-state non-interaction by state feedback.
b) Inclusion of direct compensation.
c) Complete realisation of the control scheme (example of paragraph 5.5.1.B)

We can then find the simplest structure for the compensator by taking:

$$\tilde{a}_1 = \tilde{a}_2 = -4 \qquad \tilde{b}_1 = \tilde{b}_2 = 1$$

whence

$$C_0 = \begin{bmatrix} -(1 - \frac{2}{s}) & 0 \\ -(1 - \frac{2}{s}) & -(1 + \frac{3}{s}) \end{bmatrix}$$

We then deduce the matrices R and L from

$$R^*_{11} = B^{-1*}_{11}(\alpha^*_{11} - A^*_{11}) = \begin{bmatrix} 7 & -4 \\ -11 & 4 \end{bmatrix}$$

$$R^*_{12} = -B^{*-1}_{11} A^*_{12} = \begin{bmatrix} -2 \\ 3 \end{bmatrix}$$

whence

$$R^* = \begin{bmatrix} 7 & -4 & -2 \\ -11 & 4 & 3 \end{bmatrix}$$

and

$$L^* = \begin{bmatrix} -2 & 1 \\ 3 & -1 \end{bmatrix}$$

In the original coordinates, we shall have

$$L = L^* \qquad R = R^* T^{-1} = \begin{bmatrix} 7 & -2 & -4 \\ -11 & 3 & 4 \end{bmatrix}$$

The control scheme finally obtained is illustrated in figure 5.16c). In practice, it should not be forgotten that the constraints imposed on the form of the closed-loop transfer function matrix lead to a system of non-minimum phase and that the response will thus have undesirable transient behaviour.

<u>Remarks</u>: If we compare this solution with the previous one (paragraph

A), it would seem that compensation has been achieved with the system order augmented by only two units (corresponding to the order of the compensator C_0) whereas three were necessary before. In fact, since the original system had three states and two outputs, the use of the feedback compensator R requires in practice an observer of order 1, which reconciles the two solutions in regard to their respective dimensions.

The advantage of this method is to be found:

- essentially in a simplification of the design procedures, following from the partial relaxation of the constraints imposed on the structure of the closed-loop matrices;

- in the simplification of the compensators: the compensator obtained is of the proportional-integral type frequently utilised in industry.

C - Case where the number of inputs is different from the number of outputs

In the preceding paragraphs, we assumed the process to have the same number of inputs and outputs. The attention given to this case arises from the fact that, at the expense of certain manipulations, it is possible to reduce other situations to it.

Let us consider, in fact, a process having m inputs and p outputs, with $m \neq p$. Quite clearly, the number of independent outputs is necessarily less than or equal to the number of inputs, and so we shall assume $m \geqslant p$. (We saw in paragraph 5.2.1.A that this condition is necessary for the existence of a right inverse and for reproducibility). The transfer function matrix of the process is a rectangular matrix (obviously not invertible) of dimensions (p, m), which will be written in the form:

$$Z = \begin{bmatrix} Z_a & Z_c \end{bmatrix}$$

where Z_a is a matrix of dimensions (p, p), the dimensions of Z_c being (p, m-p). The elements of the matrix Z can be treated more systematically if it is extended so as to make it square (1). This amounts to

(1) Method proposed by H. Freeman, ref. 5.10.

introducing m-p supplementary outputs, referred to as "virtual outputs" and designated by y_b, the actual outputs being y_a. We thus associate with Z a matrix Z^* of dimensions (m, m):

$$Z^* = \begin{bmatrix} Z_a & Z_c \\ Z_d & Z_b \end{bmatrix}$$

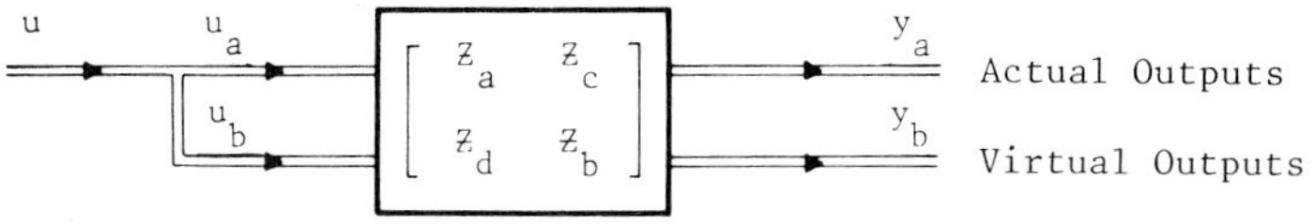

Fig. 5.17 Inputs and fictitious outputs in the case of a rectangular transfer function matrix.

Correspondingly, the input vector u will be decomposed into two components: u_a, of dimension p, and u_b, of dimension m-p, by setting (cf. figure 5.17):

$$\begin{bmatrix} y_a \\ y_b \end{bmatrix} = \begin{bmatrix} Z_a & Z_c \\ Z_d & Z_b \end{bmatrix} \begin{bmatrix} u_a \\ u_b \end{bmatrix}$$

Since the outputs y_b are fictitious, we can choose Z_d and Z_b so as to facilitate the subsequent calculations. In particular, we can arrange that (1)

$$y_b = u_b$$

i.e. take

$$Z_d = 0 \qquad Z_b = I \ .$$

Let C be the precompensator placed in the forward path upstream of Z. If we decompose C like Z^*, the open-loop matrix ZC can be written:

(1) Method proposed by H. Freeman, ref. 5.10

$$ZC = N = \begin{bmatrix} Z_a C_a + Z_c C_d & Z_a C_c + Z_c C_b \\ C_d & C_b \end{bmatrix}$$

and the non-interaction conditions, N = diagonal (N_a, N_b), take the form:

$$Z_a C_a = \text{diagonal } N_a, \qquad C_d = 0 \tag{5.49}$$

$$C_b = \text{diagonal } N_b, \qquad Z_a C_c + Z_c C_b = 0 . \tag{5.50}$$

The second equation (5.50) in particular implies a fundamental relation between the process and the compensator, whose importance we shall see later. Given that it is possible to satisfy this condition, the compensator is determined from the desired matrices by:

$$C_b = N_b \qquad C_a = Z_a^{-1} N_a \qquad C_d = 0$$

with C_c such that:

$$Z_a C_c + Z_c N_b = 0 .$$

The configuration of the system is then as indicated in figure 5.18. Evidently, the analogy between the defining equation for C_a and that for C_0 in the case m = p, entails for N_a the same constraints as were discussed in subsection 5.4.2.

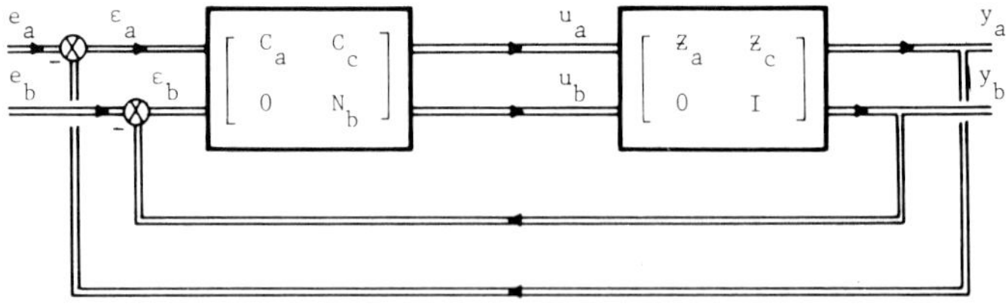

Fig. 5.18 Fictitious inputs and outputs.
Structure of the compensator.

Remark 1: We could treat in the same way the case of non-unity feedback (cf. figure 5.19). Decomposing the matrix R into

$$R = \begin{bmatrix} R_a & R_c \\ R_d & R_b \end{bmatrix}$$

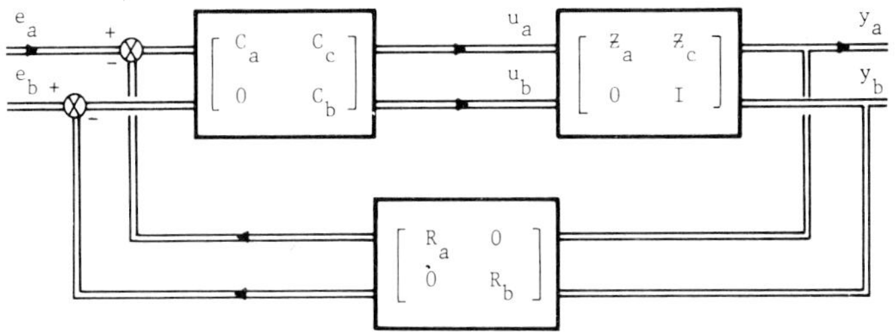

Fig. 5.19 Fictitious inputs and outputs.
Case of non-unity feedback.

we can, without unduly limiting the possible choices at the synthesis level, suppress the coupling between the primary and secondary (fictitious output) loops at the level of the feedback network. This amounts to taking:

$$R_c = 0 \qquad R_d = 0 .$$

The defining relations are then of the form:

$$Z_a\, C_a\, R_a = \mathrm{diag}\left[N_a\right] \qquad C_d = 0 \tag{5.51}$$

$$C_b\, R_b = \mathrm{diag}\left[N_b\right] \qquad Z_a\, C_c + Z_c\, C_b = 0 . \tag{5.52}$$

As before, the second equation (5.52) implies a fundamental relation between process and compensator.

Remark 2: The only extra problem connected with the existence of Z_c is created by the possible occurrence of unstable poles in its elements. Indeed, as no loop is closed around Z_c, the instabilities due to such modes will not, in general, be capable of elimination.

5.5.2 Problems of inversion

One of the problems arising in the realisation of a non-interactive system, apart from that of stability which has already been discussed, is the inversion of the process transfer function matrix in the case where operational methods are used. This problem has in fact been partially treated in subsection 5.2.1, where the notion of a right inverse was defined and its minimal structure specified. From the point of view taken here, this method, which assumes that a state-space representation of the process has been found to begin with, can however be laborious to employ in practice: it would be preferable to find directly a simple method of inversion using transfer function matrices.

We shall review here the technique of inversion by the methods of graph theory, which in most cases reduces the complexity of the calculations and whose use in problems of non-interaction has been advocated, in particular, by R.A. Mathias (cf. ref. 5.23).

A - Graphical structural representation of the system to be controlled

Let us consider the representation by a signal-flow graph of a system composed of linear elements with transfer functions F, G, H,... P, as in figure 5.20, where the u_i denote input variables, the y_i are physical output variables to be controlled, and the w_i are variables needed for specifying the state of the system. Such a graph can always be reduced to the general P-canonical structure of Mesarovic by suppressing the intermediate nodes. However, such a procedure considerably complicates the transfer functions and is harmful to the synthesis of possible controllers and the study of sensitivity to variation of system parameters. Furthermore, special reasons may prevail in the choice of one structure rather than another, even to the extent of imposing it.

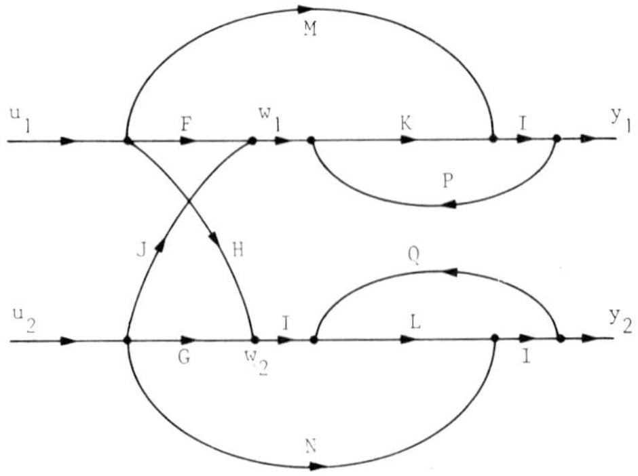

Fig. 5.20 General signal-flow graph.

R.A.Mathias (ref. 5.23) suggests, for example, that:

- if the elements F, G, H, I are subject to significant variations, they should be grouped together in a subgraph;

- the intermediate variables w can be chosen to form a non-interactive vector, even if the output signals do not define a non-interactive basis.

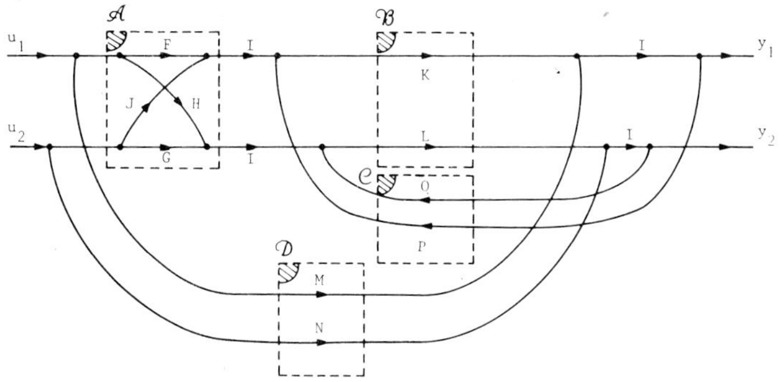

Fig. 5.21 Decomposition into subgraphs.

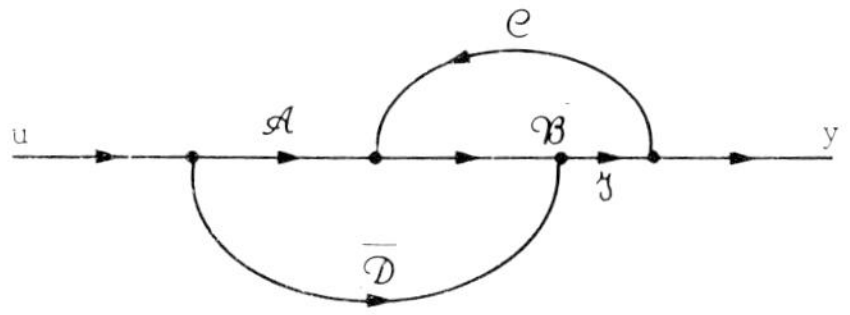

Fig. 5.22 Synthetic graph.

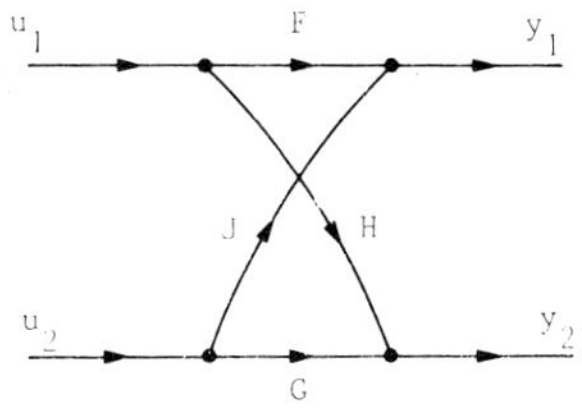

Fig. 5.23 Graph of P-canonical structure.

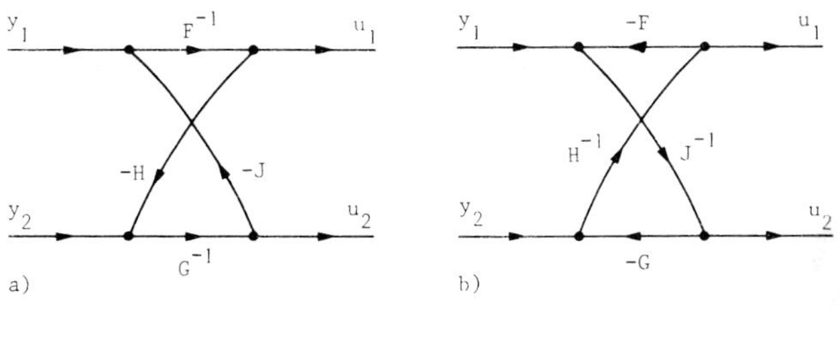

Fig. 5.24 Graphs for the inverse of the previous structure.

Thus, figure 5.21 shows the subdivision of the graph into four sub-graphs possessing a classical P-canonical structure, which facilities the matrix operations. The four corresponding matrices $\mathcal{A}, \mathcal{B}, \mathcal{C}, \mathcal{D}$ are all of rank 2, and hence invertible a priori. In graphical form, the representation of the system is as shown in figure 5.22 and Mason's rule gives directly:

$$Y = ZU$$

with

$$Z = \left[I - \mathcal{B}\mathcal{C}\right]^{-1} \left[\mathcal{B}\mathcal{A} + \mathcal{D}\right] .$$

B - Rules for inversion of graphs

1 - Application of the classical rules for inversion of a graph enables us to relate, in a simple topological manner, the P-canonical and V-canonical structures.

Let us consider, for this purpose, the P-canonical structure in figure 5.23. The inversion of this graph consists of the following steps:-

1) Choose a path leading from a source node to a sink node without passing twice through the same node. Adjoin to the nodes of the reversed path, intermediate nodes connected to the former by unit transmittances.

2) Reverse the directions along the chosen path and invert the corresponding transmittances.

3) Change the signs of the transfer functions on the branches which terminate on the chosen path, keeping their directions unaltered. Leave the rest of the graph unchanged.

2 - In the case of figure 5.23, we can reverse the paths F and G or alternatively the paths J and H, and we end up with two equivalent V-canonical structures, illustrated in figure 5.24. It will thus be noticed that the inverse of a P-canonical structure is a V-canonical structure and vice-versa. This method, which generalises to graphs of matrices, has the

advantage of facilitating matrix inversion and can be beneficially used in the choice of a controller for the purpose of making a multivariable system non-interactive.

3 - Suppose we wish to invert the structure of figure 5.20. With the system decomposed into subgraphs as indicated above, we can have the two structures represented schematically in figure 5.25, according to the paths of inversion utilised ($\mathcal{A}\mathcal{B}$ or $\mathcal{D}$). The corresponding graphs in complete detail are shown in figure 5.26a) and b). In addition to these representations, we can evidently find others which have disappeared as a result of the division into subgraphs; e.g. by using directly the paths GL and M, we obtain the diagram of figure 5.26c).

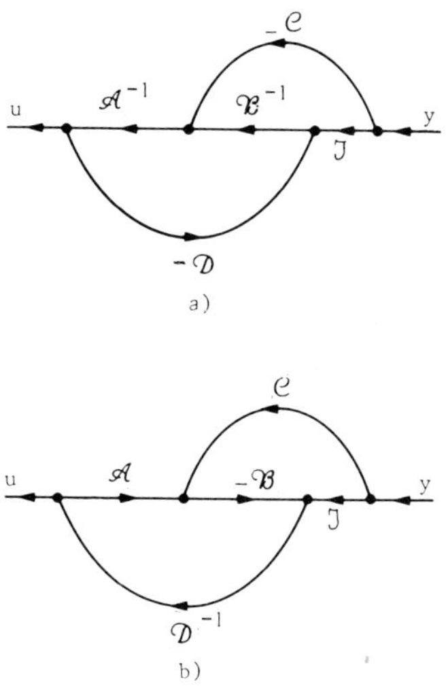

Fig. 5.25 Inverse structures for the structure of fig. 5.22.

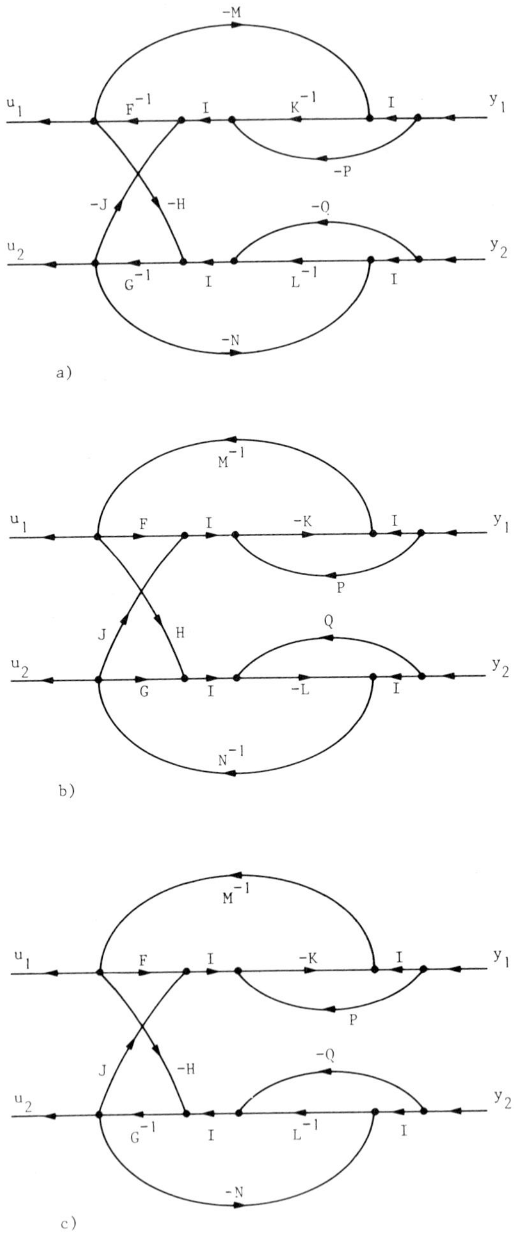

Fig. 5.26 Detailed graphs of various possible inverses.

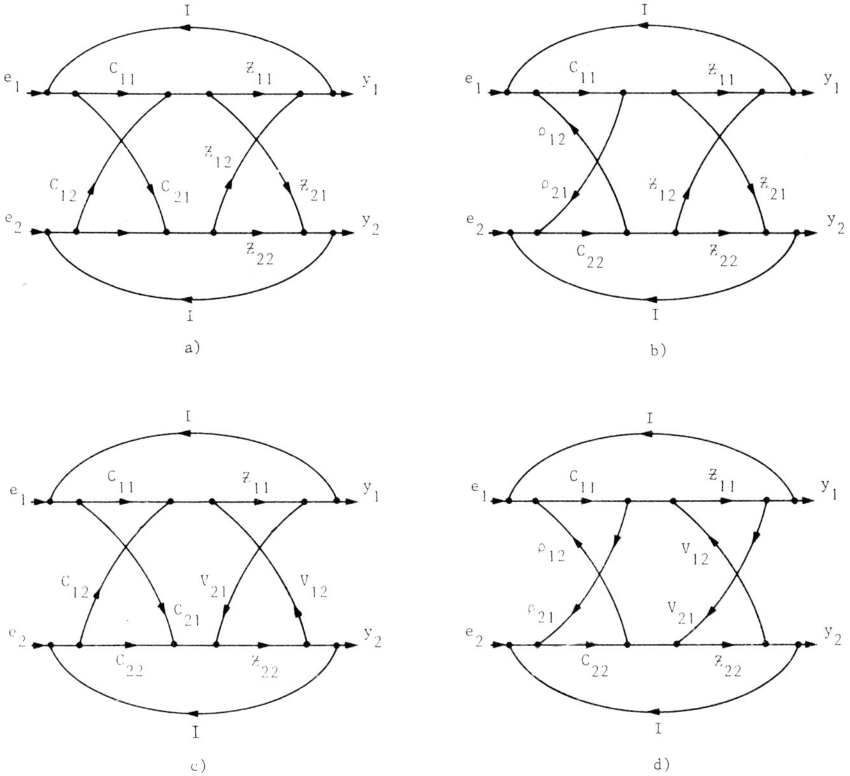

Fig. 5.27 Decoupling by controller in cascade.

C - Application to the diagonalisation of a control system

By direct application of Mason's rule, the graph method enables us, given the structure of the compensator and of the system to be controlled, to determine immediately the conditions required for non-interaction.

In the case of a cascade controller, e.g., we can envisage four possibilities, as the structures of the system to be controlled and of the compensator can each be chosen either P-canonical or V-canonical (cf. figure 5.27). Since the paths connecting the nodes e_1, e_2 to y_1, y_2 have at least one point in common with all the loops of the graph, it is sufficient, in order to determine the non-interaction conditions, to suppress the transmittances connecting the nodes e_1, e_2 to y_2, y_1

respectively.

We have immediately:-

Case a) Compensator and system P-canonical:

$$Z_{11}C_{12} + Z_{12}C_{22} = 0 \qquad Z_{22}C_{21} + Z_{21}C_{11} = 0$$

Case b) Compensator V-canonical, system P-canonical:

$$Z_{11}C_{11}\rho_{12}C_{22} + Z_{12}C_{22} = 0 \qquad Z_{21}C_{11} + Z_{22}C_{22}\rho_{21}C_{11} = 0$$

Case c) Compensator P-canonical, system V-canonical:

$$Z_{11}C_{12} + Z_{12}V_{12}Z_{22}C_{22} = 0 \qquad Z_{22}C_{21} + Z_{22}V_{21}Z_{11}C_{11} = 0$$

Case d) Compensator and system V-canonical:

$$Z_{11}C_{11}\rho_{12}C_{22} + Z_{11}V_{12}Z_{22}C_{22} = 0$$

$$Z_{22}V_{21}Z_{11}C_{11} + Z_{22}C_{22}\rho_{21}C_{11} = 0 .$$

5.5.3 Problems of compensation and physical realisation

A - Compensation

We have been concerned up to now with the problem of non-interaction as such, i.e. with the structure of compensators designed to attain this end and with the particular forms for the elements of open-loop and closed-loop transfer function matrices, which guarantee the absence of hidden instabilities. It is obvious that, in a practical problem, the non-interaction condition will not be sufficient: the compensator will also have to be designed so as to satisfy certain performance requirements.

The problem is, however, quite straightforward from here on, since, when non-interaction has been achieved, the compensation becomes a question of classical single-variable theory. We shall thus not dwell on this subject, in which the familiar techniques (frequency-response curves, Bode asymptotic diagrams, root-loci) can be utilised.

By way of example, we can refer to the work of Chen, Mathias and Sauter (ref. 5.3).

B - Physical realisation problems

1) Matrix methods: As already indicated, the very principle of design of the compensator:

$$c_{ij} = \left[Z_{ij}^{-1}\right] n_{jj}$$

has the consequence that, if n_{jj} is chosen only on the basis of the desired specifications for the open-loop matrix N, the compensator may not be physically realisable. Indeed, in the case that the elements of the inverse matrix Z^{-1} have, as usual, more zeroes than poles, the same may be true for C if N does not have sufficient dynamics. More to the point, the problem will arise if, as is often the case in industrial control, the process contains pure delays. In this case, in fact, the inverse matrix contains pure predictor elements, exp(Ts), which cannot be physically realised.

By way of example, consider a process defined by the transfer function matrix:

$$Z = \begin{bmatrix} \frac{2}{s+5} & \frac{1}{s+1} \\ \frac{2}{s+6} & \frac{1}{s+2} \end{bmatrix}$$

whose determinant is equal to:

$$\frac{-8}{(s+1)(s+2)(s+5)(s+6)}$$

We have:

$$Z^{-1} = \frac{-(s+1)(s+2)(s+5)(s+6)}{8} \begin{bmatrix} \frac{1}{s+2} & \frac{-1}{s+1} \\ \frac{-2}{s+6} & \frac{2}{s+5} \end{bmatrix}$$

and let us suppose that we wish to obtain the open-loop matrix:

$$N = \frac{5}{s+3} \quad I$$

The compensator thus defined would have as its transfer function matrix:

$$C = \frac{-5}{8}\begin{bmatrix} \frac{(s+1)(s+5)(s+6)}{s+3} & \frac{-(s+2)(s+5)(s+6)}{s+3} \\ \frac{-2(s+1)(s+2)(s+5)}{s+3} & \frac{2(s+1)(s+2)(s+6)}{s+3} \end{bmatrix}$$

which is manifestly unrealisable.

One solution would consist of putting in n_{jj} enough poles to make the compensator realisable. In this case, it is, however, important to ensure that, by so doing, we do not change the dynamical characteristics of the loop. This means that it will, in general, be impossible to adjoin these extra poles in such a way as to compensate directly those appearing in the compensator originally obtained.

Thus, if we adjoin to n_{ii} a factor $1/(s+5)(s+6)$, the compensator certainly becomes realisable, but the dynamics of the system are completely altered.

It is preferable in this case to adjoin the supplementary poles in a region where the gain of N is sufficiently low, e.g. -12 dB (cf. ref. 5.3). Then, indeed, they will not seriously affect the response, unless too many are put in the same place. If, in fact, they are placed an octave apart, this procedure will be satisfactory. We could, for example, in the present case, take N in the form:

$$N = \frac{5}{(s+3)(s+20)(s+40)} \quad I$$

which leads to a realisable compensator while preserving the original dynamics.

2) <u>Graph methods</u>: In general, the inversion of a structure increases the degrees of the numerators of transfer functions and makes their physical realisation impossible. The introduction of a diagonal matrix N containing lag elements, however, allows us to reduce the degrees of the numerators compared with those of the denominators.

Denoting by q the common element of this matrix N, the structure representing the product $Z^{-1}N$ is obtained by cutting the inverse graph with a vertical line and performing on the branches of the cut graph, the following operations:

- for the cut branches belonging to a direct path from the input to the output, multiply the transmittances by q;
- for the cut branches belonging to a feedback loop and not to any direct path, divide the transmittances by q.

We also wish to preserve the same loop transmittances and to multiply by q the transmittances of the direct paths.

This method can be applied to the graph of figure 5.20 with the paths M and N reversed. Figure 5.28 shows the inverse structure after multiplication by q.

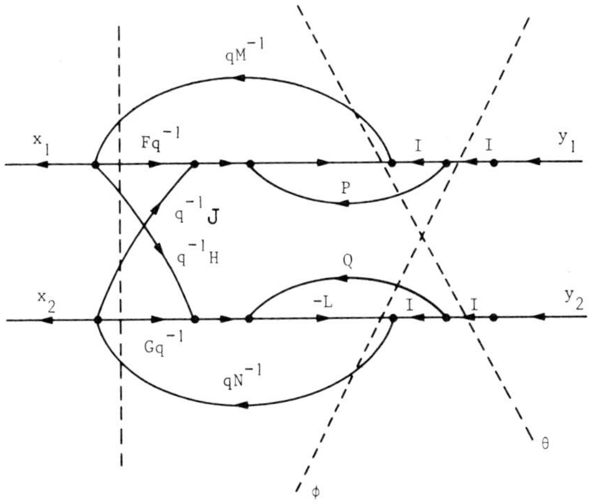

Fig. 5.28 Inversion of the graph of fig. 5.20 by the method of intersections.

Denoting by X_d the difference between the denominator and numerator degrees of a transfer function X, the realisability conditions become:

$$(F_d, G_d, H_d, J_d) \geqslant q_d \geqslant (M_d, N_d).$$

If this condition cannot be satisfied, we may be able, by using other cuts (θ, ϕ), to satisfy simply the required conditions.

Example: If

$$M = \frac{1}{1+T_1 s} \quad , \quad K = \frac{1}{(1+2s)(1+3s)} \quad ,$$

$$N = \frac{1}{1+T_2 s} \quad , \quad L = \frac{1}{(1+4s)^2} \quad ,$$

we have:

$$\begin{matrix} K_d \geqslant \theta_d \geqslant M_d \\ L_d \geqslant \phi_d \geqslant N_d \end{matrix} \qquad \text{so that} \qquad \begin{cases} 2 \geqslant \theta_d \geqslant 1 \\ 2 \geqslant \phi_d \geqslant 1 \end{cases}$$

and we can take

$$\theta = \frac{k}{s} \quad , \quad \phi = \frac{1}{1+Ts}$$

C - Non-interaction with regard to inputs and to perturbations

In the preceding paragraphs, we have primarily been concerned with the problem of realisation of a system which is non-interactive in regard to its inputs. We could also look at the problem from a "regulator" viewpoint, i.e. with regard to perturbations. The problem then consists of ensuring a suitable response of the complete system for particular "inputs", usually not well known, while leaving open the possibility of improving the response to a true control signal, profiting from the interaction with respect to inputs, and forming the control signal by means of suitable filters. It is important, in fact, to note that, contrary to what has been suggested by certain authors, non-interation with regard to inputs does not imply non-interaction with regard to perturbations, and conversely. Let us consider, for example, the system of figure 5.29, where γ represents the perturbations introduced into the loop, and Z_1, Z_2, Z_3 are transfer function matrices of the process.

We have

$$Y = \left[I + Z_2\, Z_1\, C\right]^{-1} Z_2\, Z_1\, CE + \left[I + Z_2\, Z_1\, C\right]^{-1} Z_2\, Z_3\, \Gamma \; .$$

Let us suppose that the desired characteristics of the total system with respect to control signals e and perturbations γ are represented by matrices H and H´:

$$Y = HE + H' \Gamma .$$

We have, by equating coefficients:

$$\left[I + Z_2 Z_1 C\right]^{-1} Z_2 Z_1 C = H ,$$

$$\left[I + Z_2 Z_1 C\right]^{-1} Z_2 Z_3 = H' .$$

From the last equation, it follows that:

$$C = Z_1^{-1} \left[Z_3 H'^{-1} - Z_2^{-1}\right] .$$

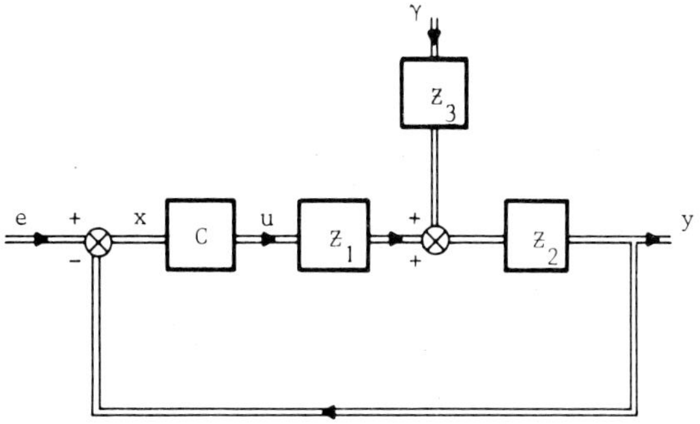

Fig. 5.29 Introduction of perturbations.

This solution is unique if the matrices Z_1, Z_2, Z_3 are of the same dimensions. This will, in general, be the case in this type of problem. Substituting the value thus obtained for C into the equation for H, we get:

$$H = I - H' Z_3^{-1} Z_2^{-1} .$$

If, then, we desire the system to be non-interactive in regard to perturbations, H´ should be diagonal. The last equation shows that the system can then be non-interactive with regard to inputs only if $Z_2 Z_3$ is diagonal.

The problem evidently arises in the same way if the perturbations enter at the level of the process inputs u. In fact, in this case:

$$Z_3 = Z_1 = I , \quad Z_2 = Z$$

whence

$$C = H'^{-1} - Z^{-1}, \quad H = I - H'Z^{-1} .$$

Simultaneous non-interaction with respect to inputs and perturbations is thus possible only in the very special case that Z is itself diagonal. In fact, this property is directly linked with the concept of structure.

D - Example

Let us consider a system whose dynamics are defined by the equations:

$$A\ddot{\phi} + E\phi - G\dot{\psi} = u + C_{p\phi}$$
$$A\ddot{\psi} + E\psi + G\dot{\phi} = v + C_{p\psi} \quad .$$

These equations could, e.g., represent the linearised equations of a satellite during the period of accurate alignment. In view of these equations, the system is best represented by a V-canonical structure, which takes account of the physical interaction of the outputs on one another (cf. figure 5.30).

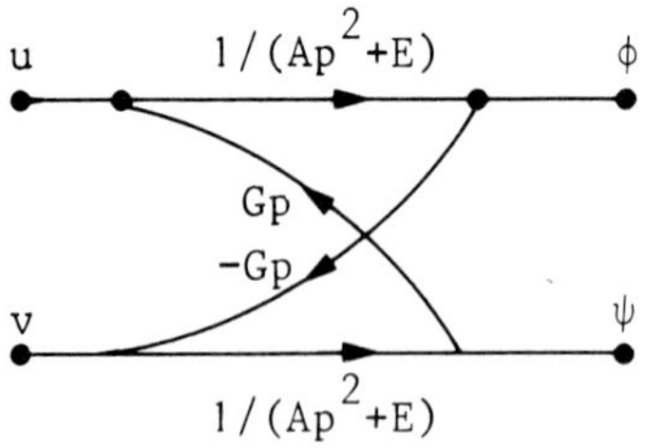

Fig. 5.30 Simplified diagram of satellite dynamics.

On the other hand, the control system to be designed should be envisaged from a "regulator" point of view, i.e. it will be required to maintain ϕ and ψ null, despite the perturbing torques $C_{p\phi}$ and $C_{p\psi}$. We should, all the same, leave open the possibility of having to control, e.g., a deviation of ϕ or ψ.

We note in passing that the normal modes of the system, given by the equation $(As^2 + E)^2 + G^2 s^2 = 0$, are oscillatory. Consequently, there is no question of realising, at least directly, a system which is non-interactive with regard to the inputs, since the response to perturbations would be unacceptable.

Let us, to begin with, try to compensate the system with a controller D in parallel. This compensation is brought about by the following considerations: the realisation of D, and the establishment in the resulting matrix H of rigid couplings, with a V-canonical structure for H allowing the choice of a series compensator.

Assuming a P-canonical structure for the compensator D, we take

$$D = \begin{bmatrix} d_{11} & d_{12} \\ d_{21} & d_{22} \end{bmatrix} \qquad \text{with} \begin{cases} d_{11} = d_{22} = k_2 s + k_3 - E, \\ d_{12} = -d_{21} = Gs - k_1 . \end{cases}$$

(This assumes that it is possible to measure the angular velocities $\dot{\phi}$ and $\dot{\psi}$ accurately by the use of gyroscopes).

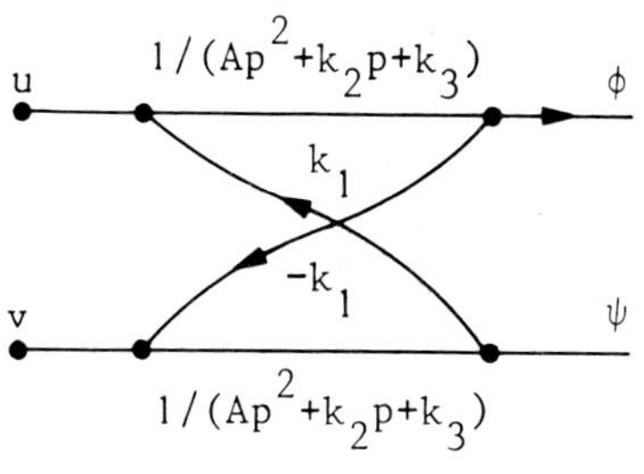

Fig. 5.31 The compensated system.

The system thus compensated can be put in the form of figure 5.31. The coefficients k_1, k_2, k_3 are determined so as to ensure that the resultant system H is stable. The coupling terms k_1 and $-k_1$ being now fixed, the conditions for non-interaction with regard to perturbations reduce simply (with the notation of figure 5.32) to:

$$C_{12} = k_1, \; C_{21} = -k_1.$$

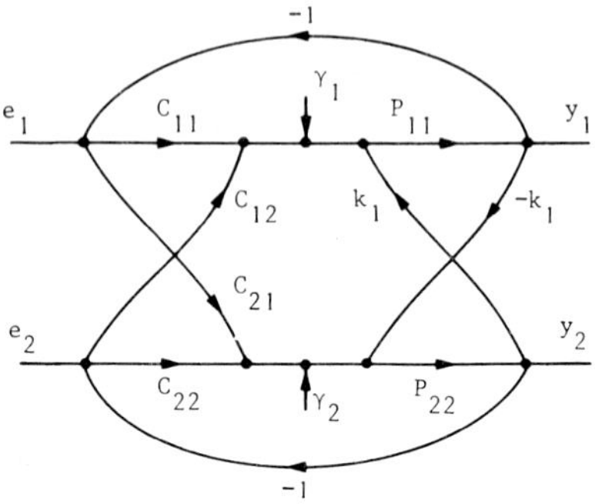

$$\begin{bmatrix} 1 + P_{11} C_{11} & P_{11}(C_{12} - k_1) \\ P_{22}(C_{21} + k_1) & 1 + P_{22} C_{22} \end{bmatrix} \begin{bmatrix} y_1 \\ y_2 \end{bmatrix} = \begin{bmatrix} P_{11} C_{11} & P_{11} C_{12} \\ P_{22} C_{21} & P_{22} C_{22} \end{bmatrix} \begin{bmatrix} e_1 \\ e_2 \end{bmatrix} + \begin{bmatrix} P_{11} & 0 \\ 0 & P_{22} \end{bmatrix} \begin{bmatrix} \gamma_1 \\ \gamma_2 \end{bmatrix}$$

Fig. 5.32 Structure.

As regards the compensators C_{11} and C_{22}, it will be advantageous to incorporate integral action into them, so as to suppress the steady-state error after a perturbation $C_{p\phi}$ or $C_{p\psi}$. If we take

$$C_{11} = C_{22} = k_4 \frac{1 + \tau s}{s(1 + \tau' s)}$$

the complete system will be represented by the diagram of figure 5.33. The system thus constituted is non-interactive with respect to perturbations; on the other hand, it is interactive with respect to the inputs e_1, e_2, since we have

$$\phi = \frac{P_{11}}{1 + P_{11}C_{11}} C_{p\phi} + \frac{P_{11}C_{11}}{1 + P_{11}C_{11}} (E_1 + C_{12} C_{11}^{-1} E_2).$$

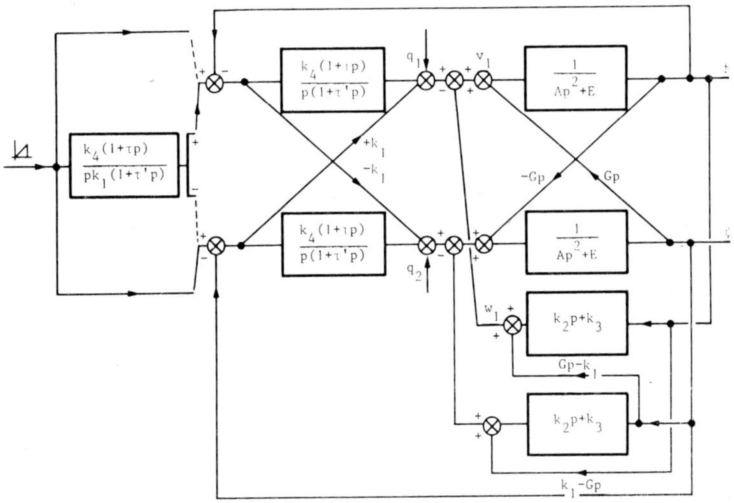

Fig. 5.33 Complete loop.

In view of the forms of C_{12} and C_{11}, we see that it is possible to control a deviation of ϕ without changing ψ, by maintaining a certain relation between e_1 and e_2. Explicitly, we must have

$$E_2(s) = \frac{k_1 s(1 + \tau' s)}{k_4(1 + \tau s)} \; E_1(s)$$

a realisation of which is indicated in figure 5.34. With the relay in position 2, $\Delta\phi$ is controlled with ψ constant; for position 1, ϕ is held constant and $\Delta\psi$ controlled. The duration of the input signal depends on the deviation to be controlled.

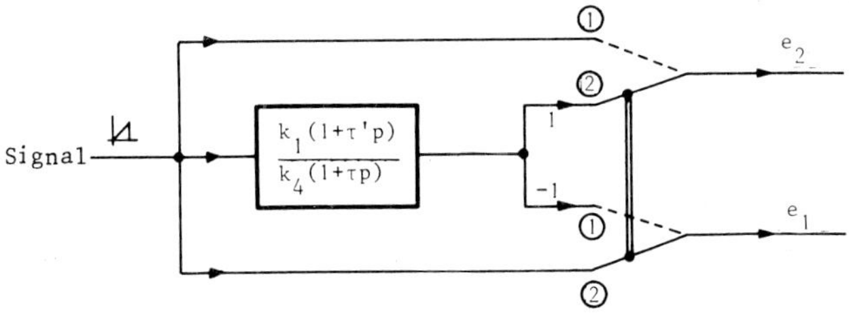

Fig. 5.34 Types of control.

Appendix

Numerical-calculation programs

1 LEVER : Transformation from state-space representation to transfer function matrix.

2 INDHOU : Test of linear independence.

3 DEMOE : Decomposition into subsystems controllable from a single input.

4 DEMOS : Decomposition into single-output observable subsystems.

5 NONIN : Non-interactive control.

A.1 SUB-PROGRAM FOR TRANSFORMATION FROM STATE-SPACE REPRESENTATION TO TRANSFER FUNCTION MATRIX : LEVER

This program implements the transformation from a state-space representation for a linear constant system of order n:

$$\dot{x} = Ax + Bu, \quad y = Cx$$

to its transfer function matrix representation:

$$Y(s) = Z(s)U(s).$$

The theoretical considerations relevant to this transformation will be found in subsection 2.1.4.B.

Notation

The matrices A,B,C are of respective dimensions (n,n), (n,m), (p,n). Leverrier's algorithm enables us to express the matrix $(sI-A)^{-1}$ in the form

$$(sI-A)^{-1} = \psi^{-1}(s)R(s)$$

with

$$\psi(s) = s^n + a_{n-1}s^{n-1} + \ldots + a_1 s + a_0$$

$$R(s) = Is^{n-1} + R_1 s^{n-2} + \ldots + R_{n-1}$$

whence

$$Z(s) = C(sI-A)^{-1}B$$

$$= \psi^{-1}(s)\left[Z_0 s^{n-1} + Z_1 s^{n-2} + \ldots + Z_{n-1}\right]$$

where

$$Z_i = CR_iB.$$

Input data

The data are:

- The number of states n(N), the number of inputs m(M), the number of outputs p(IP).
- The matrices A,B,C.

Output

The results comprise:

- The matrices Z_i $(i = 0,\ldots n-1)$
- The coefficients V(i) of the characteristic polynomial, printed in order of decreasing powers of s.

Storage

If $n > 10$ or $p > 5$, the matrices are re-dimensioned as follows: R(n,n), Z(p,n), V(n).

FLOWCHART

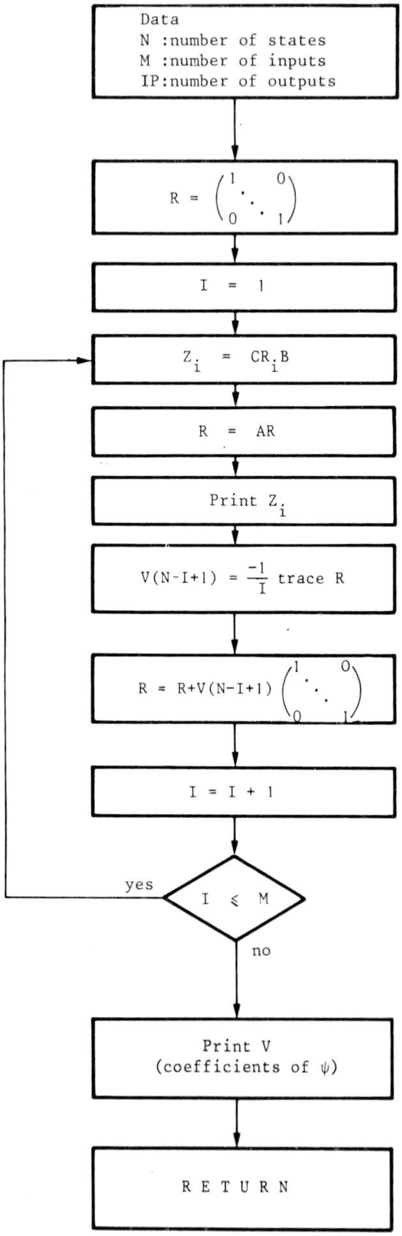
Data
N :number of states
M :number of inputs
IP:number of outputs
R = (1 ... 0; 0 ... 1)
I = 1
Z i = CR i B
R = AR
Print Z i
V(N-I+1) = -1/I trace R
R = R+V(N-I+1) (1 ... 0; 0 ... 1)
I = I + 1
yes
I ≤ M
no
Print V
(coefficients of ψ)
R E T U R N

```
      SUBROUTINE LEVER(A,B,C,N,M,IP,LN,LM)
C     CALCULATION OF Z(S) FROM A,B,C
C     LN=MAXIMUM NUMBER OF STATES
C     LM=MAXIMUM NUMBER OF INPUTS OR OUTPUTS
C     SUBROUTINE USED PRODI
      DOUBLE PRECISION A(LN,1),B(LN,1),C(LM,1),R(16,16),Z(5,16),V(16)
  100 FORMAT(8X16F7.1)
      DO1 I=1,N
      DO1 J=1,N
    1 R(I,J)=0.
      DO3 I=1,N
    3 R(I,I)=1.
      DO2 I=1,N
      CALL PRODI(IP,N,N,C,R,Z,LM,LN,LM)
      CALL PRODI(IP,N,M,Z,B,Z,LM,LN,LM)
      CALL PRODI(N,N,N,A,R,R,LN,LN,LN)
      IM=I-1
      NMI=N-I
      PRINT 101,IM,NMI
  101 FORMAT(///10X'MATRIX Z'I2,'(MULTIPLE OF S'I2,')'///)
      DO6 J=1,IP
    6 PRINT 100,(Z(J,K),K=1,M)
      V(N-I+1)=0.
      DO4 J=1,N
    4 V(N-I+1)=V(N-I+1)-R(J,J)/I
      DO2 J=1,N
    2 R(J,J)=R(J,J)+V(N-I+1)
      PRINT 102
  102 FORMAT(///10X'COEFFICIENTS OF CHARACTERISTIC POLYNOMIAL
    1 IN ORDER OF DECREASING POWERS OF S'//)
      PRINT 103,(V(N-I+1),I=1,N)
  103 FORMAT(10X5F16.3/)
      RETURN
      END
```

A.2 SUB-PROGRAM FOR TEST OF LINEAR INDEPENDENCE : INDHOU

This program allows the determination, in a sequence of vectors V, of those vectors which are linearly dependent on the preceding ones, in such a way that the coefficients of dependence are identified.

Method used

a) The method employed is Householder's, which is more stable than that of Gram-Schmidt.

It is based on the existence of an elementary transformation $\mathcal{P}$ which enables us to associate, with any vector $V = [v_1,\ldots v_n]^T$, a vector w whose last n-1 components are null while the first has magnitude equal to the norm of V, i.e.

$$w = [\pm\|V\|, 0,\ldots 0]^T.$$

$\mathcal{P}$ is thus such that

$$\mathcal{P}V = \|V\| e^1$$

and can be calculated in the form

$$\mathcal{P} = I - 2WW^T$$

where W is a vector of unit norm. We have, in fact:

$$\begin{aligned} V &= (I-2WW^T)V \\ &= V - 2W(W^TV) \\ &= [\pm \|V\|, 0,\ldots 0]^T \end{aligned}$$

whence, by equating components and setting $W^TV = K$, we get the n equations

$$2Kw_1 = v_1 \mp \|V\|$$

$$2Kw_i = v_i \qquad (i = 2,\ldots n).$$

By summing the squares of these equations, it follows that

$$2K^2\|W\|^2 = 2K^2 = \|V\|^2 \mp v_1\|V\|.$$

We take the + or - sign according as

$$|(v_1 \mp \|V\|)| = |v_1| + \|V\| .$$

b) With this in mind, let us suppose we have a sequence of vectors

$$V^1, V^2, \ldots V^n.$$

For the first vector V^1, there exists, in view of the above analysis, a transformation $\mathcal{P}_1$ such that

$$\mathcal{P}_1 V^1 = e^1 \|V^1\| .$$

The same transformation will convert V^2 into $\mathcal{P}_1 V^2$ which we shall write, separating the last n-1 components from the first one, in the form

$$\mathcal{P}_1 V^2 = \begin{bmatrix} \omega^1 \\ Z^2 \end{bmatrix}$$

The vector Z^2, of dimension n-1, can be transformed in its turn, as before, by an elementary transformation $\mathcal{P}_2^*$ of dimensions (n-1, n-1):

$$\mathcal{P}_2^* Z^2 = [\pm \|Z^2\| , 0, \ldots 0]^T.$$

If we define a matrix $\mathcal{P}_2$ by

$$\mathcal{P}_2 = \begin{bmatrix} 1 & 0 \\ 0 & \mathcal{P}_2^* \end{bmatrix}$$

the combination of transformations $\mathcal{P}_1$ and $\mathcal{P}_2$ will transform the first two vectors of the sequence into the form

$$\mathcal{P}_2 \mathcal{P}_1 [V^1, V^2] = \begin{bmatrix} \pm\|V^1\| & \times \\ 0 & \pm\|Z^2\| \\ \vdots & \vdots \\ 0 & 0 \end{bmatrix}$$

Successive application of this method enables us to triangularise the sequence of vectors. If one of the vectors is linearly dependent on the others, a zero norm will appear. The triangular structure obtained also allows us, by elimination, to obtain the coefficients of the linear dependence.

Subroutine INDHOU

The calculations corresponding to the above method are performed by the subroutine INDHOU.

Input data

The data are:

- The vectors V of the sequence to be tested, read in succession, and the dimension N of these vectors.

- Two indices, MODEX which causes printing of the orthonormalised vectors when it is equal to 1, IND1 which causes, when equal to 1, the calculation of the coefficients of linear dependence if this is detected (IND2 = 1).

Output

If MODEX = 1, the vectors of the transformed sequence are printed. The coefficients of any linear dependence relations detected are arranged in ALF and re-transmitted to the calling program.

FLOWCHART

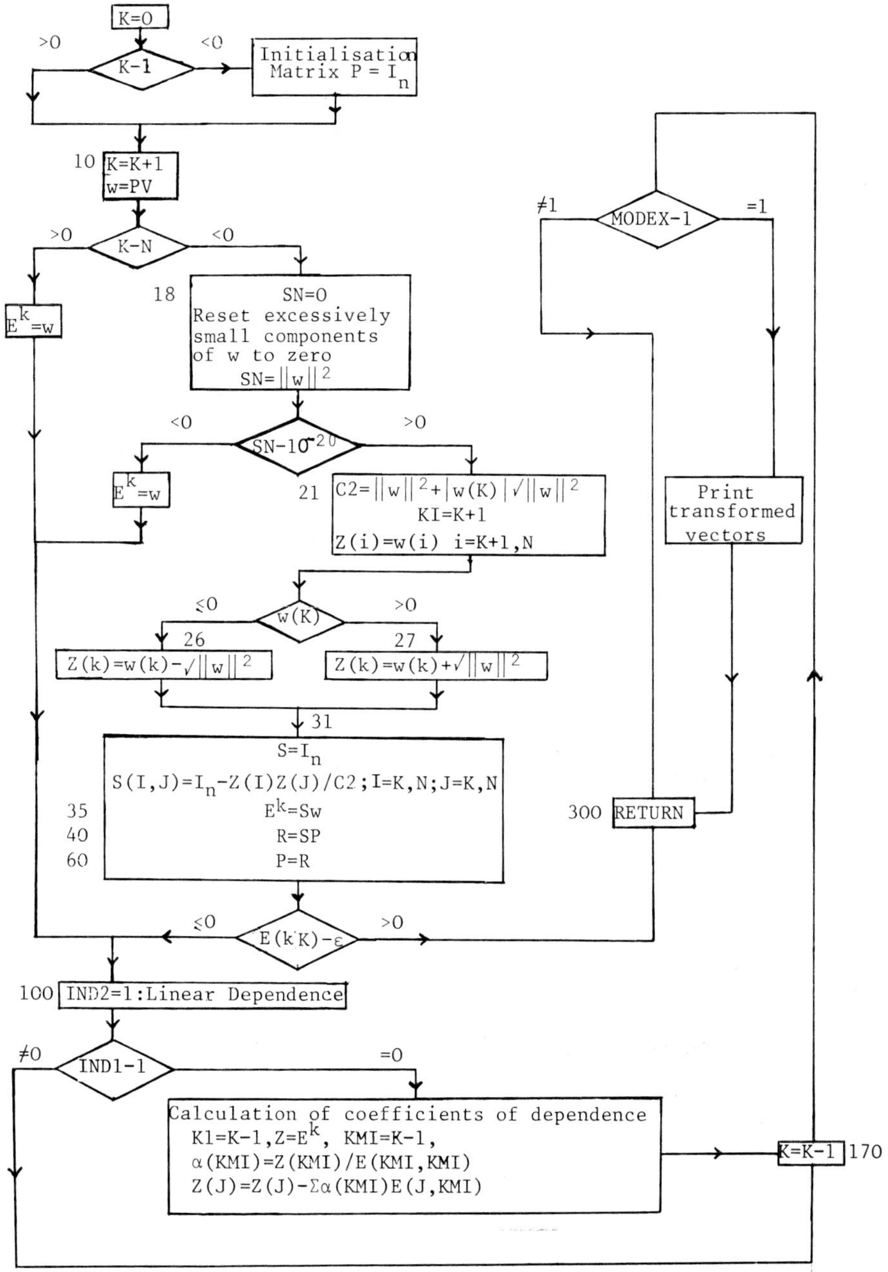

K=0
>0
<0
K-1
Initialisation Matrix P = I_n
10 K=K+1 w=PV
≠1
MODEX-1
=1
>0
K-N
<0
E^k=w
18 SN=0 Reset excessively small components of w to zero SN=$||w||^2$
<0
SN-10^{-20}
>0
E^k=w
21 C2=$||w||^2$+|w(K)|√$||w||^2$ KI=K+1 Z(i)=w(i) i=K+1,N
Print transformed vectors
≤0
w(K)
>0
26
27
Z(k)=w(k)-√$||w||^2$
Z(k)=w(k)+√$||w||^2$
31
S=I_n
S(I,J)=I_n-Z(I)Z(J)/C2;I=K,N;J=K,N
35 E^k=Sw
40 R=SP
60 P=R
300 RETURN
≤0
E(k,K)-ε
>0
100 IND2=1:Linear Dependence
≠0
IND1-1
=0
Calculation of coefficients of dependence
K1=K-1,Z=E^k, KMI=K-1,
α(KMI)=Z(KMI)/E(KMI,KMI)
Z(J)=Z(J)-Σα(KMI)E(J,KMI)
K=K-1 170

```
      SUBROUTINE INDHOU(V,N,K,ALF,MODEX,IND1,IND2)
C     V VECTOR OF THE SEQUENCE
C     N DIMENSION OF VECTORS
C     K NUMBER OF INDEPENDENT VECTORS, INITIALISED
C     AT 0 IN THE CALLING PROGRAM
C     MODEX=1 PRINTS TRANSFORMED VECTORS
C     IND1=1 CALCULATES COEFFICIENTS OF DEPENDENCE
C     IND2=1 LINEAR DEPENDENCE
C     ALF COEFFICIENTS OF LINEAR DEPENDENCE
      IMPLICIT REAL*8(A-H,O-Z)
      DIMENSION V(1),P(22,22),E(23,23),ALF(1),W(22),S(22,22),Z(22),R(22,
     122)
      EPS=0.1E-04
      IND2=0
      IF(K.GE.1) GO TO 10
      DO5 I=1,N
      DO4 J=1,N
    4 P(I,J)=0.
    5 P(I,I)=1.
   10 K=K+1
      DO15 I=1,N
      W(I)=0.
      DO15 J=1,N
   15 W(I)=W(I)+(P(I,J)*V(J))
      IF(K.LE.N) GO TO 18
      DO17 I=1,N
   17 E(I,K)=W(I)
      GOTO 100
   18 CONTINUE
      SN=0.
      DO20 I=K,N
      IF(DABS(W(I)).LT.0.1E-20) W(I)=0.
   20 SN=SN+(W(I)*W(I))
      IF(SN.GT.0.1E-20) GO TO 21
      DO22 I=1,N
   22 E(I,K)=W(I)
      GOTO 100
   21 CONTINUE
      C2=SN+(DABS(W(K))*DSQRT(SN))
      KI=K+1
      DO25 I=KI,N
   25 Z(I)=W(I)
      IF(W(K))26,26,27
   26 Z(K)=W(K)-DSQRT(SN)
      GO TO 31
   27 Z(K)=W(K)+DSQRT(SN)
   31 CONTINUE
      DO 28 I=1,N
      DO 29 J=1,N
```

```
   29 S(I,J)=0.
   28 S(I,I)=1.
      DO 30 I=K,N
      DO30 J=K,N
   30 S(I,J)=S(I,J)-(Z(I)*Z(J)/C2)
      DO 35 I=1,N
      E(I,K)=0.
      DO 35 J=1,N
   35 E(I,K)=E(I,K)+(S(I,J)*W(J))
      DO 40 I=1,N
      DO 40 J=1,N
      R(I,J)=0.
      DO 40 KK=1,N
   40 R(I,J)=R(I,J)+(S(I,KK)*P(KK,J))
      DO 60 I=1,N
      DO 60 J=1,N
   60 P(I,J)=R(I,J)
      IF(DABS(E(K,K)).GT.EPS) GO TO 300
  100 IND2=1
      IF(IND1.NE.1) GO TO 170
C     CALCULATION OF COEFFICIENTS OF DEPENDENCE
      K1=K-1
      DO 150 J=1,K1
  150 Z(J)=E(J,K)
      DO 160 I=1,K1
      KMI=K-I
      ALF(KMI)=Z(KMI)/E(KMI,KMI)
      DO 165 J=1,KMI
  165 Z(J)=Z(J)-ALF(KMI)*E(J,KMI)
  160 CONTINUE
  170 K=K-1
  180 CONTINUE
      IF(MODEX.NE.1) GO TO 300
      WRITE(6,210)
      DO 200 I=1,N
  200 WRITE(6,220)(E(I,J),J=1,K)
  210 FORMAT(//10X 'TRANSFORMED VECTORS'//)
  220 FORMAT(10E12.4)
  300 RETURN
      END
```

A.3 SUB-PROGRAM FOR DECOMPOSITION INTO SUBSYSTEMS CONTROLLABLE FROM A SINGLE INPUT: DEMOE

This program corresponds to the decomposition of a completely controllable linear constant system, of order n, defined by its state-space equations:

$$\dot{x} = Ax + Bu, \quad y = Cx$$

into a set of r single-input controllable subsystems of the form

$$\dot{\tilde{x}}_i = \tilde{A}_{ii}\tilde{x}_i + \sum_{k=i+1}^{r} \tilde{A}_{ik}\tilde{x}_k + {}^{i}Bu_i + \beta(i)u^*$$

(cf. subsection 3.2.1).

Notation

The matrices A,B,C are of respective dimensions (n,n), (n,m), (p,n). The columns of B are denoted by $[{}^1B, {}^2B, \ldots {}^mB]$, and those of M by

$$M = [M(r), \ldots M(1)] \text{ with } M(i) = [{}^1M(i), \ldots {}^{n_i}M(i)].$$

Procedure

a) The vectors ${}^1B$, $A\,{}^1B, \ldots A^{n-1}\,{}^1B$, are generated in succession. The first dependent vector of this sequence, $A^{n_1}\,{}^1B$, together with the coefficients α_i (a_i^1) of its linear dependence on the preceding ones, are determined by calling the subroutine INDHOU; n_1 and the α_i are returned to DEMOE as K and ALF. The order of the first subsystem (n_1) and the last row of the corresponding companion matrix (ALF) are thus known. This procedure is extended to the remaining subsystems as indicated in subsection 3.2.1.

b) The block M(i) of the transformation matrix, corresponding to the ith subsystem, is calculated, its n_i columns being given by the equations

$${}^{n_i}M(i) = {}^iB$$

$${}^jM(i) = \left[A^{n_i-j} + \sum_{k=1}^{n_i-j} A^{k-1}\, a^i_{k+j-1}\right]{}^iB, \quad (j=1,\ldots n_i-1).$$

Input data

The data are:

- The number of states n(IQ), the number of inputs m(M), the number of outputs p(IP).
- The matrices A,B,C, stored in the form:

$$F(I,J) = A, \quad G(I,J) = B, \quad C(I,J) = C.$$

Output

The results printed consist of:

- The data: system order (n), number of inputs (m).
- The matrices A,B, stored as F,G.
- Each time a subsystem is identified: its number, its order, the last row of its companion matrix.
- After all subsystems have been identified: the transformation matrix and its inverse, the matrices $\tilde{A},\tilde{B},\tilde{C}$.

Storage

If n > 10, the vectors and matrices are re-dimensioned as follows: F(n,1), G(n,1), V(n), TEMPO(n), ALF(n), XM(n,n).

Tests

The program makes provision for detecting the following cases:

- the number of inputs can be reduced;
- the system is uncontrollable;
- insufficient accuracy leads to the generation of a space with dimension greater than n.

FLOWCHART

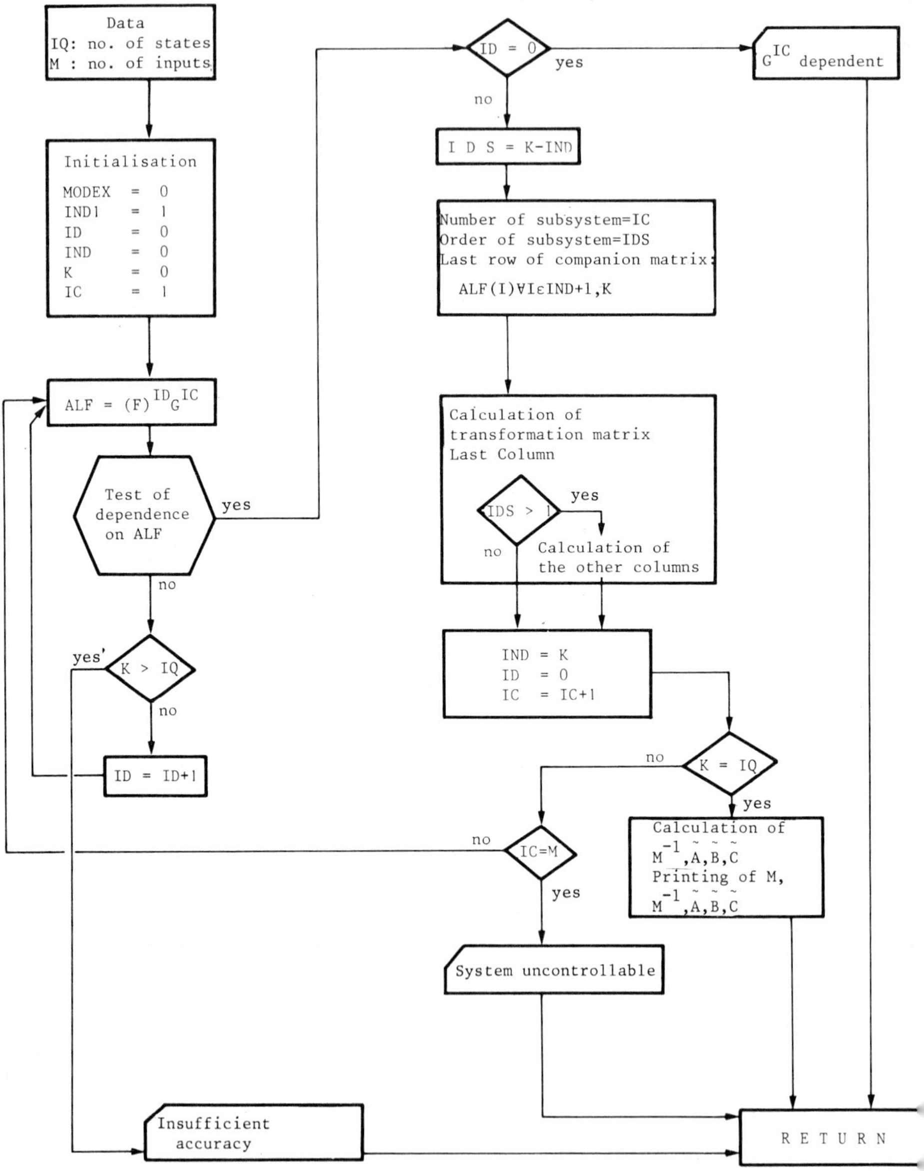
Data
IQ: no. of states
M : no. of inputs
Initialisation
MODEX = 0
IND1 = 1
ID = 0
IND = 0
K = 0
IC = 1
ALF = (F)^ID G^IC
Test of dependence on ALF
yes
no
K > IQ
yes
no
ID = ID+1
ID = 0
yes
no
G^IC dependent
I D S = K-IND
Number of subsystem=IC
Order of subsystem=IDS
Last row of companion matrix:
ALF(I)∀I∈IND+1,K
Calculation of transformation matrix
Last Column
IDS > 1
yes
no
Calculation of the other columns
IND = K
ID = 0
IC = IC+1
K = IQ
no
yes
Calculation of M^-1, Ã, B̃, C̃
Printing of M, M^-1, Ã, B̃, C̃
IC=M
no
yes
System uncontrollable
Insufficient accuracy
R E T U R N

```
      SUBROUTINE DEMOE(F,G,C,IQ,M,IP,LN,LM)
C     SUBROUTINE FOR DECOMPOSITION INTO SINGLE-INPUT
C     CONTROLLABLE SUBSYSTEMS
C     LUENBERGER'S METHOD
C     IQ=SYSTEM ORDER
C     M=NUMBER OF INPUTS
C     IP=NUMBER OF OUTPUTS
C     LN=NUMBER OF ROWS OF F AND G (MAXIMUM NUMBER OF STATES)
C     LM=NUMBER OF ROWS OF C (MAXIMUM NUMBER OF OUTPUTS)
C     NON=0 FOR USE WITH SUBROUTINE NONIN
C     SUBROUTINES USED ASLI
C                      INDHOU
C                      PRODI
C     PROGRAM LIMITED TO (16X16)MATRICES
      DOUBLE PRECISION F(LN,1),G(LN,1),C(LM,1),V(16),ALF(16),XM(16,16)
  103 FORMAT(10X10F11.3)
      WRITE(6,102) IQ,M
  102 FORMAT(//10X'ORDER OF SYSTEM'I2//10X'NUMBER OF INPUTS'I2)
      PRINT 112
  112 FORMAT(//30X'MATRIX A'///)
      DO7 I=1,IQ
    7 PRINT 103,(F(I,J),J=1,IQ)
      PRINT 113
  113 FORMAT(//30'MATRIX B'///)
      DO10 I=1,IQ
   10 PRINT 103,(G(I,J),J=1,M)
C     THE MATRICES F AND G ARE PRINTED BY ROWS
C     INITIALISATION
      ID=0
      IND=0
      K=0
      IC=1
      DO22 I=1,IQ
      DO22 J=1,IQ
   22 XM(I,J)=0
   21 DO23 I=1,IQ
   23 ALF(I)=G(I,IC)
      GO TO 16
    8 CALL PRODI(IQ,IQ,1,F,ALF,LN,LN,LN)
   16 CALL INDHOU(ALF,IQ,0,ALF,0,1,IND2)
      IF(IND2-1) 50,92,50
   50 IF(K-IQ) 63,63,54
   54 WRITE(6,104)
  104 FORMAT(//10X' INSUFFICIENT ACCURACY'//)
      GO TO 300
   63 ID=ID+1
      GO TO 8
C     DIMENSION AND CHARACTERISTICS OF SUBSYSTEMS
   92 IF(ID) 15,15,14
   15 WRITE(6,111) IC
```

```
111    FORMAT(10X 'THE VECTOR G('I4,')IS DEPENDENT'/10X 'THE
      1 CORRESPONDING SUBSYSTEM IS ABSENT'/)
       GO TO 5
 14    IDS=K-IND
       INE=IND+1
       WRITE(6,205) IC,IDS, (ALF (I), I=INE,K)
205    FORMAT(/10X'THE SUBSYSTEM 'I4,2X' HAS ORDER 'I4//10X'
      1 THE LAST ROW OF ITS COMPANION MATRIX IS'//2(10X10F11.3/))
C      IC=NUMBER OF SUBSYSTEM, IDS=ORDER OF SUBSYSTEM
C      CALCULATION OF TRANSFORMATION MATRIX
       DO1 I=1,IQ
       V(I)=G(I,IC)
  1    XM(I,IQ-IND)=G(I,IC)
       IF(IDS-1) 5,5,2
  2    IDSS=IDS-1
       DO17 I=1,IDS
       IF(I-1) 3,3,4
  4    CALL PRODI(IQ,IQ,1,F,V,V,LN,LN,LN)
  3    DO17 J=1,IDSS
       IF(I+J-IDS-1) 18,19,17
 18    DO25 L=1, IQ
 25    XM(L,IQ-K+J)=XM(L,IQ-K+J)-ALF(IND+I+J)*V(L)
       GO TO 17
 19    DO26 L=1,IQ
 26    XM(L,IQ-K+J)=XM(L,IQ-K+J)+V(L)
 17    CONTINUE
C      END OF CALCULATION OF TRANSFORMATION MATRIX
  5    IND=K
       ID=0
       IC=IC+1
       IF(K-IQ) 9,6,6
  9    IF(IC-M) 21,21,11
C      PRINTING OF TRANSFORMATION MATRIX
  6    WRITE(6,106)
106    FORMAT(//20X'TRANSFORMATION MATRIX'///)
       DO12 I=1,IQ
 12    PRINT 103,(XM(I,J),J=1,IQ)
C      CALCULATION OF A TILDE
 33    CALL PRODI(IQ,IQ,IQ,F,XM,F,LN,LN,LN)
C      CALCULATION OF C TILDE
       CALL PRODI(IP,IQ,IQ,C,XM,C,LM,LN,LM)
C      CALCULATION OF INVERSE TRANSFORMATION MATRIX
       CALL ASLI(XM,IQ,LN)
       WRITE(6,108)
108    FORMAT(//20X'INVERSE MATRIX'///)
       DO13 I=1,IQ
 13    PRINT 103,(XM(I,J),J=1,IQ)
C      CALCULATION OF A TILDE
       CALL PRODI(IQ,IQ,M,XM,F,F,LN,LN,LN)
C      CALCULATION OF B TILDE
       CALL PRODI(IQ,IQ,M,XM,G,G,LN,LN ,LN)
       PRINT 107
```

```
107     FORMAT(//20X'NEW MATRIX A'///)
        DO30 I=1,10
 30     PRINT 103,(F(I,J),J=1,IQ)
        PRINT 109
109     FORMAT(//20X'NEW MATRIX B'///)
        DO31 I=1, IQ
 31     PRINT 103,(G(I,J),J=M)
        PRINT 114
114     FORMAT(//20X'NEW MATRIX C'///)
        DO32 I=1,IP
 32     PRINT 103,(C(I,J),J=1,IQ)
        GO TO 300
 11     WRITE(6,110)
110     FORMAT(10X'SYSTEM UNCONTROLLABLE'/)
300     RETURN
        END
```

A.4 SUB-PROGRAM FOR DECOMPOSITION INTO SINGLE-OUTPUT OBSERVABLE SUBSYSTEMS: DEMOS

This program corresponds to the decomposition of a completely observable linear constant system, of order n, defined by its state equations

$$\dot{x} = Ax + Bu, \quad y = Cx$$

into a set of p single-output observable subsystems of dimensions n_i (cf. subsection 3.2.3).

Notation

The matrices A,B,C are of dimensions (n,n), (n,m), (p,n) respectively. We define

$$H = C^T, \quad \psi = \phi^T.$$

The columns of H are denoted by $[{}^1H, \ldots {}^pH]$; those of ψ are designated, first in blocks by

$$\psi = [\psi(1), \ldots \psi(p)],$$

then inside each block by

$$\psi(i) = [{}^1\psi(i), \ldots {}^{n_i}\psi(i)].$$

Procedure

a) We first look for n independent vectors in the sequence

$$\Sigma_1 : \; {}^1H, \; {}^2H, \; \ldots {}^pH, \; A^T \, {}^1H, \ldots A^T \, {}^pH, \ldots (A^T)^{n-1} \, {}^pH$$

taking account of the fact that, if $(A^T)^j \, {}^iH$ is dependent, so also are all vectors of the form $(A^T)^{j+k} \, {}^iH$, which can thus be rejected. Every time that a vector of the form $(A^T)^{n_i} \, {}^iH$ is found to satisfy the test for linear dependence, we have identified the subsystem, of order n_i, associated with the ith output.

By elimination of the dependent vectors mentioned above, the subroutine for linear independence INDHOU (called by the present program) displays the linear dependence relations in the form

$$(A^T)^{n_i} \, {}^iH = \sum_{j=1}^{k} \alpha_j \, {}^jV$$

where the vectors ${}^jV$ are the k independent vectors already found in Σ_1; to determine the coefficients ξ_k^j (i) in

$$(A^T)^{n_i} \, {}^iH = \sum_{k=0}^{n_i} \sum_{j=1}^{p} (A^T)^k \, \xi_k^j(i) \, {}^jH$$

it will thus be important to know the rejected vectors, for which the corresponding $\xi_k^j(i)$ are zero.

b) The calculation of the companion matrices A(i,i) and the matrix ψ is performed using the equations of paragraph 3.2.3.C, and the matrices θ and $P = M^{-1}$ are calculated from the equations of paragraphs 3.2.3.D and 3.2.3.E.

Input data

The data are:

- The number of states n(IQ), of inputs m(M), and of outputs p(N), on one card in the format (3I2).

- The matrices A,C,B, punched in sequence, row by row for the first two and column by column for B, $({}_1A,\ldots{}_nA,\ {}_1C\ldots{}_pC,\ {}^1B,\ldots{}^mB)$, in the format (8F10.3), A and C being transposed as they are read. The matrices are thus stored in the form

$$\mathrm{AFT(I,J)} = F = A^T$$

$$\mathrm{CHT(I,J)} = H = C^T$$

$$\mathrm{B(I,J)} = B$$

with

$$\mathrm{AFT(I,J)} = A_{ji} .$$

Output

The results printed are:

- The data: system order, number of outputs, matrices A,B,C.
- Each time a subsystem is identified: its number (i.e. the corresponding output), its order, the transpose of the first column of the associated companion matrix, the corresponding block of the matrix ϕ.
- After identification of all the subsystems: the matrix θ, the transformation matrix and its inverse, the matrix $\tilde{B}$ (stored in place of B).

Storage

If $n > 10$ or $p > 5$, the matrices and vectors are re-dimensioned as follows: AFT(n,1), CHT(n,1), B(n,1), XM(n,n), V(n,n), TEMPO(n), ALF(n), W(n), SKI(n,p,p), SPI(p,n), IPSEX(p+1),IDEX(p).

FLOWCHART

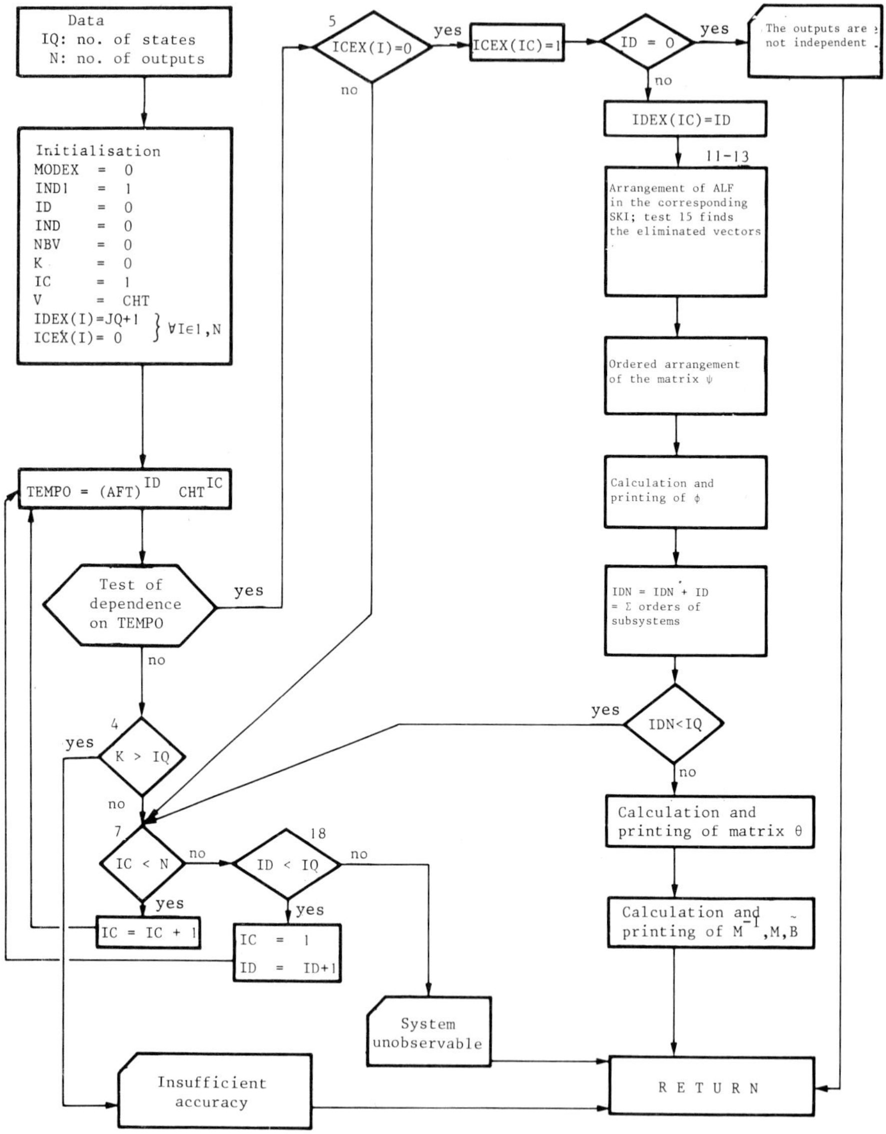

Data
IQ: no. of states
N: no. of outputs
Initialisation
MODEX = 0
IND1 = 1
ID = 0
IND = 0
NBV = 0
K = 0
IC = 1
V = CHT
IDEX(I)=JQ+1
ICEX(I)= 0
∀I∈1,N
TEMPO = (AFT)^ID CHT^IC
Test of dependence on TEMPO
yes
no
4
K > IQ
yes
no
7
IC < N
no
yes
IC = IC + 1
18
ID < IQ
no
yes
IC = 1
ID = ID+1
System unobservable
Insufficient accuracy
5
ICEX(I)=0
yes
no
ICEX(IC)=1
ID = 0
yes
The outputs are not independent
no
IDEX(IC)=ID
11-13
Arrangement of ALF in the corresponding SKI; test 15 finds the eliminated vectors
Ordered arrangement of the matrix ψ
Calculation and printing of φ
IDN = IDN + ID = Σ orders of subsystems
IDN<IQ
yes
no
Calculation and printing of matrix θ
Calculation and printing of M^-1,M,B
R E T U R N

```
      SUBROUTINE DEMOS(AFT,CHT,B,IQ,M,N,LN,LM)
C     SUBROUTINE FOR DECOMPOSITION INTO SINGLE-OUTPUT SUBSYSTEMS
C     LUENBERGER'S METHOD
C     IQ=SYSTEM ORDER
C     M=NUMBER OF INPUTS
C     N=NUMBER OF OUTPUTS
C     LN=MAXIMUM NUMBER OF STATES
C     SUBROUTINES USED ASL1
                       INDHOU
                       PRODI
C     .........................................
C     THE MATRICES F AND H ARE TRANSPOSED ON READING
C     .........................................
C     PROGRAM LIMITED TO (16X16) MATRICES
      DOUBLE PRECISION AFT(LN,1),CHT(LN,1),B(LN,1),XM(16,16),V(16,16),
     1 ALF(16),W(16),TEMPO(16)
      DIMENSION SKI(16,5,5),PSI(5,16),IPSEX(6),IDEX(5),ICEX(5)
  101 FORMAT(10X10F8.3)
      PRINT 100,IQ,N
  100 FORMAT(10X'ORDER OF SYSTEM'I2/10X'NUMBER OF OUTPUTS'I2//)
      PRINT 117
  117 FORMAT(30X'MATRIX A'///)
      DO1 I=1,IQ
    1 PRINT 101,(AFT(J,I),J=1,IQ)
      PRINT 102
  102 FORMAT(//30X'MATRIX B'///)
      DO70 I=1,IQ
   70 PRINT 101,(B(I,J),J=1,M)
      PRINT 118
  118 FORMAT(//30X'MATRIX C'///)
      DO2 I=1,N
    2 PRINT 101,(CHT(J,I),J=1,IQ)
C     INITIALISATION
      ID=0
      IND-0
      K=0
      IC=1
      NBV=0
      DO49 I=1,N
      IDEX(I)=IQ+1
      DO66 J=1,IQ
   66 V(J,I)=CHT(J,I)
   49 ICEX(I)=0
      GO TO 8
   57 CALL PRODI(IQ,IQ,N,AFT,V,V,LN,LN,LN)
    8 DO3 I=1,IQ
    3 TEMPO(I)=V(I,IC)
      CALL INDHOU(TEMPO,IK,K,ALF,0,1,IND2)
      IF(IND2-1) 4,5,4
    4 IF(K-IQ) 7,7,6
    6 PRINT 103
```

```
  103 FORMAT(10X'INSUFFICIENT ACCURACY'/)
      GO TO 1000
    7 IF(IC-V) 17,18,18
   17 IC=IC+1
      GO TO 8
   18 IF(ID-IQ) 16,19,19
   19 PRINT 105
  105 FORMAT(10X'SYSTEM UNOBSERVABLE'/)
      GO TO 1000
   16 IC=1
      ID=ID+1
      GO TO 57
C     ORDERS AND CHARACTERISTICS OF SUBSYSTEMS
    5 IF(ICEX(IC)) 7,21,7
   21 ICEX(IC)=1
      IF(ID) 9,9,10
    9 PRINT 104
  104 FORMAT(10X'OUTPUTS NOT INDEPENDENT'/)
      GO TO 1000
   10 PRINT 106,IC,ID
  106 FORMAT(//10X'SUBSYSTEM' I4, 3X'HAS ORDER'I4//)
      IDEX(IC)=ID
C     STORAGE OF SKI
      DO11 I=1,IQ
      DO11 J=1,N
   11 SKI(I,J,IC)=0.
      NN=1
      DO14 I=1,IQ
      DO14 J=1,N
      IF(N*(I-ID-1)+J-IC) 15,20,20
   15 IF(IDEX(J)-I+1) 14,14,13
   13 SKI(I,J,IC)=ALF(NN)
      NN=NN+1
   14 CONTINUE
C     CALCULATION OF COMPANION MATRICES
   20 DO22 I=1,ID
      L=ID-I+1
   22 ALF(I)=SKI(L,IC,IC)
      PRINT 107
  107 FORMAT(10X'THE TRANSPOSE OF THE FIRST COLUMN OF ITS
     1 COMPANION MATRIX IS'//)
      PRINT 101,(ALF(I),I=1,ID)
C     REARRANGEMENT OF THE MATRIX PSI
      NBV=NBV+1
      IPSEX(NBV)=IC
      IDEXL=0
      DO51 L=1,N
      IF(IPSEX(L)-IC) 51,52,52
   52 DO53 LL=L,N
      LNL=N+L-LL
   53 IPSEX(LNL)=IPSEX(LNL-1)
      IPSEX(L)=IC
      IDEXN=IDEXL+1+ID
      DO54 LL=IDEXN,IQ
      LNL=IDEXN+IQ-LL
      DO54 LLL=1,N
```

```
 54 PSI(LLL,LNL)=PSI(LLL,LNL-ID)
    GO TO 55
 51 IDEXL=IDEXL+IDEX(IPSEX(L))
C   CALCULATION AND PRINTING OF BLOCKS OF PHI
 55 PRINT 109
109 FORMAT(//30X'CORRESPONDING BLOCK OF MATRIX PHI'///)
    DO24 J=1,ID
    ICD=ID-J+1
    DO25 KF=1,N
    IF(KF-IC)40,41,42
 40 TEMPO(KF)=SKI(ICD,KF,IC)+SKI(ID+1,KF,IC)*SKI(ICD,IC,IC)
    GO TO 25
 41 TEMPO(KF)=0.
    GO TO 25
 42 TEMPO(KF)=SKI(ICD,KF,IC)
 25 CONTINUE
    PRINT 101, (TEMPO(I),I=1,N)
    DO24 I=1,N
    PSI(I,IDEXL+J)=TEMPO(I)
 24 CONTINUE
    IDN=IDN+ID
    IF(IDN-IQ) 7,26,26
C   CALCULATION OF MATRIX THETA
 26 PRINT 111
111 FORMAT(//30X'MATRIX THETA'///)
    DO31 I=1,N
    DO27 J=1,N
    IF(J-I) 28,29,30
 30 TEMPO(J)=0.
    GO TO 27
 29 TEMPO(J)=1.
    GO TO 27
 28 TEMPO(J)=SKI(IDEX(I)+1,J,I)
 27 CONTINUE
    PRINT 101, (TEMPO(J),J=1,N)
 31 CONTINUE
C   CALCULATION OF INVERSE TRANSFORMATION MATRIX
    DO32 I=1,N
    IM=I-1
    IQG=1
    IF(I-1) 33,33,34
 34 DO35 J=1,IM
 35 IQG=IQG+IDEX(J)
    DO36 J=1,IQ
    XM(J,IQG)=CHT(J,I)
    DO36 L=1,IM
 36 XM(J,IQG)=XM(J,IQG)-CHT(J,L)*SKI(IDEX(I)+1,L,I)
    GO TO 38
 33 DO 37 J=1,IQ
 37 XM(J,IQG)=CHT(J,I)
 38 IF(IDEX(I)-1) 32,32,56
 56 IDEXI=IDEX(I)
    DO46 L=1,IQ
```

```
   46 W(L)=XM(L,IQG)
      DO32 J=2,IDEXI
      JM=J-1
      DO50 L=1,IQ
      DO58 LL=1,IQ
   58 V(L,LL)=AFT(L,LL)
   50 ALF(L)=0.
      DO43 KI=1,JM
      KM=KI-1
      JMK=J-KI
      DO44 L=1,IQ
      TEMPO(L)=0.
      DO39 LL=1,N
   39 TEMPO(L)=TEMPO(L)+CHT(L,LL)*PSI(LL,JMK)
   44 TEMPO(L)=TEMPO(L)+SKI(IDEX(I)-JMK+1,I,I)*XM(L,IQG)
      IF(KM-1) 59,60,61
   61 CALL PRODI(IQ,IQ,IQ,AFT,V,V,LN,LN,LN)
   60 CALL PRODI(IQ,IQ,1,V,TEMPO,TEMPO,LN,LN,LN)
   59 DO43 K=1,IQ
   43 ALF(L)=ALF(L)+TEMPO(L)
      CALL PRODI(IQ,IQ,1,AFT,W,W,LN,LN,LN)
      DO32 L=1,IQ
      XM(L,IQG+JM)=W(L)-ALF(L)
   32 CONTINUE
      PRINT 115
  115 FORMAT(//20X'INVERSE MATRIX '///)
      DO47 I=1,IQ
   47 PRINT 101,(XM(J,I),J=1,IQ)
C     CALCULATION OF B TILDE
      DO68 J=1,M
      DO67 I=1,IQ
      W(I)=0.
      DO67 K=1,IQ
   67 W(I)=W(I)+XM(K,I)*B(K,J)
      DO68 K=1,IQ
   68 B(K,J)=W(K)
C     CALCULATION OF TRANSFORMATION MATRIX
      CALL ASLI(XM,IQ,LN)
      PRINT 113
  113 FORMAT(//20X'TRANSFORMATION MATRIX '///)
      DO48 J=1,IQ
   48 PRINT 101,(XM(I,J),I=1,IQ)
      PRINT 108
  108 FORMAT(//30X'NEW MATRIX  B'///)
      DO69 I=1,IQ
   69 PRINT 101,(B(I,J),J=1,M)
1 000 RETURN
      END
```

A.5 SUB-PROGRAM FOR NON-INTERACTIVE CONTROL: NONIN

This program corresponds to the non-interactive compensation of a linear constant system, of order n, defined by the state equations

$$\dot{x} = Ax + Bu, \quad y = Cx$$

with a control law of the form

$$u = Le + Rx.$$

The analysis relevant to this type of control may be found in chapter 5 (section 5.3).

Input data

The data are:

- The number of states n(N), the number of inputs and outputs m(M).
- The matrices A,B,C, stored as A(I,J), B(I,J), C(I,J).

Output

The results printed comprise:

- The data: system order, number of inputs and outputs, matrices A,B,C.
- Each time a subsystem is identified: its number and order.
- When all subsystems have been identified: the matrices T, $\tilde{A}$, $\tilde{B}$, $\tilde{C}$.
- After determination of each of the rows of $\tilde{R}$: the corresponding $\nu(i)$ and M(i), the matrices $\tilde{R}$, $\tilde{L}$, R, L.

Storage

If n > 10 or m > 5, the matrices are re-dimensioned as follows: Z(n,n), T(n,n), BTI (n,n), XI(m,m), AST(m,n), V(n), W(n).

Tests

The program provides several tests to detect cases where insufficient accuracy would lead to the generation of a space with dimension greater than n.

FLOWCHART

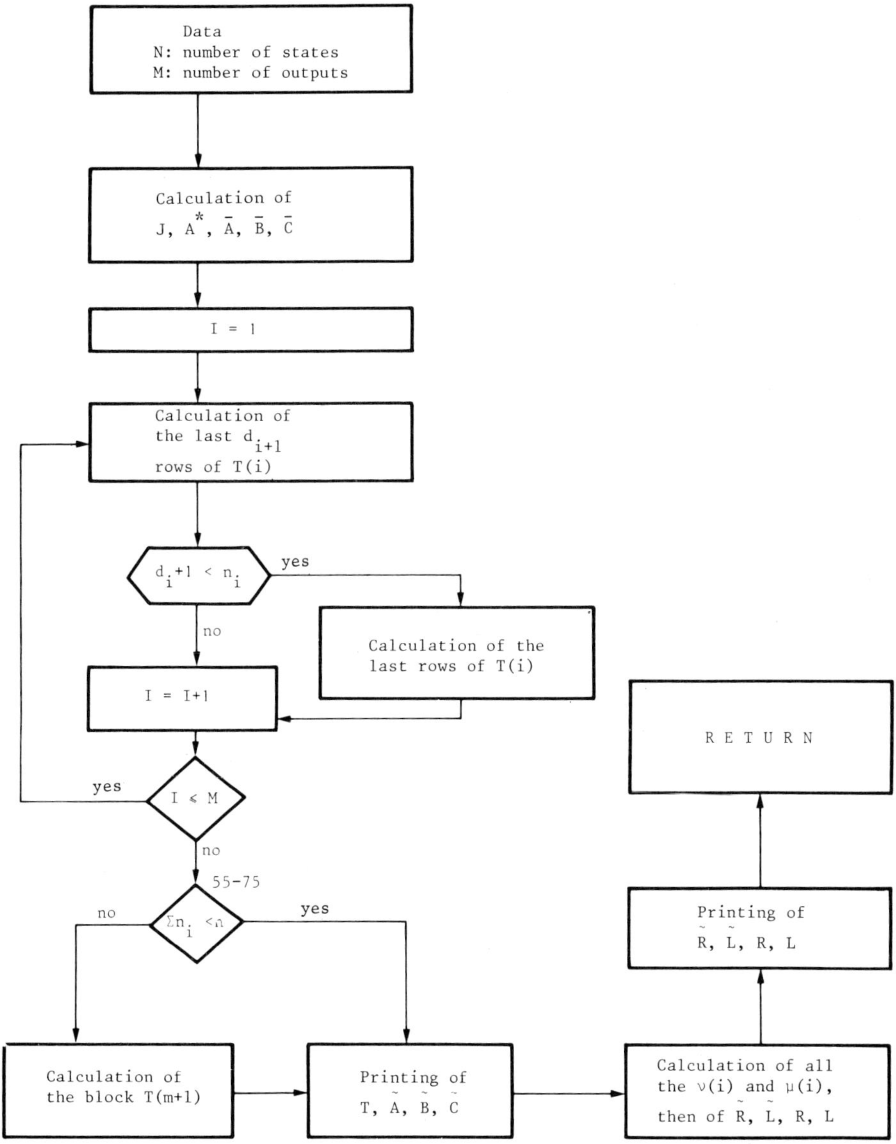
Data
N: number of states
M: number of outputs
Calculation of
J, A*, Ā, B̄, C̄
I = 1
Calculation of
the last d_{i+1}
rows of T(i)
d_i+1 < n_i
yes
no
Calculation of the
last rows of T(i)
I = I+1
yes
I ≤ M
no
55-75
no
Σn_i < n
yes
Calculation of
the block T(m+1)
Printing of
T, Ã, B̃, C̃
Calculation of all
the ν(i) and μ(i),
then of R̃, L̃, R, L
Printing of
R̃, L̃, R, L
R E T U R N

```
      SUBROUTINE NONIN(A,B,C,N,M,R,XL,LN,LM)
C     SUBROUTINE FOR NON-INTERACTIVE CONTROL BY STATE-SPACE METHOD
C     N=SYSTEM ORDER
C     M=NUMBER OF INPUTS AND OUTPUTS
C     LN=NUMBER OF ROWS OF A AND B (MAXIMUM NUMBER OF STATES)
C     LM=NUMBER OF ROWS OF C (MAXIMUM NUMBER OF OUTPUTS)
C     SUBROUTINES USED PRODI
C                      ASLI
C                      INDHOU
C     PROGRAM LIMITED TO (16X16) MATRICES
      DOUBLE PRECISION A(LN,1),B(LN,1),C(LM,1),R(LM,1),XL(LM,1),Z(16,16)
     1,T(16,16),XI(5,5),BTI(16,16),AST(5,16),V(16),W(16),ACC
      DIMENSION ID(5),NI(5)
  100 FORMAT(10X10F11.3)
      PRINT 101,N,M
  101 FORMAT(10X'SYSTEM ORDER='I2//10X'NUMBER OF INPUTS=
      NUMBER OF OUTPUTS='I2//)
      PRINT 102
  102 FORMAT(30X'MATRIX A'//)
      DO10 I=1,N
   10 PRINT 100,(A(I,J),J=1,N)
      PRINT 103
  103 FORMAT(///30X'MATRIX B'//)
      DO20 I=1,N
   20 PRINT 100,(B(I,J),J=1,M)
      PRINT 104
  104 FORMAT(///30X'MATRIX C'//)
      DO30 I=1,M
   30 PRINT 100,(C(I,J),J=1,N)
C     CALCULATION OF J,A*
      DO2 I=1,M
      IDEX=0
      DO60 J=1,N
   60 V(J)=C(I,J)
      GO TO 1
   70 CALL PRODI(1,N,N,V,A,V,1,LN,1)
    1 CALL PRODI(1,N,N,V,B,W,1,LN,1)
      ACC=0.
      DO21 J=1,M
   21 ACC=ACC+DABS(W(J))/M
      DO31 J=1,M
      IF(DABS(W(J))-0.00001*ACC) 31,31,51
   31 CONTINUE
      IDEX=IDEX+1
      GO TO 70
   51 CALL PRODI(1,N,N,V,A,V,1,LN,1)
      DO61 J=1,M
   61 XI(I,J)=W(J)
      DO71 J=1,N
   71 AST(I,J)=V(J)
    2 ID(I)=IDEX
      IDEX=M
      DO12 J=1,M
```

```
 12 IDEX=IDEX+ID(J)
    IF(N-IDEX) 59,32,32
 32 CALL ASLI(XI,M,LM)
C   CALCULATION OF A BAR AND B BAR
 42 CALL PRODI(N,M,M,B,XI,B,LN,LM,LN)
    CALL PRODI(N,M,N,B,AST,AST,LN,LM,LM)
    DO52 L=1,N
    DO52 LL=1,N
 52 A(L,LL)=A(L,LL)-AST(L,LL)
C   CALCULATION OF T
    DO55 I=1,M
    K=0
    ACC=0.
    DO62 L=1,N
    DO62 LL=1,M
 62 BTI(L,LL)=B(L,LL)
    GO TO 3
 82 CALL PRODI(N,N,M,A,BTI,BTI,LN,LN,LN)
  3 DO53 J=1,M
    IF(J-I) 13,53,13
 13 DO33 L=1,N
 33 V(L)=BTI(L,J)
    CALL INDHOU (V,N,K,W,0,0,IND2)
    IF(IND2-1) 43,23,43
 43 IF(N-K) 59,53,53
 53 CONTINUE
    GO TO 82
 23 NI(I)=N-K
    PRINT 114,I,NI(I)
114 FORMAT(//10X'SUBSYSTEM ORDER 'I2,'='I2//)
    IF(I-1) 63,63,73
 63 NIT=1
    GO TO 83
 73 NIT=NIT+NI(I-1)
 83 DO93 J=1,N
 93 V(J)=C(I,J)
    IDEX=NIT
    GO TO 24
  4 IDEX=IDEX+1
    CALL PRODI(1,N,N,V,A,V,1,LN,1)
 24 DO44 J=1,N
    ACC=ACC+DABS(V(J))/(ID(I)+1)/N
 44 T(IDEX,J)=V(J)
    CALL INDHOU (V,N,K,W,0,0,IND2)
    IF(IND2-1) 34,79,34
 34 IF(ID(I)+NIT-IDEX) 74,74,4
 74 IF(NI(I)-ID(I)-1) 55,55,64
 64 DO35 J=1,N
    DO84 L=1,N
 84 V(L)=0.
    V(J)=1.
    CALL INDHOU (V,N,K,W,0,0,IND2)
    IF(IND2-1) 5,35,5
```

```
  5 IDEX=IDEX+1
    DO15 L=1,N
 15 T(IDEX,L)=W(L)*ACC
    IF(IDEX-NI(I)-NIT+1) 35,55,55
 35 CONTINUE
 55 CONTINUE
    IF(IDEX-N) 75,96,96
 75 K=0
    ACC=0.
    DO6 J=1,IDEX
    DO95 L=1,N
    V(L)=T(J,L)
 95 ACC=ACC+DABS(V(L))/IDEX/N
    CALL INDHOU (V,N,K,W,O,O,IND2)
    IF(IND2-1) 6,89,6
  6 CONTINUE
    DO76 J=1,N
    DO36 L=1,N
 36 V(L)=0.
    V(J)=1.
    CALL INDHOU (V,N,K,W,O,O,IND2)
    IF(IND2-1) 46,76,46
 46 DO56 L=1,N
 56 T(K,L)=W(L)*ACC
    IF(N-K) 96,96,76
 76 CONTINUE
C   PRINTING OF T
 96 PRINT 105
105 FORMAT(///30X'MATRIX  T'//)
    DO86 I=1,N
 86 PRINT 100,(T(I,J),J=1,N)
    DO66 L=1,N
    DO66 LL=1,N
 66 BTI(L,LL)=T(L,LL)
    CALL ASLI(BTI,N,LN)
C   T CONTAINS T, XI CONTAINS J INVERSE, BTI CONTAINS T INVERSE
C   CALCULATION AND PRINTING OF A TILDE. B TILDE, C TILDE
  7 CALL PRODI(N,N,N,T,A,A,LN,LN,LN)
    CALL PRODI(N,N,N,A,BTI,A,LN,LN,LN)
    CALL PRODI(N,N,M,T,B,B,LN,LN,LN)
    CALL PRODI(M,N,N,C,BTI,C,LM,LN,LM)
    PRINT 106
106 FORMAT(///30X'MATRIX  A TILDE'//)
    DO17 I=1,N
 17 PRINT 100,(A(I,J),J=1,N)
    PRINT 107
107 FORMAT(///30X'MATRIX  B TILDE'//)
    DO27 I=1,N
 27 PRINT 100,(B(I,J),J=1,M)
    PRINT 108
108 FORMAT(///30X'MATRIX  C TILDE'//)
    DO37 I=1,M
 37 PRINT 100,(C(I,J),J=1,N)
    DO48 I=1,M
    DO47 J=1,N
```

```
 47 R(I,J)=0.
    DO48 J=1,M
 48 XL(I,J)=0.
    IDEX=0
    DO38 I=1,M
    NII=NI(I)
    IDP=IDEX+1
    IDPP=IDEX+NII
    WRITE(4,121) I,NI(I)
121 FORMAT(1X'COEFFICIENTS(IN ORDER OF INCREASING POWERS OF S)
  1 OF SUBSYSTEM 'I2,'WITH ORDER'I2)
    READ(4,115) (R(I,J),J=IDP,IDPP)
115 FORMAT(10F8.3)
    DO18 L=1,NII
    DO18 LL=1,NII
 18 Z(L,LL)=A(IDEX+L,IDEX+LL)
    K=0
    DO228 L=1,NII
    DO228 J=1,NII
228 BTI(L,J)=0.
    DO28 L=1,NII
 28 W(L)=B(IDEX+L,I)
    GO TO 328
528 CALL PRODI(NII,NII,1,Z,W,W,LN,LN,LN)
328 CALL INDHOU (W,NII,K,W,0,1,IND2)
    IF(IND2-1) 728,88,728
728 IF(K-NII) 528,528,59
 88 DO388 L=1,K
388 R(I,IDEX+L)=R(I,IDEX+L)+W(L)
    PRINT 222,I,(W(L),L=1,K)
222 FORMAT(///10X' VECTOR  NU ('I2,')'//2(10X10F11.3/))
    DO588 L=1,NII
    V(L)=B(IDEX+L,I)
588 BTI(L,NII)=B(IDEX+L,I)
    IF(K-1) 398,398,788
786 KM=K-V
    DO98 L=1,K
    IF(L-1) 988,988,988
888 CALL PRODI(NII,NII,1,Z,V,V,LN,LN,LN)
988 DO98 J=1,KM
    IF(L+J-K-1) 198,298,98
198 DO688 LL=1,NII
688 BTI(LL,NII-K+J)=BTI(LL,NII-K+J)-W(L+J)*V(LL)
    GO TO 98
298 DO488 LL=1,NII
486 BTI(LL,NII-K+J)=BTI(LL,NII-K+J)+V(LL)
 98 CONTINUE
398 PRINT 223
223 FORMAT(//20X'TRANSFORMATION MATRIX '///)
    DO498 L=1,NII
```

```
498 PRINT 100,(BTI(L,J),J=1,NII)
    CALL ASLI(BTI,NII,LN)
    DO698 L=1,NII
    V(L)=0.
    DO698 LL=1,NII
698 V(L)=V(L)-R(I,IDEX+LL)*BTI(LL,L)
    DO898 L=1,NII
898 R(I,IDEX+L)=V(L)
 38 IDEX=IDEX+NI(I)
    WRITE(4,116) M
116 FORMAT(1X'DIAGONAL OF MATRIX L WITH DIMENSION'I2)
    READ(4,115)(XL(I,I),I=1,M)
    PRINT 117
117 FORMAT (///30X' MATRIX  R TILDE'//)
    DO58 I=1,M
 58 PRINT 100,(R(I,J),J=1,N)
    PRINT 118
118 FORMAT(///30X' MATRIX  L TILDE'//)
    DO68 I=1,M
 68 PRINT 100,(XL(I,J),J=1,M)
C   CALCULATION OF R AND L
    CALL PRODI(M,N,N,R,T,R,LM,LN,LM)
    DO78 L=1,M
    DO78 LL=1,N
 78 R(L,LL)=R(L,LL)-AST(L,LL)
    CALL PRODI(M,M,N,XI,R,R,LM,LM,LM)
    CALL PRODI(M,M,M,XI,XL,XL,LM,LM,LM)
    PRINT 119
119 FORMAT(///30X' MATRIX  R'//)
    DO19 I=1,M
 19 PRINT 100,(R(I,J),J=1,N)
    PRINT 120
120 FORMAT(///30X' MATRIX  L'//)
    DO29 I=1,M
 29 PRINT 100,(XL(I,J),J=1,M)
    GO TO 99
 59 PRINT 109
109 FORMAT(//10X'PRECISION INSUFFICIENT ACCURACY'//)
    GO TO 99
 79 PRINT 111
111 FORMAT(//10X'ONE ROW OF T(I) IS DEPENDENT ON THE COLUMNS B,...,
  1 (A)(N-1)B'//)
    GO TO 99
 89 PRINT 112
112 FORMAT(//10X' ROWS OF T NOT INDEPENDENT'//)
 99 RETURN
    END
```

Bibliography

1.1 L.G. BIRTA : "Formal Approach to Concepts of Interaction". S.R.C. 81, A, 65, 31.

1.2 L.G. BIRTA, I.H. MUFTI : "Some Results on an Inverse Problem in Multivariable Systems". I.E.E.E. on Auto. Control, February, 1967.

1.3 R. BROCKETT, M.D. MESAROVIC : "The Reproducibility of Multivariable Systems". Joint Automatic Control Conference, Stanford 1964, pp. 481-486.

1.4 C.A. DESOER : "Controllability and Observability of Feedback Systems". I.E.E.E. Trans. on Auto. Control, Aug. 1967, pp. 474-475.

1.5 A. FOSSARD, M. GAUVRIT, C. GUEGUEN : "Comments on an Inverse Problem in Multivariable Systems". I.E.E.E. Trans. on Auto. Control, April 1968.

1.6 F.R. GANTMACHER : "Matrix Theory". Vol. 1, Chelsea Publishing Company, New York 1960.

1.7 M. GAUVRIT, F. LEORAT, M. BROSSON : "Controllability of Multivariable Systems using a Frequency Domain Criterion". I.E.E.E. Trans. on Auto. Control, February 1968.

1.8 E.G. GILBERT : "Controllability and Observability in Multivariable Control Systems". S.I.A.M. Journal on Control, Serie A, Vol. 1, n° 2, 1963.

1.9 C.J. GUEGUEN, A. FOSSARD, M. GAUVRIT : "Une représentation intermédiaire des systèmes Multidimensionnels". I.F.A.C. Symposium on Multivariable Control Systems, Düsseldorf 1968.

1.10 R.E. KALMAN, Y.C. HO, K.S. NARENDRA : "Controllability of Linear Dynamical Systems" in : "Contributions to differential equations. Wiley and Sons, New York 1963, pp. 189-197.

1.11 R.E. KALMAN, L. WEISS : "Contribution to Linear System Theories". R.I.A.S. Technical report 64.9, April 1964.

1.12 R.E. KALMAN : "Linear dynamical Systems". S.I.A.M. Journal on Control, Serie A, Vol. 1 n° 2, 1963.

1.13 E. KREINDLER, P.E. SARACHIK : "On the Concepts of Controllability and Observability of Linear Systems". I.E.E.E. Trans. on Auto. Control, April 1964.

1.14 E.C. TITCHMARSH : "Introduction to the theory of Fourier integrals". Clarendon Press, 1948.

1.15 L.A. ZADEH, C.A. DESOER : "Linear System Theory". Mac Graw Hill, New York 1963, pp. 1-39.

1.16 L.A. ZADEH, E. POLAK : "System Theory". Mac Graw Hill, New York 1969, pp. 3-41.

2.1 F.M BROWN : "A new way to construct state equation". Dept E.E., Air Force Inst. Tech. Wright Patterson, Ohio.

2.2 A. FOSSARD, J.J. ELTGEN : "Simulation et formulation des systèmes multidimensionnels". Revue des annales de l'A.I.C.A., Vol. XI (1), pp. 21-26, January 1969.

2.3 F.R. GANTMACHER : "Matrix Theory". Chelsea Publ. Company, New York 1960,

2.4 E.G. GILBERT : "Controllability and Observability in Multivariable Control Systems". S.I.A.M. Journal on Control. Series A, Vol. 1 no 2, 1963, pp. 128-151.

2.5 M. GAUVRIT, A. FOSSARD : "Construction de réalisations irréductibles dans le cas de pôles multiples complexes". Automatisme Vol. XIII (5), pp. 203-207, May 1968.

2.6 M. GAUVRIT, C. GUEGUEN, A. FOSSARD : "Effective Construction of Irreducible Realization", First I.F.A.C. Symposium on multivariable-Control Systems, Düsseldorf, October 1968

2.7 C. GUEGUEN, A. FOSSARD, M. GAUVRIT : "Une représentation intermédiaire des systèmes multidimensionnels". First I.F.A.C. Symposium on multivariable Control Systems, Düsseldorf, October 1968, (paper 1.2).

2.8 C. GUEGUEN, E. TOUMIRE : "Mise sous forme de variables d'état d'un système donné par sa matrice de transfert". Automatisme Vol. XIII (5), pp. 109-203, May 1968.

2.9 R.E. KALMAN : "Irreductible Realizations and the Degree of a Rational Matrix", S.I.A.M. Journal on Control, Series A, Vol. 1, no 2. 1963.

2.10 R.E. KALMAN : "On structural properties of linear constant multivariable systems", Proceeding 3d Inter. I.F.A.C. Congress, London 1966, Butterworths Scientific Publications.

2.11 S. PERLIS : "Theory of Matrices", Addison and Wesley, Cambridge, 1942.

2.12 E. POLAK : "An Algorithm for Reducing a Linear time-invariant Differential System to State Form". Trans. I.E.E.E. AC, Vol. 11, pp. 577-570.

3.1 E. BIONDI, L. DIVIETI, C. ROVEDA, R. SCHMID : "On the Optimal Implementation of Multivariable Discrete Linear Systems". 4th Int. I.F.A.C.Congress, Varsovie 1969 (Preprint 67.5).

3.2 J.J. BONGIORNO et C.C. YOULA : "On Observers in Multivariable Control Systems". International Journal of Control, Vol. 8, n° 3, Oct. 1968.

3.3 M.R. CHIDAMBARA et E.J. DAVISON : On "A Method for Simplifying Linear Dynamic Systems". I.E.E.E. Trans. on Auto. Control, Vol. AC-12, February 1967, pp. 119-121.

3.4 P.R. CHIDAMBARA et E.J. DAVISON : "Further Remarks on Simplifying linear Dynamic Systems". I.E.E.E. Trans. on Auto. Control, Vol. AC-12, April 1967, pp. 213-214.

3.5 M.R. CHIDAMBARA et E.J. DAVISON : "Further Comments on "A method for Simplifying Linear Dynamic Systems". I.E.E.E. Trans. on Auto. Control, December 1967, pp. 799-800.

3.6 E.J. DAVISON : "A new Method for Simplifying Large Linear Dynamic Systems". I.E.E.E. Trans. on Auto. Control, Vol. AC-12, April 1968, pp. 214-225.

3.7 J.E. DAVISON : "A Method for Simplifying Linear Dynamic Systems I.E.E.E. Trans. on Auto. Control, Vol. AC-12. February 1967, pp. 119-121.

3.8 FALB and WOLOVICH : On the Structure of MV, Systems. S.I.A.M. Journal Control, U.S.A. 1969, 7 n° 3, pp. 437-51.

3.9 A.J. FOSSARD : Editorial "Revue des annales de l'A.I.C.A.". Vol. X, no 4, October 1968.

3.10 A.J. FOSSARD, J.J. ELTGEN : "Simulation et formulation des systèmes multidimensionnels". Revue des annales de l'A.I.C.A., Vol. X n° 1, January 1969.

3.11 A.J. FOSSARD : "Sur la simplification des modèles linéaires". Automatisme, Vol. XV, no 4, April 1970.

3.12 A.J. FOSSARD : "A Method for Simplifying Linear Dynamic Systems" I.E.E.E. Trans. on Auto. Control, Vol. AC-15, n° 2.

3.13 A. FOSSARD, C. GUEGUEN : "Structures identifiables des systèmes multivariables et leurs applications à la commande des processus". Rapport d'avancement : Convention 66.00.227 D.G.R.S.T. December 1968, Chapter VIII.

3.14 F.R. GANTMACHER : The Theory of Matrices. Vol. 1 (Chelsea Publ. Cy. 1960).

3.15 J.C. GILLE, M. PELEGRIN, P. DECAULNE : "Théorie et Techniques des Asservissements". Dunod 1965, p. 31.

3.16 P. GODIN, E. IRVING : "Identification et régulation multivariable par le filtre de Kalman". Communication Journée A.F.C.E.T. "Les modèles et leur Emploi", 17 April 1970.

3.17 B.L. HO, R.E. KALMAN : "Effective Construction for Linear State - variable Models from Input - output Functions". Regelungstechnik 14 Jahrg. Heft 12, 1966, pp. 545-48.

3.18 IRVING : "Algorithmes de détermination des systèmes d'équations différentielles linéaires à partir des données d'entrée et de sortie". E.D.F. Bulletin de la Direction des Etudes et Recherches ; Série C, Mathématiques n° 2, 1969, pp. 5-28.

3.19 R.E. KALMAN : "Mathematical Description of Linear dynamical Systems". Journal S.I.A.M., Series A on Control, Vol. 1, no 2, 1963.

3.20 D.G. LUENBERGER : "Canonical Forms for Linear Multivariable Systems". I.E.E.E. Trans. on Auto. Cont. Vol. AC-12, n° 3, Juin 1967.

3.21 D.G. LUENBERGER : "Observing the State of a Linear System". I.E.E.E. Trans. Mil. élect., Vol. MIL 8, n° 2, April 1964.

3.22 L.E. Mc BRIDE, K.S. NARENDRA : "An Expanded Matrix Representation for Multivariable Systems". I.E.E.E. Trans. on Auto. Cont. Vol. AC-8 n° 3, July 1963.

3.23 S.A. MARSCHALL : "An Approximate Method for Reducing the Order of a Linear System". December 1966, Control.

3.24 P.D. MESAROVIC : "The Control of Multivariable Systems". John Whiley and Sons, 1960.

3.25 T. MITRA : "The Reduction of Complexity of Linear, Time-invariant Dynamical Systems". 4th Int. I.F.A.C. Congress, Varsovie 1969, Preprint 67.2.

3.26 K.S. NARENDRA, L.E. Mc BRIDE : "Comparison of different Modes of Representation of Multivariable Systems". Cruft Laboratory TR 356, Harvard University, March 1962.

3.27 D. PEGADO : Compensation des systèmes linéaires multivariables par la technique du retour d'état. Thèse Maîtrise E.N.S.A., 1967.

3.28 B. PORTER : "Synthesis of dynamical Systems". Nelson, 1969.

3.29 H.H. ROSENBROCK, CK STOREY : "Mathematics of dynamical Systems". Nelson, 1969.

3.30 R. SINGER, P. FROST : "New canonical Forms of general multivariable linear Systems with Application to Estimation and Control", Preprints 1st I.F.A.C. Symposium ou Multivariable Control Systems, October 1968.

3.31 WILKINSON : "The algebraic eigenvalue Problem". Clarendon, 1965.

4.1 J.J. BONGIORNO, D.C. YOULA : "On Observers in Multivariable Control Systems". International Journal of Control, Vol. 8, n° 3, pp. 221-243.

4.2 R. BROCKETT, M.D. MESAROVIC : "The Reproducibility of multivariable Systems". J.A.A.C. Stanford, 1964, pp. 481-486.

4.3 E.J. DAVISON, R.W. GOLDBERG : "A Design Technique for the Incomplete State Feedback Problem in Multivariable Control Systems", Automatica, Vol. 5, n° 3, 1969.

4.4 F.R. GANTMACHER : "Théorie des matrices", (traduit du russe), Vol. 1 (Chelsea Publ. Cy. 1960).

4.5 E.G. GILBERT : "Controllability and Observability in Multivariable Control Systems". S.I.A.M. Journal Control, Series A, Vol. 1, no 2, pp. 128-151, 1963.

4.6 R.E. KALMAN : "Mathematical Description of Linear Dynamical Systems". Journal S.I.A.M., Series A on Control, Vol. 1, n° 2, pp. 152-192, 1963.

4.7 E. KREINDLER, PE.E. SARACHIK : "On the Concepts of Controllability and Observability of Linear Systems". I.E.E.E. Trans. on Auto. Control, April, 1964.

4.8 D.G. LUENBERGER : "Observers for Multivariable Systems". I.E.E.E. Trans. P.G.A.C., Vol. 11, n° 2, pp. 190-197, April 1966.

4.9 D.G. LUENBERGER : "Observing the State of a Linear System". I.E.E.E. Trans. on Military Electronics. Vol. MIL-8, n° 2, April 1964.

4.10 J.B. PEARSON : "Compensator Design for Multivariable Linear Systems". I.E.E.E. Trans. on Auto. Control, Vol. AC-14, no 2, April 1969.

4.11 D. PEGADO : "Compensation des systèmes linéaires multivariables par la technique du retour d'état". Thèse maîtrise E.N.S.A., 1967.

4.12 B. PORTER : "Synthesis of dynamical Systems". Nelson, 1969.

4.13 H.H. ROSENBROCK : "On the Design of linear multivariable Control systems". 3rd I.F.A.C. Congress London 1966. Session 1 Paper IA.

4.14 H.H. ROSENBROCK, C. STOREY : "Mathematics of dynamical Systems", Nelson, 1969.

4.15 W.M. WONHAM : "On Pole Assignment in Multi-input Controllable Linear Systems". I.E.E.E. Trans. (Auto. Control), Vol. P.G.A.C.-12, n° 6, pp. 660-665.

5.1 A.S. BOKSENBOM and R. HOOD : "General Algebraic Method Applied to Control Analysis of Complex Engines Types". N.A.C.A. T. R. 980, pp. 1-12, 1960.

5.2 H.K. CHATTERJEE : "Multivariable Process Control". Proceedings of the first intern. I.F.A.C. Congress, Moscow 1960. Edited by Butterworth 1961, Vol. I, pp. 132-141.

5.3 K. CHEN, R.A. MATHIAS and D.M. SAUTER : "Design of Non-interacting Controller Systems Using Bode Diagrams". A.I.E.E. Trans. Appl. and Industry, no 58, January 1962.

5.4 I.L. CHIEN, E.I. ERGIN and C. LING : "The Non-interacting Controls for a Steam Generating System". Control Engineering, Vol. 5, pp. 95-101, October 1958.

5.5 R.T. DARCY and D.K. FORBES : "The Theory, Error Analysis and Practical Problems Encountered in the Design of Non-interacting Control Systems". M.S. Thesis, Univ. of Michigan, May 1958.

5.6 E.J. ERGIN and C. LING : "Developement of a Non-interacting Controller for Boiler". Proceedings of the first International I.F.A.C. Congress, part III. Edited by Butterworth, London. pp. 1261-1265, 1961.

5.7 P.L. FALB and W.A. WOLOVICH : "Decoupling in the Design Synthesis of Multivariable Control Systems". I.E.E.E. Trans. on Auto. Control, Vol. AC-12, n° 6, pp. 651-659, December 1967.

5.8 P.L. FALB and W.A. WOLOVICH : "On the Decoupling of Multivariable Systems". Proceedings of the Joint Auto. Control, Conf. Philadelphia, pp. 791-796, June 1967.

5.9 H. FREEMAN : "A Synthesis Method for Multiple Control Systems". A.I.E.E. Trans. Part II, Vol, 76, pp. 28-31, March 1957.

5.10 H. FREEMAN : "Stability, Physical Realisability Consideration in Synthesis of Multiple Control Systems". A.I.E.E. Trans. Part II, Vol. 77, pp. 1-5, March 1958.

5.11 E.G. GILBERT : "Controllability and Observability in Multivariable Control System". S.I.A.M. Journal on Control, Series A, Vol. 1, no 2, pp. 128-151, 1963.

5.12 E.G. GILBERT : "The Decoupling of Multivariable Systems by State Feedback". S.I.A.M. Journal Control, Vol. 7, n° 1, pp. 50-63, February 1969.

5.13 C.F. HAUGH : "Synthesis of Controllers for Linear Multivariable Systems with Considerations of Physical Realizability, Sensitivity and Interaction". Ph. D. Thesis, Univers. of Illinois, 1961.

5.14 I.H. HOROWITZ : "Synthesis of Linear Multivariable Feedback Control Systems". I.R.E. Trans. on Auto. Control, Vol. AC-7, n° 3, pp. 47-57, April 1962.

5.15 I.H. HOROWITZ : Synthesis of Feedback Systems, Chapter 10, Academic Press Inc., New York 1963.

5.16 A. JEFFREY : "The Stability of Interacting Control System". Proceedings of the first intern I.F.A.C. Congress, Moscow 1960, Edited by Butterworth, 1961, pp. 813-817.

5.17 R.E. KALMAN : "The Continuous Linear Systems : on Structural Properties of Linear Constant, Multivariable Systems". Preprints of the 3rd Intern. I.F.A.C. Congress, London 1966, paper 6 A.

5.18 R.E. KALMAN : "Mathematical Description of Linear Dynamical Systems". S.I.A.M. Journal on Control, Series A, Vol. 1, no 2, pp. 152-192, 1963.

5.19 R.J. KAVANAGH : "Multivariable Control System Synthesis". A.I.E.E., part II, Vol. 77, pp. 425-429, November 1958.

5.20 R.J. KAVANAGH : "Non-Interacting Controls in Linear Multivariable Systems". A.I.E.E. Trans., part II, Vol. 76, pp. 95-100, May 1957.

5.21 J. MARVIN and H. HAMMOND Jr. : "Multivariable Systems", Master Thesis Kansas State University, 1962.

5.22 J.L. MASSEY and M.K. SAIN : "Inverses of Linear Sequential Circuits". I.E.E.E. Trans. on computers, Vol. 17, n° 4, pp. , April 1968.

5.23 R.A. MATHIAS : "Multivariable Controller Design by Signal Flow Graph Techniques" I.E.E.E. Trans. on Auto. Control, Vol. AC-9, n° 5, pp. 283-292, October 1963.

5.24 M.D. MESAROVIC and L. BIRTA : "Synthesis of Interaction in Multivariable Control Systems". Automatica, Vol. 2, n° 1, pp. 15-39, Pergamon Press, 1964.

5.25 M.D. MESAROVIC : "Measure of Interaction in a System and its Application to Control Problems". Case Inst. Techn., S.R.C. 4. A, 62.4, January 1962.

5.26 B.S. MORGAN : "The Synthesis of Linear Multivariable Systems by State Variable feedback". Trans. I.E.E.E. on Auto. Control, Vol. AC-9, n° 5, pp. 468-472, October 1964.

5.27 V.T. MOROZOVSKI : "Discussion de la communication de H.K.C. CHATTERJEE, cf. ref. 5.2.

5.28 J.A. PLANCHARD and V.J. LAW : "The Application of Non-interacting Control Theory to a Continuous Multivariate System". Preprints 4th Intern. I.F.A.C. Congress, Varsovie, 1969, Session 67, paper 3.

5.29 Z.V. REKASIUS : "Decoupling of Multivariable Systems by Means of State Variable Feedback". Proc. 3rd Ann. Allerton Conf. on Circuit and Syst. Theory, pp. 439-447, 1965.

5.30 M.M. ROSENBROCK : "On the Design of Linear Multivariable Control Systems". Preprints of the 3rd Intern. I.F.A.C. Congress, London 1966, paper 1 A, pp. 1A1-1A16.

5.31 M.M. ROSENBROCK : "On the Design of Linear Multivariable Control Systems". Preprints 3rd I.F.A.C. Congress, London 1966, Paper IA, pp. 1A-16.

5.32 M.K. SAIN and J.L. MASSEY : "Invertibility of Linear Time Invariant Dynamical Systems". I.E.E.E. Trans. Auto. Control, Vol. AC-14, n° 2, pp. 141-149, April 1969.

5.33 M. SOBRAL : "Extension of the Optimum Design Through Digital Compensation for Linear Multivariable Sampled-date Control Systems", A.I.E.E. Trans., Appl. and Ind. n° 63, pp. 253-255, nov. 1962.

5.34 H.TOKUMARU and Z. IWAI : "On the Non-interacting Control of Linear Invariant Multivariable Systems" : Mem. Fac. Eng. Kyoto University, 1969, 30, n° 4, pp. 592-613.

Index of cited authors

(The numbers between parentheses refer to the Bibliography arranged by chapter.)